Evolutionary Algorithms in Engineering Design Optimization

Evolutionary Algorithms in Engineering Design Optimization

Editors

David Greiner
António Gaspar-Cunha
Daniel Hernández-Sosa
Edmondo Minisci
Aleš Zamuda

MDPI • Basel • Beijing • Wuhan • Barcelona • Belgrade • Manchester • Tokyo • Cluj • Tianjin

Editors
David Greiner
Universidad de Las Palmas de Gran Canaria
Spain

António Gaspar-Cunha
University of Minho
Portugal

Daniel Hernández-Sosa
Universidad de Las Palmas de Gran Canaria
Spain

Edmondo Minisci
University of Strathclyde
UK

Aleš Zamuda
University of Maribor
Slovenia

Editorial Office
MDPI
St. Alban-Anlage 66
4052 Basel, Switzerland

This is a reprint of articles from the Special Issue published online in the open access journal *Mathematics* (ISSN 2227-7390) (available at: https://www.mdpi.com/journal/mathematics/special_issues/Evolutionary_Algorithms_Engineering_Design_Optimization).

For citation purposes, cite each article independently as indicated on the article page online and as indicated below:

LastName, A.A.; LastName, B.B.; LastName, C.C. Article Title. *Journal Name* **Year**, *Volume Number*, Page Range.

ISBN 978-3-0365-2714-7 (Hbk)
ISBN 978-3-0365-2715-4 (PDF)

Cover image courtesy of David Greiner

© 2022 by the authors. Articles in this book are Open Access and distributed under the Creative Commons Attribution (CC BY) license, which allows users to download, copy and build upon published articles, as long as the author and publisher are properly credited, which ensures maximum dissemination and a wider impact of our publications.

The book as a whole is distributed by MDPI under the terms and conditions of the Creative Commons license CC BY-NC-ND.

Contents

About the Editors vii

Preface to "Evolutionary Algorithms in Engineering Design Optimization" ix

Alberto Pajares, Xavier Blasco, Juan Manuel Herrero and Miguel A. Martínez
A Comparison of Archiving Strategies for Characterization of Nearly Optimal Solutions under Multi-Objective Optimization
Reprinted from: *Mathematics* **2021**, *9*, 999, doi:10.3390/math9090999 1

Xavier Blasco, Gilberto Reynoso-Meza, Enrique A. Sánchez-Pérez, Juan Vicente Sánchez-Pérez and Natalia Jornard-Pérez
A Simple Proposal for Including Designer Preferences in Multi-Objective Optimization Problems
Reprinted from: *Mathematics* **2021**, *9*, 991, doi:10.3390/math9090991 29

Alejandra Ríos, Eusebio E. Hernández and S. Ivvan Valdez
A Two-Stage Mono- and Multi-Objective Method for the Optimization of General UPS Parallel Manipulators
Reprinted from: *Mathematics* **2021**, *9*, 543, doi:10.3390/math9050543 49

Sarah Stirrat, Mohammed Afsar and Edmondo Minisci
Assessment of Optimization Methods for Aeroacoustic Prediction of Trailing-Edge Interaction Noise in Axisymmetric Jets
Reprinted from: *Mathematics* **2021**, *9*, 998, doi:10.3390/math9090998 69

Francisco Badea, Jesus Angel Perez and Jose Luis Olazagoitia
Detailed Study on the Behavior of Improved Beam T-Junctions Modeling for the Characterization of Tubular Structures, Based on Artificial Neural Networks Trained with Finite Element Models
Reprinted from: *Mathematics* **2021**, *9*, 943, doi:10.3390/math9090943 87

Margarita Antoniou and Gregor Papa
Differential Evolution with Estimation of Distribution for Worst-Case Scenario Optimization
Reprinted from: *Mathematics* **2021**, *9*, 2137, doi:10.3390/math9172137 105

Valentin Koblar and Bogdan Filipič
Evolutionary Design of a System for Online Surface Roughness Measurements
Reprinted from: *Mathematics* **2021**, *9*, 1904, doi:10.3390/math9161904 127

Mihailo Micev, Martin Ćalasan and Diego Oliva
Fractional Order PID Controller Design for an AVR System Using Chaotic Yellow Saddle Goatfish Algorithm
Reprinted from: *Mathematics* **2020**, *8*, 1182, doi:10.3390/math8071182 145

Francesco Marchetti and Edmondo Minisci
Genetic Programming Guidance Control System for a Reentry Vehicle under Uncertainties
Reprinted from: *Mathematics* **2021**, *9*, 1868, doi:10.3390/math9161868 167

Lorenzo A. Ricciardi, Christie Maddock and Massimiliano Vasile
Multi-Objective Optimisation under Uncertainty with Unscented Temporal Finite Elements
Reprinted from: *Mathematics* **2021**, *9*, 3010, doi:10.3390/math9233010 187

António Gaspar-Cunha, Paulo Costa, Wagner de Campos Galuppo, João Miguel Nóbrega, Fernando Duarte and Lino Costa
Multi-Objective Optimization of Plastics Thermoforming
Reprinted from: *Mathematics* **2021**, *9*, 1760, doi:10.3390/math9151760 209

Andrés Cacereño, David Greiner and Blas J. Galván
Multi-Objective Optimum Design and Maintenance of Safety Systems: An In-Depth Comparison Study Including Encoding and Scheduling Aspects with NSGA-II
Reprinted from: *Mathematics* **2021**, *9*, 1751, doi:10.3390/math9151751 229

Nabeel Al-Milli, Amjad Hudaib and Nadim Obeid
Population Diversity Control of Genetic Algorithm Using a Novel Injection Method for Bankruptcy Prediction Problem
Reprinted from: *Mathematics* **2021**, *9*, 823, doi:10.3390/ math9080823 269

Petr Bujok
The Real-Life Application of Differential Evolution with a Distance-Based Mutation-Selection
Reprinted from: *Mathematics* **2021**, *9*, 1909, doi:10.3390/math9161909 289

About the Editors

David Greiner, Associate Professor. He belongs to Civil Engineering Department and Instituto Universitario de Sistemas Inteligentes y Aplicaciones Numéricas en Ingeniería (SIANI), of Universidad de Las Palmas de Gran Canaria (ULPGC), Spain. His research is focused mainly on optimum design in engineering applications using evolutionary algorithms, covering more than 120 works including JCR-indexed journals, edited books, book chapters, international and national conferences; he has been involved in research projects and contracts. He has been a visiting scholar at Rensselaer Polytechnic Institute, USA and at University of Jyväskylä, Finland. He has served as reviewer in WOS-indexed JCR journals (Elsevier, Springer, IEEE, Taylor & Francis, MDPI, Emerald), and as organizer and/or scientific/programme committee member of international conferences (EUROGEN, CMN, GECCO, IEEE-CEC, Civil-Comp). He is editorial board member for the Mathematics MDPI and Mathematical Problems in Engineering, journals.

António Gaspar-Cunha received a PhD degree in Optimization and Modelling of Single Screw Extrusion from the University of Minho, Portugal, in 2000. He is currently an Auxiliary Professor of Polymer Processing at the University of Minho. His main areas of scientific activity are modelling of polymer extrusion based processes and multi-objective optimization. He is author or co-author of more than 170 works, including books edited, book chapters, papers published in international refereed journals, and more published in proceedings of international conferences. In 2015, he was the general chair of the 8th International Conference on Evolutionary Multi-Criterion Optimization (EMO2015) and in 2019 was the general chair of the EUROGEN 2019 international conference.

Daniel Hernández-Sosa, Associate Professor. He belongs to Informatics and Systems Department and Instituto Universitario de Sistemas Inteligentes y Aplicaciones Numéricas en Ingeniería (SIANI), of Universidad de Las Palmas de Gran Canaria (ULPGC), Spain. His main research interests are autonomous robotics and computer vision; more specifically he is focused on path planning optimization for autonomous surface and underwater vehicles. He has co-authored more than 75 papers, including JCR journals, book chapters and major conferences (ICRA, GECCO, CEC or ICPR); he has been involved in research projects and contracts. He has been a visiting scholar for short periods at different universities and research centers: CIIMAR (Madeira), Tomas Bata (Zlín) or AGH (Krakow). He has served as reviewer in several WOS-indexed JCR journals, and as organizer and/or scientific/programme committee member of national and international conferences (IROS, OCEANS, IbPRIA).

Edmondo Minisci, Associate Professor. He received MSc and PhD degrees from the Technical University of Turin and currently is the director of the Intelligent Computational Engineering Laboratory (ICELab) at the University of Strathclyde. He has 20+ years of experience in the development and application of analysis and design optimization methods for engineering systems, documented in nearly 130 publications, including international journals, edited books, book chapters, and national and international conferences, and is currently involved in projects regarding the design optimization of high lift devices and high-aspect ratio wings, the multidisciplinary analysis and design optimization under uncertainties of wind turbines, and the use of machine learning techniques for data analytics, meta-modelling, and intelligent control. He has served as program committee member for several international events, and was the chair/co-chair of EUROGEN 2015, EUROPT 2019, and BIOMA 2020.

Aleš Zamuda is an Associate Professor and Senior Researcher at University of Maribor (UM), Slovenia. He received Ph.D. (2012), M.Sc. (2008), and B.Sc. (2006) degrees in computer science from UM. He is UM project part leader for EU H2020 RIA project DAPHNE: Integrated Data Analysis Pipelines for Large-Scale Data Management, HPC, and Machine Learning. He was also MC member for Slovenia at COST at actions CA15140 (ImAppNIO) and IC1406 (cHiPSet). He is IEEE Senior Member and chaired several IEEE positions at chapter, section, and CIS society levels, and also an ACM SIGEVO member. He is also an associate editor for Swarm and Evolutionary Computation. His areas of interest include differential evolution, multiobjective optimization, evolutionary robotics, artificial life, and cloud computing. He was awarded IEEE R8 SPC award and Danubuius Young Scientist Award, and 1% top reviewer at Publons Peer Review Awards, including reviews for dozens of journals, conferences, and research projects.

Preface to "Evolutionary Algorithms in Engineering Design Optimization"

Evolutionary algorithms (EAs) are population-based global optimizers, which, due to their characteristics, have allowed us to solve, in a straightforward way, many real-world optimization problems in the last three decades, particularly in engineering fields. Their main advantages are the following: they do not require any requisite to the objective/fitness evaluation function (continuity, derivability, convexity, etc.); they are not limited by the appearance of discrete and/or mixed variables or by the requirement of uncertainty quantification in the search. Moreover, they can deal with more than one objective function simultaneously through the use of evolutionary multi-objective optimization algorithms. This set of advantages, and the continuously increased computing capability of modern computers, has enhanced their application in research and industry.

To help address and resolve these engineering optimization problems, this book comprises 14 chapters that present a series of contributions in the field. The chapters collect the papers included in the "Evolutionary Algorithms in Engineering Design Optimization" Special Issue of the Mathematics MDPI journal, 2020, first decile of the JCR 2020 in the Mathematics category.

The manuscripts cover a wide spectrum in terms of type of problems, methodologies and applications. Type of problems: single-objective and multi-objective optimization (among them, analysis of archiving strategies in evolutionary multi-objective algorithms, and preference directions in multi-objective optimization problems). Methods: genetic programming, genetic algorithms, particle swarm optimization, differential evolution, estimation of distribution algorithms, memetic algorithms, among others. Applications: Identification of thermal systems, plastics thermoforming, reliability (maintenance) and design of systems, multi-objective design of general universal–prismatic–spherical Gough–Stewart structure platforms, aero-acoustical trailing-edge noise problem, surrogate modelling of beam T-junctions for characterization of tubular structures, vibration absorber, online surface roughness measurement of automobile components, daily diet design problem, bankruptcy prediction problem, optimal tuning of a fractional order proportional–integral-derivative controller for an automatic voltage regulator system, control system for an aerospace re-entry vehicle, and design of descent trajectories for spaceplane-based two-stage launch systems.

We would like to thank the MDPI publishing editorial team, the scientific peer reviewers and the 41 authors who have contributed to this volume. We hope that the manuscripts are of value to researchers, academics and professionals of multidisciplinary engineering, mathematicians and computer scientists involved in the resolution and optimization of real world engineering problems.

David Greiner, António Gaspar-Cunha, Daniel Hernández-Sosa, Edmondo Minisci, Aleš Zamuda
Editors

Article

A Comparison of Archiving Strategies for Characterization of Nearly Optimal Solutions under Multi-Objective Optimization

Alberto Pajares *, Xavier Blasco, Juan Manuel Herrero and Miguel A. Martínez

Instituto Universitario de Automática e Informática Industrial, Universitat Politècnica de València, 46022 Valencia, Spain; xblasco@upv.es (X.B.); juaherdu@upv.es (J.M.H.); mmiranzo@upv.es (M.A.M.)
* Correspondence: alpafer1@upv.es

Abstract: In a multi-objective optimization problem, in addition to optimal solutions, multimodal and/or nearly optimal alternatives can also provide additional useful information for the decision maker. However, obtaining all nearly optimal solutions entails an excessive number of alternatives. Therefore, to consider the nearly optimal solutions, it is convenient to obtain a reduced set, putting the focus on the potentially useful alternatives. These solutions are the alternatives that are close to the optimal solutions in objective space, but which differ significantly in the decision space. To characterize this set, it is essential to simultaneously analyze the decision and objective spaces. One of the crucial points in an evolutionary multi-objective optimization algorithm is the archiving strategy. This is in charge of keeping the solution set, called the archive, updated during the optimization process. The motivation of this work is to analyze the three existing archiving strategies proposed in the literature ($ArchiveUpdateP_{Q,\epsilon}D_{xy}$, $Archive\_nevMOGA$, and $targetSelect$) that aim to characterize the potentially useful solutions. The archivers are evaluated on two benchmarks and in a real engineering example. The contribution clearly shows the main differences between the three archivers. This analysis is useful for the design of evolutionary algorithms that consider nearly optimal solutions.

Keywords: multi-objective optimization; nearly optimal solutions; archiving strategy; evolutionary algorithm; non-linear parametric identification

1. Introduction

Many real-world applications pose different objectives (usually in conflict) to optimize [1–3]. This leads to the proposal of a multi-objective optimization approach (MOOP—multi-objective optimization problem) [4–7]. In a posteriori multi-objective approach [8], after the MOOP definition and the optimization stage, a set of Pareto optimal solutions [9] is generated. The decision maker (DM) can then analyze, at the decision-making stage, the trade-off of the optimal alternatives for each design objective. This enables a better understanding of the problem and a better-informed final decision.

For the DM, it is useful to have a diverse set of solutions. Traditionally, diversity is sought in the objective space. However, obtaining a diverse set in the decision space also offers advantages [10]: (1) it enables the DM to obtain different (even significantly different) alternatives before the final decision; (2) it helps speed up the search, improving exploration, and preventing premature convergence towards a non-global minimum. In addition, the best solutions are sometimes too sensitive to disturbances, or are not feasible in practice [11–14]. In this scenario, the multimodal solutions or the nearly optimal solution set (also called approximate or ϵ-efficient solutions) plays a key role in enhancing the diversity of solutions. Two multimodal solutions are those that, being optimal, obtain the same performance. Nearly optimal solutions are those that have similar performance to optimal solutions. Generalizing, it can be considered that multimodal solutions are included in nearly optimal solutions. Nearly optimal solutions have been studied by many authors in the bibliography [15–19], have similar performance to optimal solutions and can

sometimes be more adequate according to DM preferences (for instance, more robust [14])
Therefore, an additional challenge then arises: to obtain a set of solutions that, in addition
to good performance in the design objectives, offer the greatest possible diversity.

However, considering all the nearly optimal solutions requires obtaining and analyzing a great number of alternatives and this causes two problems:

1. It slows down the optimization process. In evolutionary algorithms, an *archive* (a set to store solutions during the execution) is required. The computational cost of the optimization process largely depends on the archive size. This is because to check for the inclusion of a new candidate solution in the archive, it is necessary to check the dominance (or ϵ−dominance) for each solution in the current archive. Many new candidate solutions are analyzed in an optimization process. Therefore, a large archive results in a significantly higher computational cost.
2. The decision stage is made more difficult. The designer must choose the final solution from a much larger number of alternatives.

Therefore, it is necessary to reduce the set of optimal solutions obtained by the designer. In the literature, there are different algorithms aimed at finding nearly optimal solutions in multi-objective optimization problems. The multimodal multi-objective evolutionary algorithms (MMEAs [20]) are intended for multimodal optimization problems. Some of the MMEAs take into account nearly optimal solutions in the optimization process, but most of them do not provide these solutions to the DM. Furthermore, evolutionary algorithms with an unbounded external archive [21] can also be interesting to analyze these solutions. These unbounded external archives can be analyzed to obtain the relevant nearly optimal solutions.

One of the crucial points in an evolutionary multi-objective optimization algorithm is the archiving strategy (or archiver). An archiving strategy is the strategy that selects and updates a solution set, called the archive, during the evolutionary process. Some archivers have been studied previously [19,22–26]. In this paper, we address the problem of discretization of the potentially useful alternatives. For this purpose, we compare different archiving strategies that aim to obtain the set of potentially useful nearly optimal solutions. An archiving strategy must take into account the decision space to ensure that the potentially useful nearly optimal solutions are not discarded.

For the comparison of the results in this real problem, we have chosen to embed the archiver in a basic evolutionary algorithm. First, to observe the impact of each archiver when incorporated into an evolutionary mechanism. In addition, second, because the computational cost associated with the objective functions of the real problem does not allow simulations on large numbers of points. Therefore, it is not feasible to test each archiver with a random or exhaustive search as has been done with the benchmarks.

These archivers have not been compared in the literature, so this work is useful for future designs of evolutionary algorithms that consider nearly optimal solutions or even to modify the archivers of the old evolutionary algorithms considering such solutions. Therefore, the purpose of the paper is: (1) to understand the properties of these archivers, (2) to provide an analysis for choosing one of these archivers and (3) to give ideas for designing new archivers. The design of these algorithms are currently open issues in this research area [18,20,27].

This work is structured as follows. In Section 2, a small state of the art on potentially useful nearly optimal solutions is introduced. In Section 3 some basic multi-objective backgrounds are presented. In Section 4, different archiving strategies to characterize the optimal and nearly optimal set are described. In Section 5 the MOOPs and the archivers comparison procedure are presented. The results obtained on the archivers are shown in Section 6. Finally, the conclusions are given in Section 7.

2. State of the Art

As discussed in the previous section, obtaining all nearly optimal solutions leads to problems. Considering only the most relevant solutions largely avoids the problems mentioned above. Not all nearly optimal solutions are equally useful to the DM. Therefore, if we manage to discard those that are less useful, we will reduce both mentioned problems. Let us see a graphic example to illustrate what we consider as potentially useful solutions. Suppose we have a MOOP with two design objectives and two decision variables (see Figure 1). Three solutions x^1, x^2 and x^3 are selected. x^1 is an optimal solution (member of the Pareto set), and it slightly dominates the nearly optimal solutions x^2 and x^3. x^1 and x^2 are very similar alternatives in their parameters (both belong to $neighborhood_1$, the same area in the parameter space), while x^3 is significantly different (it belongs to $neighborhood_2$). In this scenario, x^2 does not provide new relevant information to the DM. This solution is similar to x^1 but with a worse performance in design objectives. Predictably, both will have similar characteristics, therefore the DM will choose x^1 since it obtains a better performance in the design objectives. However, x^3 does provide useful new information to the DM because it has a similar performance to the optimal ones and is in a different neighborhood. The solutions in $neighborhood_1$ could be, for example, not very robust or not feasible in practice. In this context, x^3 (and the solutions in $neighborhood_2$) could be a potentially useful solution due to their significantly different characteristics. It is possible, and often common, for the DM to analyze in the decision stage additional indicators/objectives not included in the optimization phase. Thus, the DM can assume a small loss of performance in the design objectives in exchange for an improvement in a new feature not contemplated in the optimization process. This analysis can decide the final choice in one way or another. In short, including solutions of $neighborhood_2$ increases the diversity with useful solutions and enables the DM to make a better-informed final decision.

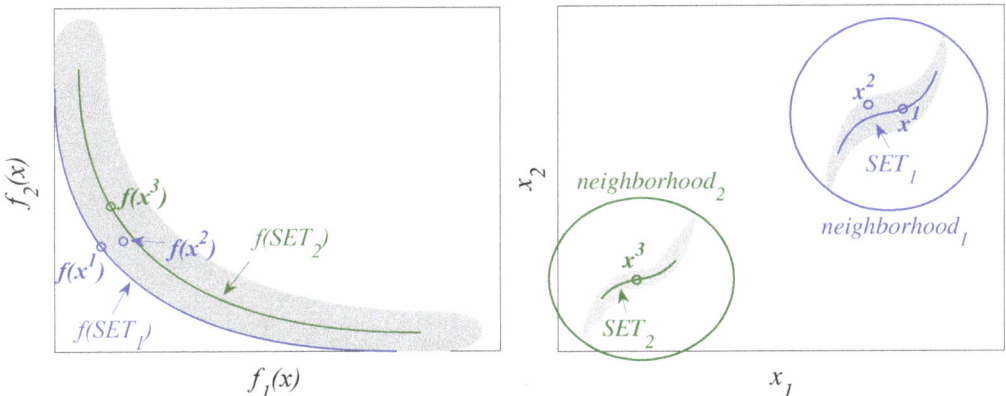

Figure 1. A MOOP example. On the left, the objective space is shown, and on the right, the decision space is shown. SET_1 is the Pareto optimal set and SET_2 is a potentially useful nearly optimal set.

Therefore, the potentially useful nearly optimal solutions are those nearly optimal alternatives that differ significantly in the parameter space [28–30]. Thus, the new set must: (1) not neglect the diversity existing in the set of nearly optimal alternatives; (2) obtain the least number of solutions. To achieve both aims, it is necessary to employ an evolutionary algorithm that characterizes the set of solutions by means of a discretization which takes into account both the decision space and the objective space, simultaneously. A discretization that takes into account only the objective space can lead to the loss of significantly different nearly optimal alternatives in the decision space. This loss is a drawback because, as we have previously discussed, these alternatives are potentially useful. On the other hand, a discretization that takes into account only the decision space can lead to archives with a huge number of solutions [19] and cause the two problems previously mentioned.

3. Background

A multi-objective optimization problem can be defined as follows:

$$\min_{x \in Q} f(x) \qquad (1)$$

where $x = [x_1, ..., x_k]$ is defined as a decision vector in the domain $Q \subset \Re^k$ and $f: Q \to \Re^m$ is defined as the vector of objective functions $f(x) = [f_1(x), ..., f_m(x)]$. A maximization problem can be converted into a minimization problem. For each objective to be maximized $\max f_i(x) = -\min(-f_i(x))$ will be performed. The domain Q is defined by the set of constraints on x. For instance (but not limited to –any other constraints could be introduced in a general MOOP–):

$$\underline{x_i} \leq x_i \leq \overline{x_i}, \ i = [1, ..., k] \qquad (2)$$

where $\underline{x_i}$ and $\overline{x_i}$ are the lower and upper bounds of x components.

Consequently, the MOOP obtains a Pareto set P_Q (see Definition 2). This set has solutions non-dominated by any other solution (see Definition 1) in Q.

Definition 1 (Dominance [31]). *A decision vector x^1 is dominated by another decision vector x^2 if $f_i(x^2) \leq f_i(x^1)$ for all $i \in [1, ..., m]$ and $f_j(x^2) < f_j(x^1)$ for at least one $j, j \in [1, ..., m]$. This is denoted as $x^2 \preceq x^1$.*

Definition 2. *(Pareto set P_Q): is the set of solutions in Q that is non-dominated by any other solution in Q: $P_Q := \{x \in Q | \nexists x' \in Q : x' \preceq x\}$*

Definition 3. *(Pareto front $f(P_Q)$): given a set of Pareto optimal solutions P_Q, the Pareto front is defined as:*

$$f(P_Q) := \{f(x) | x \in P_Q\}$$

In any MOOP, there is a set of solutions with objective values close to the Pareto front. These solutions receive several names in the bibliography: nearly optimal, approximate, or ϵ-efficient solutions. To formalize the treatment of the nearly optimal solutions, the following definitions are used:

Definition 4. *($-\epsilon$-dominance [32]): define $\epsilon = [\epsilon_1, ..., \epsilon_m]$ as the maximum acceptable degradation. A decision vector x^1 is $-\epsilon$-dominated by another decision vector x^2 if $f_i(x^2) + \epsilon_i \leq f_i(x^1)$ for all $i \in [1, ..., m]$ and $f_j(x^2) + \epsilon_j < f_j(x^1)$ for at least one $j, j \in [1, ..., m]$. This is denoted by $x^2 \preceq_{-\epsilon} x^1$.*

Definition 5. *(Set of nearly optimal solutions, $P_{Q,\epsilon}$ [30]): is the set of solutions in Q which are not $-\epsilon$-dominated by another solution in Q:*

$$P_{Q,\epsilon} := \{x \in Q | \nexists x' \in Q : x' \preceq_{-\epsilon} x\}$$

The sets defined P_Q and $P_{Q,\epsilon}$ usually contain a great, or even infinite, number of solutions. Optimization algorithms try to characterize these sets using a discrete approximation P_Q^* and $P_{Q,\epsilon}^*$. In general, if such an approach has a limited set of solutions, the computational cost falls. However, the number of solutions must be sufficient to obtain a good characterization of these sets.

To compare the archiving strategies, it is useful to use a metric. Different metrics are used in the literature to measure the convergence of the outcome set. An example of these is the Hausdorff distance (d_H [33–35] see Equation (3)).

$$d_H(A, B) := max(dist(A, B), dist(B, A))$$
$$dist(A, B) := \sup_{u \in A} dist(u, B) \qquad (3)$$
$$dist(u, B) := \inf_{v \in B} ||u - v||$$

This metric is a measure of the distance between two sets. Therefore, d_H can be used to measure convergence between the outcome set (or final archive) $f(A)$ to the target set $f(H)$ of a given MOOP (or archiving strategy). However, d_H only penalizes the largest outlier of the candidate set. Thus, a high value of $d_H(A, H)$ can indicate both that A is a bad approximation of H and that A is a good approximation but contains at least one outlier. The d_H is used by the archiver $ArchiveUpdateP_{Q,\epsilon}D_{xy}$.

To avoid this problem, a new indicator appears in the literature: the averaged Hausdorff distance Δ_p (the standard Hausdorff distance is recoverable from Δ_p by taking the limit $\lim_{p \to \infty} \Delta_p = d_H$). This metric (with $1 \leq p < \infty$) assigns a lower value to sets uniformly distributed throughout its domain. Δ_p is based on the known generational distance metrics (GD [36]) and represents how "far" $f(H)$ is from $f(A)$), and inverted generational distance (IGD [37]) represents how "far" $f(A)$ is from $f(H)$). However, these metrics are slightly modified in Δ_p (GD_p and IGD_p, see Equation (5)). This modification means that the larger archive sizes and finer discretizations of the target set do not automatically lead to better approximations under Δ_p [34]. Δ_p measures the diversity and convergence in the decision and objective spaces. In this work we use Δ_p with $p = 2$ (as in [18,38]) so that the influence of outliers is low.

$$\Delta_p(X, Y) := max(GD_p(X, Y), IGD_p(X, Y)) \qquad (4)$$

where

$$GD_p = \left(\frac{1}{n_x} \sum_{i=1}^{n_x} dist(x_i, Y)^p \right)^{1/p}$$
$$IGD_p = \left(\frac{1}{n_y} \sum_{i=1}^{n_y} dist(y_i, X)^p \right)^{1/p} \qquad (5)$$

where

$$dist(u, A) := \inf_{v \in A} ||u - v|| \qquad (6)$$

To use Δ_p it is necessary to define the target set H with which to compare the final archive A obtained by the archivers. The target set is defined in the decision space (H) and it has its representation in the objective space ($f(H)$). The definition of H is possible on the benchmarks used in this work (where the global and local optimum are known), but this definition is not trivial. $Archive\_nevMOGA$ and $ArchiveUpdateP_{Q,\epsilon}D_{xy}$ archivers discard solutions similar in both spaces at the same time. However, $Archive\_nevMOGA$, unlike $ArchiveUpdateP_{Q,\epsilon}D_{xy}$, discards dominated solutions in their neighborhood. $targetSelect$ looks for diversity in both spaces simultaneously. On the one hand, defining H as the optimal and nearly optimal solutions that are not similar in both spaces at the same time, would give the archive $ArchiveUpdateP_{Q,\epsilon}D_{xy}$ an advantage. If H is defined as the set of optimal and nearly optimal solutions non-dominated in their neighborhood, $Archive\_nevMOGA$ would be benefited. However, the archivers have a common goal: to obtain the solutions close to the optimals in the objective space but significantly different in the decision space (potentially useful solutions). Consequently, to be as "fair" as possible we must define H as the set that defines the common objective. Thus, the potentially useful solutions can be represented by the local Pareto set (see Definition 6).

Definition 6. (Local Pareto set [5,39]): is the set of solutions in Q that is non-dominated by any another neighbor solution in Q (where n is a small positive number):

$$H := \{x \in Q | \nexists y \in Q : \ ||x - y||_\infty \leq n \text{ and } y \preceq x\}$$

Figure 2 shows an example of a MOOP with two design objectives and two decision variables. Sets SET_1 and SET_2 form the global Pareto set. Both sets, together with SET_3 and SET_4 form the local Pareto set (since the global Pareto set is also a local Pareto set). No solution of a local Pareto set is dominated by a neighboring solution. Furthermore, all the solutions neighboring the local Pareto set are dominated by a neighboring solution, and therefore they are not part of this set. This can be verified by the colored areas around the sets SET_1, SET_2, SET_3 and SET_4. For example, solutions in the gray area, which are neighboring solutions to SET_3, obtain a worse objective value than SET_3. For this reason, the solutions of the gray area are dominated by neighboring solutions, and therefore are not part of the local Pareto set. Sets SET_3 and SET_4 provide the DM with alternatives potentially useful (significantly different to SET_1 and SET_2), enabling the DM to make a more informed final decision. For this work, the ϵ−dominated solutions (solutions that are not in the set $P_{Q,\epsilon}$) will not be considered to be local Pareto solutions (H) because their degradation in performance is significant for the DM.

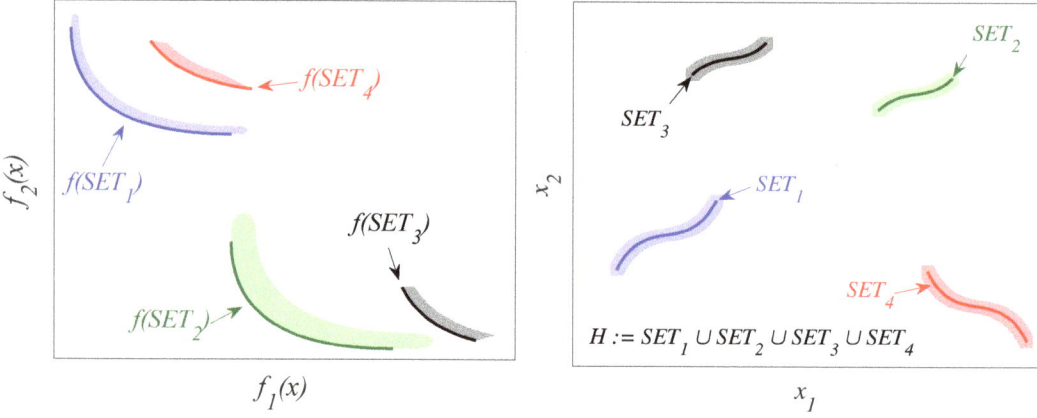

Figure 2. Visualization of an MOOP in the objective space (on the left) and decision space (on the right). The sets SET_1 and SET_2 are the optimal solutions. Both sets and SET_3 and SET_4 form the local Pareto set.

4. Description of the Compared Archivers

As already discussed above, nearly optimal solutions can be very useful. However, it is necessary to discretize this set in order to find a reduced set of solutions to avoid the problems associated with an excessive number of solutions. Furthermore, it is necessary not to neglect the potentially useful nearly optimal solutions, i.e., nearly optimal alternatives significantly different (in the decision space) to the optimal solutions. To achieve both purposes, it is essential to discretize the set of solutions taking into account the decision and objective spaces at the same time. In this section, three archivers that discretize the set $P_{Q,\epsilon}$ in both spaces are described.

There are MMEAs and algorithms that consider nearly optimal solutions that offer these solutions: $P_{Q,\epsilon}$-NSGA-II [28], $P_{Q,\epsilon}$-MOEA [30], nevMOGA [29], NϵSGA [18], DIOP [10], 4D-Miner [40,41], MNCA [42].

$P_{Q,\epsilon}$-NSGA-II [28] was one of the first algorithms aimed at finding approximate (nearly optimal) solutions. $P_{Q,\epsilon}$-NSGA-II uses the same classification strategy as the algorithm on which it is based, NSGA-II [43], and therefore, the highest pressure of the population is taken toward the Pareto set. Thus, this may result in the neighborhoods with only nearly optimal solutions not being adequately explored [30]. To avoid this problem, the algorithm $P_{Q,\epsilon}$-MOEA [30] is created. This algorithm was designed to avoid Pareto set bias. Nevertheless, $P_{Q,\epsilon}$-MOEA does not take into account the location of solutions in the decision space. $P_{Q,\epsilon}$-MOEA does not then guarantee that the potentially useful alternatives will not be discarded. To overcome this problem, the nevMOGA [29] algorithm was designed. This algorithm seeks to ensure that the potentially useful alternatives are not discarded. DIOP [10] is a set-based algorithm that can maintain dominated solutions. This algorithm simultaneously evolves two populations A and T. Population A approaches the Pareto front, and is not provided to the DM, while T is the target population that seeks to maximize diversity in the decision and objective spaces. MNCA [42] is an evolutive algorithm that simultaneously evolves multiple subpopulations. In MNCA each subpopulation converges to a different set of non-dominated solutions. Finally, 4D-Miner [40,41] is an algorithm especially designed for functional brain imaging problems.

One of the crucial points in an evolutionary multi-objective optimization algorithm is the archiving strategy. The $P_{Q,\epsilon}$-NSGA-II and $P_{Q,\epsilon}$-MOEA algorithms share the $ArchiveUpdateP_{Q,\epsilon}$ archiver. This archiver seeks to characterize all nearly optimal solutions without taking into account the decision space. In [19], different archiving strategies are compared: $ArchiveUpdateP_{Q,\epsilon}$, $ArchiveUpdateP_{Q,\epsilon}D_x$, $ArchiveUpdateP_{Q,\epsilon}D_y$ and $ArchiveUpdateP_{Q,\epsilon}D_{xy}$. On the one hand, $ArchiveUpdateP_{Q,\epsilon}$ gets an excessive number of solutions. On the other hand, $ArchiveUpdateP_{Q,\epsilon}D_x$, $ArchiveUpdateP_{Q,\epsilon}D_y$ do not discretize the decision and objective spaces simultaneously. Therefore, these archivers do not achieve the two purposes discussed above. The mentioned work concludes that archiver $ArchiveUpdateP_{Q,\epsilon}D_{xy}$ is most practical use within stochastic search algorithms. Furthermore, this archiver is the only one of the archivers compared in this paper that discretizes the decision and objective spaces simultaneously [27], a factor that we consider necessary to obtain potentially useful solutions. The archiver $ArchiveUpdateP_{Q,\epsilon}D_{xy}$ has been employed in the recent NϵSGA algorithm to maintain a well-distributed representation in the decision and objective spaces. For this reason, the present work compares the archiver $ArchiveUpdateP_{Q,\epsilon}D_{xy}$ and not $ArchiveUpdateP_{Q,\epsilon}$, $ArchiveUpdateP_{Q,\epsilon}D_x$ and $ArchiveUpdateP_{Q,\epsilon}D_y$.

The second archiver included in this comparison is the archiver of the nevMOGA algorithm ($Archive\_nevMOGA$). This archiver characterizes the set of potentially useful solutions by discretizing both spaces simultaneously. Finally, the archiver of the DIOP algorithm ($targetSelect$) is also compared in this work. $targetSelect$ seeks to find the population that maximizes an indicator that measures diversity in the decision and objective spaces simultaneously. Therefore, a metric-based archiver $targetSelect$ is compared to the distance-based archivers $ArchiveUpdateP_{Q,\epsilon}D_{xy}$ and $Archive\_nevMOGA$. The three archivers compared in this work seek to characterize the potentially useful solutions. The archiver of the MNCA algorithm has not been included in the comparison because it looks for non-dominated solutions. The archiver of the 4D-Miner algorithm has also not been included in the comparison because 4D-Miner is a very specific algorithm for functional brain imaging problems.

4.1. $ArchiveUpdateP_{Q,\epsilon}D_{xy}$

$ArchiveUpdateP_{Q,\epsilon}D_{xy}$ is the proposed archiving strategy in [19] (see Algorithm 1). As already mentioned, potentially useful solutions are those that obtain similar performance (in the design objectives) but differ significantly in their parameters (in the decision space). This archiver aims to maintain these solutions. This archiver uses, in addition to the parameter ϵ (maximum degradation acceptable to the DM, see Definition 4), the parameters Δ_x and Δ_y. Two solutions are considered similar if their distance in the decision space is less than Δ_x. Therefore, the parameter Δ_x is the maximum distance, in the decision space, between two similar solutions. Two alternative solutions obtain a similar performance if their distance in the objective space is less than Δ_y. Therefore, the parameter Δ_y is the maximum distance between two solutions to be considered similar in the objectives space. Both parameters are measured using the Hausdorff distance [34] (d_H, see Equation (3)). The archive A stores the set of obtained alternatives. A new solution p from P (new candidate solutions) will only be incorporated in A if: (1) p is a nearly optimal solution; and (2) A does not contain any solution similar to p, in the decision and objective spaces at the same time. If the new solution p is stored in archive A, the new set of optimal and nearly optimal solutions ($\hat{A}$) belonging to archive A is calculated. Thus, a solution $a \in A$ will be removed if: (1) it is not a nearly optimal solution ($p \prec_{-(\epsilon+\Delta_y)} a$); and (2) the distance to the set $\hat{A}$ fulfills the condition $dist(a, \hat{A}) \geq 2\Delta_x$.

Algorithm 1 $A := ArchiveUpdateP_{Q,\epsilon}D_{xy}(P, A_0, \epsilon, \Delta_x, \Delta_y)$

Require: population P, archive A_0, $\epsilon \in \mathbb{R}_+^m$, $\Delta_x \in \mathbb{R}_+$, $\Delta_y \in \mathbb{R}_+$
Ensure: updated archive A
1:   $A := A_0$
2:   **for all** $p \in P$ **do**
3:     **if** $(\nexists a_1 \in A : a_1 \prec_{-\epsilon} p)$ and $(\nexists a_2 : (d_H(f(a_2), f(p)) \leq \Delta_y$ and $d_H(a_2, p) \leq \Delta_x))$ **then**
4:       $A \leftarrow A \cup \{p\}$
5:       $\hat{A} = \{a_1 \in A | \nexists a_2 \in A : a_2 \prec_{-(\epsilon+\Delta_y)} a_1\}$
6:       **for all** $a \in A \setminus \hat{A}$ **do**
7:         **if** $p \prec_{-(\epsilon+\Delta_y)} a$ and $dist(a, \hat{A}) \geq 2\Delta_x$ **then**
8:           $A \leftarrow A \setminus \{a\}$
9:         **end if**
10:       **end for**
11:     **end if**
12:   **end for**

The archiver $ArchiveUpdateP_{Q,\epsilon}D_{xy}$ goes through all the set of candidate solutions $p \in P$, and in the worst case, the algorithm compares them with all the solutions $a \in A$. Thus, the complexity of the archiver is $O(|P||A|)$ [44]. Also, $ArchiveUpdateP_{Q,\epsilon}D_{xy}$ has a maximum number of solutions $|A(D_{xy})|$ [19] which is given by:

$$|A(D_{xy})| \leq |A(D_{xy}^x)||A(D_{xy}^y)| \tag{7}$$

where $|A(D_{xy}^x)|$ is the maximum number of neighborhoods that the decision space can contain (based on Δ_x) and $|A(D_{xy}^y)|$ is the maximum number of solutions that can exist in each neighborhood (based on Δ_y and ϵ), and are defined as:

$$|A(D_{xy}^x)| \leq \prod_{i=1}^{k}\left(\frac{1}{\Delta_x}+1\right)\prod_{j=1}^{k}(\overline{x_j} - \underline{x_j}) \tag{8}$$

$$|A(D_{xy}^y)| \leq \left(\frac{1}{\Delta_y}\right)^{m-1}\sum_{i=1}^{m}\left(\frac{\epsilon_i}{\Delta_y}+3\right)\prod_{\substack{i=1 \\ i \neq j}}^{m}(M_j - m_j + \Delta_y) \tag{9}$$

$\underline{x_j}$ and $\overline{x_j}$ are the bounds in the decision space ($P_{Q,\epsilon+2\Delta_y}$ is included in $[\underline{x_1}, \overline{x_1}]...[\underline{x_k}, \overline{x_k}]$) and M_j and m_j are the bounds in the objective space of the set to discretize $f(P_{Q,\epsilon+2\Delta_y})$. Also, it is assumed that any ϵ_i is greater than Δ_y.

4.2. Archive_nevMOGA

Archive_nevMOGA is the archiving strategy used by the nevMOGA evolutionary algorithm [29]. This archiver, just as $ArchiveUpdateP_{Q,\epsilon}D_{xy}$, aims to guarantee solutions that obtain similar performance to the optimals, but are significantly different in the decision space. The archiver Archive_nevMOGA uses the same three parameters as $ArchiveUpdateP_{Q,\epsilon}D_{xy}$ (ϵ, Δ_x, Δ_y). However, there are differences in the definition of some parameters in this archiver with respect to $ArchiveUpdateP_{Q,\epsilon}D_{xy}$: (1) Δ_x is a vector that contains the maximum distances (in the decision space) between similar solutions for each dimension. Thus, two individuals a and b are similar if: $|a_i - b_i| \leq \Delta_{x_i} \ \forall i \in [1, ..., k]$. (2) Δ_y is also a vector that contains the maximum distances (in the objective space) between solutions with similar performance for each dimension. Thus, two individuals a and b have a similar performance if: $|f_i(a) - f_i(b)| \leq \Delta_{y_i} \ \forall i \in [1, ..., m]$.

The archiver $Archive\_nevMOGA$ will add a new candidate solution p to the archive A if the following conditions are met simultaneously: (1) p is a nearly optimal solution; (2) there is no similar solution to p (in the decision space) $\in A$ that dominates it; and (3) there is no similar solution to p in A in both spaces at the same time (if it exists, and p dominates it, it will be replaced). If a solution p is incorporated in the archive A, it will remove from A: (1) the similar individuals (in the parameter space) that are dominated by p and (2) the individuals ϵ−dominated by p.

The complexity of the archiver $Archive\_nevMOGA$ is equivalent to the complexity of $ArchiveUpdateP_{Q,\epsilon}D_{xy}$ previously defined ($O(|P||A|)$). Moreover, $Archive\_nevMOGA$ has a maximum number of solutions $|A(nMOGA)|$ which is given by:

$$|A(nMOGA)| \leq |A(nMOGA^x)||A(nMOGA^y)| \tag{10}$$

where $|A(nMOGA)|^x$ is the maximum number of neighborhoods that the decision space can contain (based on Δ_x) and $|A(nMOGA)|^y$ is the maximum number of solutions that can exist in each neighborhood (based on Δ_y and ϵ), and are defined as:

$$|A(nMOGA^x)| \leq \prod_{i=1}^{k}\left(\frac{1}{\Delta_{x_i}}+1\right)\prod_{j=1}^{k}(\overline{x_j}-\underline{x_j}) \tag{11}$$

$$|A(nMOGA^y)| \leq \frac{\prod_{i=1}^{m} n\_box_i}{n\_box_{max}} \tag{12}$$

where $n\_box_i = (M_i - m_i)/\Delta_{y_i}$, $n\_box_{max} = \max_i n\_box_i$ and M_j and m_j are the bounds in the objective space of the set to discretize $f(P_{Q,\epsilon})$.

The archive size with respect to the decision space is equivalent for the compared archivers ($A(D_{xy}^x)$ and $A(nMOGA^x)$). However, there is a difference between $A(D_{xy}^y)$ and $A(nMOGA^y)$ (objective space). The archiver $Archive\_nevMOGA$, unlike $ArchiveUpdateP_{Q,\epsilon}D_{xy}$, discards nearly optimal solutions dominated by a similar solution in the decision space. Thus, in the worst case, $Archive\_nevMOGA$ will obtain the best solutions (non-dominated) in each neighborhood. These solutions will have a maximum number of alternatives depending on Δ_y. However, $ArchiveUpdateP_{Q,\epsilon}D_{xy}$, in the worst case, will obtain, in each neighborhood, in addition to the best solutions, additional solutions. These additional solutions are dominated by neighboring solutions, but are considered solutions with different performance (based on Δ_y). As a result, the archive of $Archive\_nevMOGA$ will have fewer solutions than the archive of $ArchiveUpdateP_{Q,\epsilon}D_{xy}$. Therefore, we can deduce that the archiver $Archive\_nevMOGA$ has a lower computational cost than $ArchiveUpdateP_{Q,\epsilon}D_{xy}$ because its archive contains fewer solutions (the candidate solutions are compared with a smaller number of solutions).

4.3. targetSelect

$targetSelect$ is the archiving strategy used by the DIOP evolutionary algorithm [10]. This archiver seeks to obtain a diverse set of solutions (keeping solutions close to the Pareto set) in the decision and objective spaces. $A = targetSelect(F, T, \mu^t, \epsilon)$ has as inputs: an approximation to the Pareto front F, the set of solutions to be analyzed T, the size of the set target μ^t, and ϵ. This archiver selects μ^t solutions from set T. The goal is to find the population A, with size μ^t, to maximize $G(T)$ (see Equation (13)). $G(T)$ is defined as the sum of the product between a metric and its respective weight.

$$G(T) := \omega_o \cdot D_o(T) + \omega_d \cdot D_d(T), |T| = \mu \text{ with } q_F(t) \leq \epsilon \ \forall t \in T, \omega_o + \omega_d = 1 \tag{13}$$

where D_o and D_d are metrics that measure diversity in the objective and decision spaces respectively, and q_F is a distance metric defined as:

$$q_F(x) := \min\{\epsilon \mid \exists y \in F : x \preceq_\epsilon y\} \tag{14}$$

D_o is an indicator that measures diversity and convergence to the Pareto front, and D_d is an indicator that measures diversity in the decision space. In this work, as in [10], D_o and D_d were specified by the hypervolume indicator [45] and the Solow–Polasky diversity measure [46], respectively. An advantage of the archiver *targetSelect* is that you can directly and arbitrarily specify the archive size.

5. Materials and Methods

In this section, the MOOPs, on which the three archivers will be compared, will be defined. In addition, the methodology for carrying out the comparison is introduced.

5.1. Definition of MOOPs

The archivers will be compared on two benchmarks and a real engineering example Benchmarks have a very low computational cost for the objective function. For this reason, it is inexpensive to obtain a target set (with a very fine discretization in the decision space) H. This discrete set is necessary for the use of the metric Δ_p (see Section 3). Furthermore, the definition of this set must be as "fair" as possible. Therefore, for the benchmarks, the target set H is defined as the local and global Pareto set (see justification in Section 3). This set is obtained by discretizing the decision space with 10,000,000 solutions in the range of the parameters. Furthermore, the target set H has its representation in the objective space. In the real engineering example, obtaining a set H would be computationally very expensive (or even unaffordable). Therefore, this set is not defined in this MOOP. By means of these problems it is possible to analyze the behavior of the archivers for different characteristics in the MOOP: multimodal solutions; local Pareto sets; or discontinuous Pareto fronts. However, there are other features of MOPs that are not analyzed in this article (such as MOPs with many objectives and/or decision variables).

5.1.1. Benchmark 1

Benchmark 1 (see Equation (15)) is a test problem called SYM-PART defined in [47] widely used in the literature [18,20,48–50] for the evaluation of algorithms that characterize nearly optimal or multimodal solutions. Benchmark 1 has the Pareto set located in a single neighborhood, and it also has eight local Pareto set that overlap in the objective space (see Equation (20) and Figure 3). Thus, this benchmark is very useful to observe if the compared archivers can adequately characterize the nine existing neighborhoods, and provide all the existing diversity to the DM.

$$\min_x f(x) = [f_1(x) \; f_2(x)] \tag{15}$$

$$\begin{aligned} f_1(x) &= (x_1 - t_1(c+2a) + a)^2 + (x_2 - t_2 b)^2 + \delta_t \\ f_2(x) &= (x_1 - t_1(c+2a) - a)^2 + (x_2 - t_2 b)^2 + \delta_t \end{aligned} \tag{16}$$

where

$$\begin{aligned} t_1 &= sgn(x_1) \min\left(\left\lceil \frac{|x_1|-a-c/2}{2a+c} \right\rceil, 1\right) \\ t_2 &= sgn(x_2) \min\left(\left\lceil \frac{|x_2|-b/2}{b} \right\rceil, 1\right) \end{aligned}$$

and

$$\delta_t = \begin{cases} 0 & \text{for } t_1 = 0 \text{ and } t_2 = 0 \\ 0.1 & \text{else} \end{cases}$$

subject to:
$$-20 \leq x_1 \leq 20$$
$$-20 \leq x_2 \leq 20 \quad (17)$$

using $a = 0.5$, $b = 5$ and $c = 5$.

This MOOP contains one global Pareto set:
$$P_{0,0} = [-0.5, 0.5] \times \{0\} = P_Q \quad (18)$$

as well as the following eight local Pareto sets:
$$\left.\begin{array}{ll} P_{-1,-1} & = [-6.5, -5.5] \times \{-5\} \\ P_{0,-1} & = [-0.5, 0.5] \times \{-5\} \\ P_{1,-1} & = [5.5, 6.5] \times \{-5\} \\ P_{-1,0} & = [-6.5, -5.5] \times \{0\} \\ P_{1,0} & = [5.5, 6.5] \times \{0\} \\ P_{-1,1} & = [-6.5, -5.5] \times \{5\} \\ P_{0,1} & = [-0.5, 0.5] \times \{5\} \\ P_{1,1} & = [5.5, 6.5] \times \{5\} \end{array}\right\} \text{Local Pareto set} \quad (19)$$

Therefore, the target set H is defined as:
$$H := P_{0,0} \cup P_{-1,-1} \cup P_{0,-1} \cup P_{1,-1} \cup P_{-1,0} \cup P_{1,0} \cup P_{-1,1} \cup P_{0,1} \cup P_{1,1} \quad (20)$$

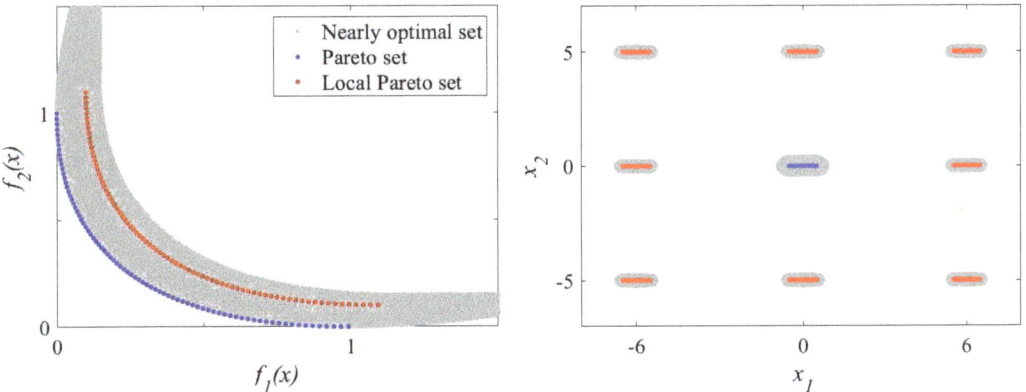

Figure 3. Target set H for the benchmark 1 formed by the global (blue) and local (red) Pareto set.

To evaluate the archivers on benchmark 1, the parameters are defined in Table 1. The parameters ϵ, Δ_x and Δ_y are defined based on prior knowledge of the problem. The *targetSelect* archiver is based on an indicator, and therefore has different parameters from the rest of the compared archives. For the choice of the parameter μ^t, based on the size of the archives obtained by the rest of the archivers, the following values have been analyzed: $\mu^t = \{100, 75, 50\}$. For the choice of the ω_o parameter, the following parameters suggested in [10] have been analyzed: $\omega_o = \{0, 0.7692, 0.9091, 0.9677, 1\}$. Among all these values, $\mu^t = 100$ and $\omega_o = 0.9677$ have been defined as the parameters that obtain the best performance, with respect to Δ_p, for the uniform dispersion. For random dispersion, $\omega_o = 0.7692$ obtain the best performance.

Table 1. Parameters for benchmark 1.

Parameters	Value
ϵ	[0.15 0.15]
ArchiveUpdate$P_{Q,\epsilon}D_{xy}$	
Δ_x	1
Δ_y	0.2
Archive_nevMOGA	
Δ_x	[1 1]
Δ_y	[0.2 0.2]
targetSelect	
μ^t	100
ω_o	0.9677 (uniform dispersion)
ω_o	0.7692 (random dispersion)

5.1.2. Benchmark 2

The benchmark 2 (see Equation (21)) is an adaptation of the modified Rastrigin benchmark [51–54]. Figure 4 show the global and local Pareto set H of benchmark 2. This benchmark has a discontinuous Pareto front made up of solutions in different neighborhoods. In addition, it also provides nearly optimal solutions significantly different from the optimal solutions (in different neighborhoods).

$$\min_x f(x) = [f_1(x)\ f_2(x)] \quad (21)$$

$$f_1(x) = -(\sum_{i=1}^{2} 10 + 9cos(2\pi \cdot k_i \cdot x_i))(1 - \sqrt{(x_1 - 0.65)^2 + (x_2 - 0.5)^2})$$

$$f_2(x) = min((x_1 - 0.65) + (x_2 - 0.5), (x_1 - 0.65) + (x_2 - 0.25), (x_1 - 0.65) + (x_2 - 0.75), (x_1 - 0.35) + (x_2 - 0.5), (x_1 - 0.35) + (x_2 - 0.75), (x_1 - 0.35) + (x_2 - 0.25)) \quad (22)$$

where $k_1 = 2$ and $k_2 = 3$, and subject to:

$$0 \leq x_1 \leq 4$$
$$0 \leq x_2 \leq 4 \quad (23)$$

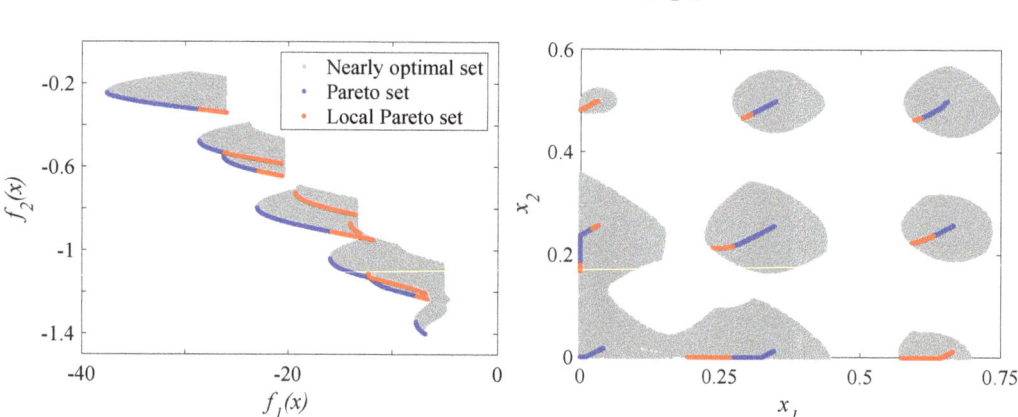

Figure 4. Target set H for the benchmark 2 formed by the global (blue) and local (red) Pareto set.

To analyze the archivers on the benchmark 2, we define the parameters of Table 2. The parameters ϵ, Δ_x and Δ_y are defined based on prior knowledge of the problem. Following the same procedure as the previous benchmark, $\mu^t = 150$ (in this case $\mu^t = \{200, 150, 100\}$ has been analyzed) and $\omega_o = 0.9677$ is defined as the parameters that obtains the best performance, with respect to Δ_p, for the uniform dispersion. For random dispersion, $\omega_o = 0.9091$ obtains the best performance.

Table 2. Parameters for benchmark 2.

Parameters	Value
ϵ	[2.5 0.15]
ArchiveUpdateP$_{Q,\epsilon}$D$_{xy}$	
Δ_x	0.15
Δ_y	0.25
Archive_nevMOGA	
Δ_x	[0.15 0.15]
Δ_y	[0.25 0.25]
targetSelect	
μ^t	150
ω_o	0.9677 (uniform dispersion)
ω_o	0.9091 (random dispersion)

5.1.3. Identification of a Thermal System

Finally, a MOOP is defined to solve a real engineering problem: identification of a thermal system. In this problem, the energy contribution inside the process is due to the power dissipated by the resistance inside it (see Figure 5). Air circulation inside the process is produced by a fan, which constantly introduces air from outside. The actuator is made up of a voltage source which is controlled by voltage. The actuator input range is [0 100] % ([0 7.5] V). Two thermocouples are used to measure the resistance temperature and the air temperature in the range [−50 250] °C. Figure 6 shows the signals that will be used in the identification process. The ambient temperature T_a is considered constant and equal to 17 °C for the entire identification test.

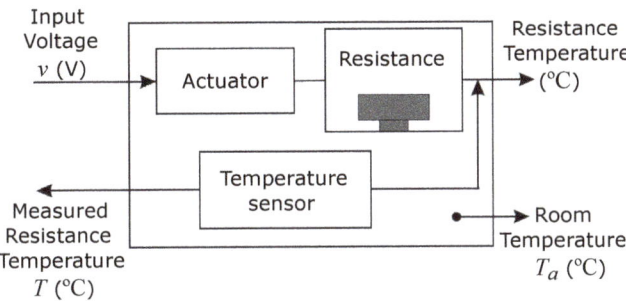

Figure 5. Block diagram of the thermal system.

Taking into account the physical laws of thermodynamics [55], the initial structure of the model can be defined using the following differential equations, where heat losses due to convection and conduction are modeled, as well as losses due to radiation:

$$\dot{T}(t) = \frac{1}{1000}\left(x_1 v(t)^2 - x_2(T(t) - T_a(t)) - x_3\left(\frac{273.0 + T(t)}{100}\right)^4\right) \quad (24)$$

where $T(t)$ is the process output temperature and state variable in °C, $v(t)$ is the input voltage to the process in volts, $T_a(t)$ is the ambient temperature in °C and $\mathbf{x} = [x_1\ x_2\ x_3]$ are the parameters of the model to estimate.

The MOOP is defined as follows:

$$\min_x f(x) = [f_1(x)\ f_2(x)] \quad (25)$$

subject to:

$$\underline{x} \leq x \leq \overline{x}$$

where:

$$f_1 = \frac{1}{\tau}\int_0^\tau |\hat{T} - T|dt \quad (26)$$

$$f_2 = \max_{i \in 1...\tau} |\hat{T}_i - T_i| \quad (27)$$

$\tau = 2500$ is the duration of the identification test, variables with circumflex accent are process outputs (experimental data), variables without circumflex accent are the model outputs, x the parameter vector:

$$x = [x_1\ x_2\ x_3] \quad (28)$$

and $\underline{x}$ and $\overline{x}$ (see Table 3) the lower and upper limits of the parameter vector x which define the decision space Q.

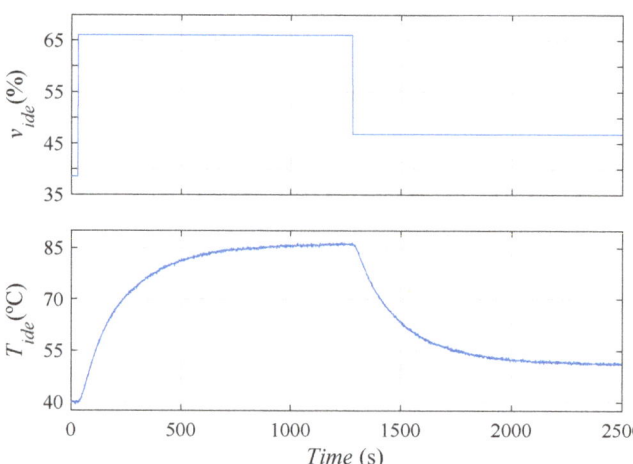

Figure 6. Identification test of the thermal system.

Table 3. Lower ($\underline{x}$) and upper ($\overline{x}$) limits of the parameters x.

Limits	θ_1	θ_2	θ_3
$\underline{x}$	0.01	2	0
$\overline{x}$	0.15	10	0.8

In this MOOP, the design objectives measure the mean and maximum error between the temperatures of the process outlet and the model. The parameters to be estimated and the design objectives have a physical meaning. This fact makes it easier to choose the ϵ, Δ_x and Δ_y (see Table 4). The ϵ parameter (maximum acceptable degradation) is the same for all archivers. Δ_y is similar for $ArchiveUpdateP_{Q,\epsilon}D_{xy}$ and $Archive\_nevMOGA$, taking into account the difference in vector size. However, Δ_x is different for $ArchiveUpdateP_{Q,\epsilon}D_{xy}$ and $Archive\_nevMOGA$. For $ArchiveUpdateP_{Q,\epsilon}D_{xy}$, $\Delta_x = 0.1$. A lower value increases significantly higher number of solutions. A higher value gives a poor approximation to the set of optimal and nearly optimal solutions. For $Archive\_nevMOGA$, $\Delta_x = [0.0015\ 0.2\ 0.1]$. In this case, Δ_x is independent for each dimension in the decision space, being different for each parameter due to its different limits (see Table 3). For $targetSelect$, $\mu^t = 78$ solutions to obtain the same number of solutions as $Archive\_nevMOGA$. In this way, both archivers will have equal conditions. Additionally, $\omega_o = 0.7692$ is defined to give greater weight to diversity in the decision space.

Table 4. Parameters for identification of thermal system.

Parameters	Value
ϵ	[0.25 0.25]
$ArchiveUpdateP_{Q,\epsilon}D_{xy}$	
Δ_x	0.1
Δ_y	0.025
$Archive\_nevMOGA$	
Δ_x	[0.0015 0.2 0.1]
Δ_y	[0.025 0.025]
$targetSelect$	
μ^t	78
ω_o	0.7692

5.2. Archivers Comparison Procedure

The procedure performed to carry out the archiver comparison is different on the benchmarks and on the real example. Benchmarks have a very low computational cost for the objective function. Therefore, it is possible to evaluate large populations that discretize the entire search space. These populations are entered in the archiver as input population. To analyze the behavior of the archivers on different types of populations, a uniform and random distribution is used to obtain these initial populations.

Because the computational cost of the objective functions of the engineering problem are significantly higher, it is not feasible to test each archiver with random or exhaustive searches as has been done with the benchmarks. Thus, the archiver has been embedded in a basic evolutionary algorithm. In addition to the reduction of computational cost, this enables observing the impact of each archiver when incorporated into an evolutionary mechanism.

5.2.1. Benchmarks

For the comparison of the archivers on the two benchmarks presented, the archivers will be fed in two ways: (1) by a uniform distribution; (2) by a random distribution of solutions in the search space. The comparison of the results will be made using the averaged Hausdorff distance Δ_p [34] (see Equation (4)). This metric measures the averaged Hausdorff distance between the outcome set (or final archive) $f(A)$ and the target set $f(H)$ (local Pareto set in this paper) of a given MOOP. Since Δ_p considers the averaged distances between the entries of $f(A)$ and $f(H)$, this indicator is very insensitive to outliers. This metric measures the diversity and convergence towards the target set. Furthermore, this metric can be applied both in the decision and objective space.

To carry out the comparison of the archiving strategies with data of a uniform dispersion, each archiver is fed with a file of 100,000 solutions uniformly distributed throughout the domain Q (generating a hypergrid). These solutions are introduced in a random order. This process is repeated with 25 different files, where each file slightly displaces the generated hypergrid vertically and/or horizontally on the search space.

To carry out the analysis with data of a random dispersion, each archiver is fed with a file of 100,000 solutions randomly distributed throughout the domain Q. This process is repeated with 25 different files, avoiding as far as possible the random component.

5.2.2. Identification of Thermal System

In this example, we are going to analyze the different archivers on a multi-objective generic optimization algorithm [23,34] (see Algorithm 2). This algorithm generates the initial population randomly with $Nind_{P_0}$ individuals, obtaining the initial archive A_0 (through the archiver). Subsequently, in each iteration, new solutions are generated and the archive A_t is updated (using the selected archiver). The $Generate()$ function generates new individuals in each iteration. To do this, two solutions are randomly selected from the current file A_{t-1}. A random number $u \in [0\ 1]$ is generated. If $u > Pcm$ (probability of crossing/mutation) a crossing is made. The crossover generates two new solutions using the simulated binary crossover (SBX [56]) technique. If $u \leq Pcm$ a mutation is performed. The mutation generates two solutions through the polynomial mutation [43]. In this way, the three archivers are compared using the same evolutionary strategy. For this example, an initial population of 500 individuals ($Nind_{P_0} = 500$), a probability of crossing/mutation of 0.2 ($Pcm = 0.2$), and 5000 generations are used. Therefore, 10,500 solutions are evaluated (500 + 2 × 5000) for each archiver.

Algorithm 2 Generic multi-objective optimization algorithm

1: $P_0 \subset Q$ ▷ Random selection
2: $A_0 := Archiver(P_0, \emptyset)$
3: **for** $t := 1$:Number of iterations **do**
4:     $P_t := Generate(A_{t-1})$
5:     $A_t := Archiver(P_t, A_{t-1})$
6: **end for**

6. Results and Discussion

This section shows the results obtained on the two benchmarks and the real example previously introduced.

6.1. Benchmark 1 with Uniform Dispersion

The archivers are tested on 25 different input populations obtained by uniform dispersion. Figure 7 shows the median results of archive A on the benchmark 1 for both decision and objective spaces with respect to $\Delta_p(A, H)$ in decision space. As can be seen, the archivers make a good approximation to the target set H, characterizing the nine neighborhoods that compose it. However, there are differences between the sets found by the archivers. First, the *Archive_nevMOGA* archiver obtains fewer solutions. The number of solutions $\mu^t = 100$ for the *targetSelect* is user-defined, but a smaller size makes Δ_p worse for both spaces. The archive A obtained by $ArchiveUpdateP_{Q,\epsilon}D_{xy}$ obtains a larger number of solutions.

The *targetSelect* archiver obtains a better approximation to the Pareto front than the other archivers. This is because the weight ω_o has a high value, giving greater weight to the D_o indicator that measures convergence and diversity for the Pareto front. The $ArchiveUpdateP_{Q,\epsilon}D_{xy}$ and *Archive_nevMOGA* archivers do not select a candidate solution p (even if p belongs to the Pareto set) if an alternative already exists in the current archive that is similar in both spaces (in *Archive_nevMOGA*, p is selected if it dominates the similar solution). These archivers could obtain a better approximation to the Pareto front by reducing the parameter Δ_y (parameter with which the degree of similarity in the objective space is decided), but it probably also implies obtaining a greater number of solutions.

Regarding the local Pareto set, the archive *Archive_nevMOGA* obtains a better approximation in the comparison. *ArchiveUpdate$P_{Q,\epsilon}D_{xy}$* and *targetSelect* obtain solutions in all neighborhoods where nearly optimal solutions exist. However, these solutions are rarely located on the lines that define the local Pareto set.

Notice that *ArchiveUpdate$P_{Q,\epsilon}D_{xy}$* and *Archive_nevMOGA* archivers obtain $\epsilon-$ dominated solutions. In *ArchiveUpdate$P_{Q,\epsilon}D_{xy}$*, it is possible that a solution in A that is no longer nearly optimal due to the apparition of a new candidate solution p. p may not be removed because it does not satisfy condition in the line 7 of Algorithm 1. In *Archive_nevMOGA*, a new candidate solution p can be added to the archive A through the condition of line 8 of Algorithm 3. In some cases, solutions that are not nearly optimal due to the appearance of p (by line 8 of Algorithm 3) are not eliminated. Therefore, the archive A obtained by both archivers may contain solutions that do not belong to the nearly optimal set.

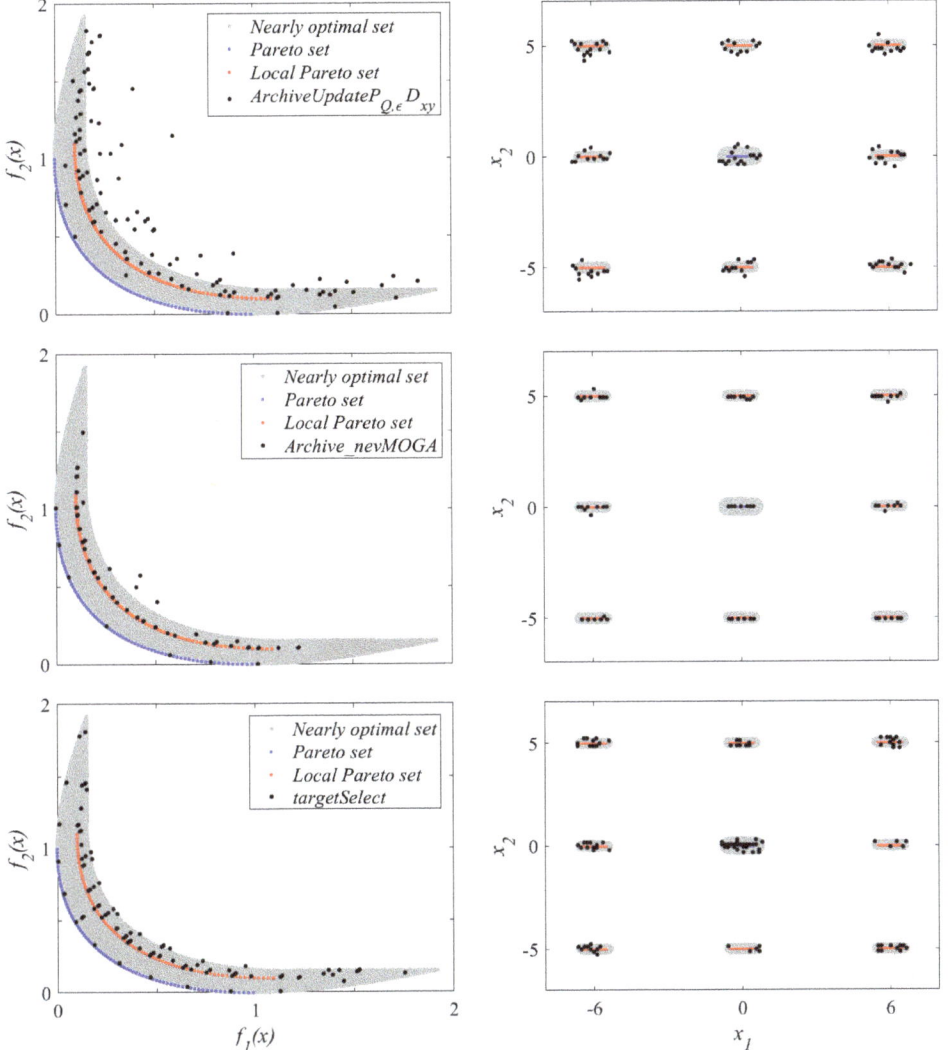

Figure 7. Median result of archive A obtained by *ArchiveUpdate$P_{Q,\epsilon}D_{xy}$*, *Archive_nevMOGA* and *targetSelect* with a uniform dispersion on benchmark 1.

Figure 8 shows the boxplot of the indicators $\Delta_p(f(A), f(H))$, $\Delta_p(A, H)$ and archive size for the 25 tests performed. *Archive_nevMOGA* achieves a better approximation in the decision and objective spaces, and obtains fewer solutions. *targetSelect* also obtains a better approximation in both spaces with fewer solutions than $ArchiveUpdateP_{Q,\epsilon} D_{xy}$. $ArchiveUpdateP_{Q,\epsilon} D_{xy}$ obtains greater variability among the 25 archives obtained. Therefore, *Archive_nevMOGA* has achieved, in a better way, the two main objectives: not neglecting the diversity of solutions (locates all nine neighborhoods) and obtains a reduced number of solutions (simplifying the optimization and decision stages).

Algorithm 3 $A := Archive\_nevMOGA(P, A_0, \epsilon, \Delta_x, \Delta_y)$

Require: population P, archive A_0, $\epsilon \in \mathbb{R}_+^m$, $\Delta_x \in \mathbb{R}_+^k$, $\Delta_y \in \mathbb{R}_+^m$
Ensure: updated archive A
1: $A := A_0$
2: **for all** $p \in P$ **do**
3:     **if** ($\nexists a_1 \in A : a_1 \prec_{-\epsilon} p$) and ($\nexists a_2 \in A : |a_2 - p| \leq \Delta_x$ and $a_2 \prec p$) and ($\nexists a_3 : |a_3 - p| \leq \Delta_x$ and $|f(a_3) - f(p)| \leq \Delta_y$) **then**
4:         $A \leftarrow A \cup p$
5:         **if** $\exists a_4 \in A : p \prec_{-\epsilon} a_4$ or $|a_4 - p| \leq \Delta_x$ and $p \prec a_4$ **then**
6:             $A \leftarrow A \setminus a_4$
7:         **end if**
8:     **else if** $\exists a_5 : |a_5 - p| \leq \Delta_x$ and $|f(a_5) - f(p)| \leq \Delta_y$ and $p \prec a_5$ **then**
9:         $a_5 := p$
10:     **end if**
11: **end for**

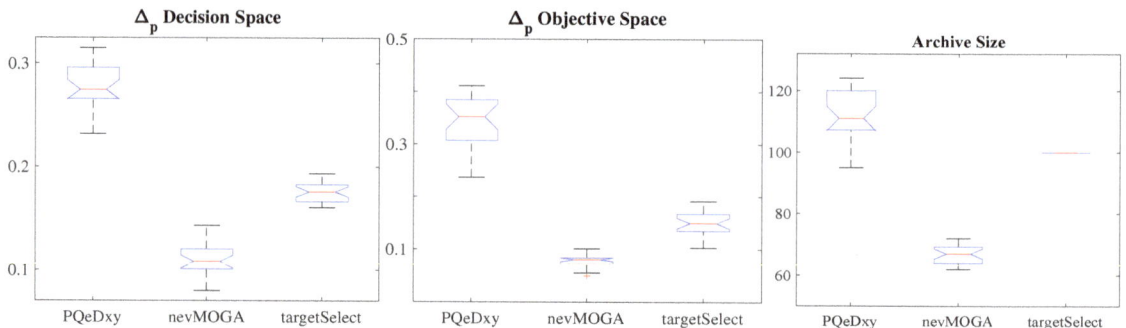

Figure 8. Boxplot of $\Delta_p(A, H)$ (decision space), $\Delta_p(f(A), f(H))$ (objective space) and archive size with a uniform dispersion on the benchmark 1.

6.2. Benchmark 1 with Random Dispersion

The archivers are tested on 25 different input populations obtained by random dispersion. Figure 9 shows the archive A, with median result for $\Delta_p(A, H)$. The archivers characterize the nine neighborhoods that form the target set H. The archive A obtained by *Archive_nevMOGA* has a smaller number of solutions. Decreasing the number of solutions $\mu^t \leq 100$ for the archive *targetSelect* makes Δ_p worse in both spaces. Comparing Figures 7 and 9, *targetSelect* in random dispersion produces a worse approximation of the Pareto front than in uniform search. This is for two reasons: (1) the lower value of the weight $\omega_o = 0.7692$ (lower weight of the metric D_o, which measures convergence in Pareto front); (2) the initial population has been obtained in a random way (meaning certain areas have not been adequately explored). Figure 10 shows the boxplot, of the 25 archives obtained in the tests. *Archive_nevMOGA* obtains better results in both spaces, also obtaining a smaller number of solutions. On the benchmark 1, the approximations obtained by the archivers in a random dispersion are slightly worse than in a uniform dispersion.

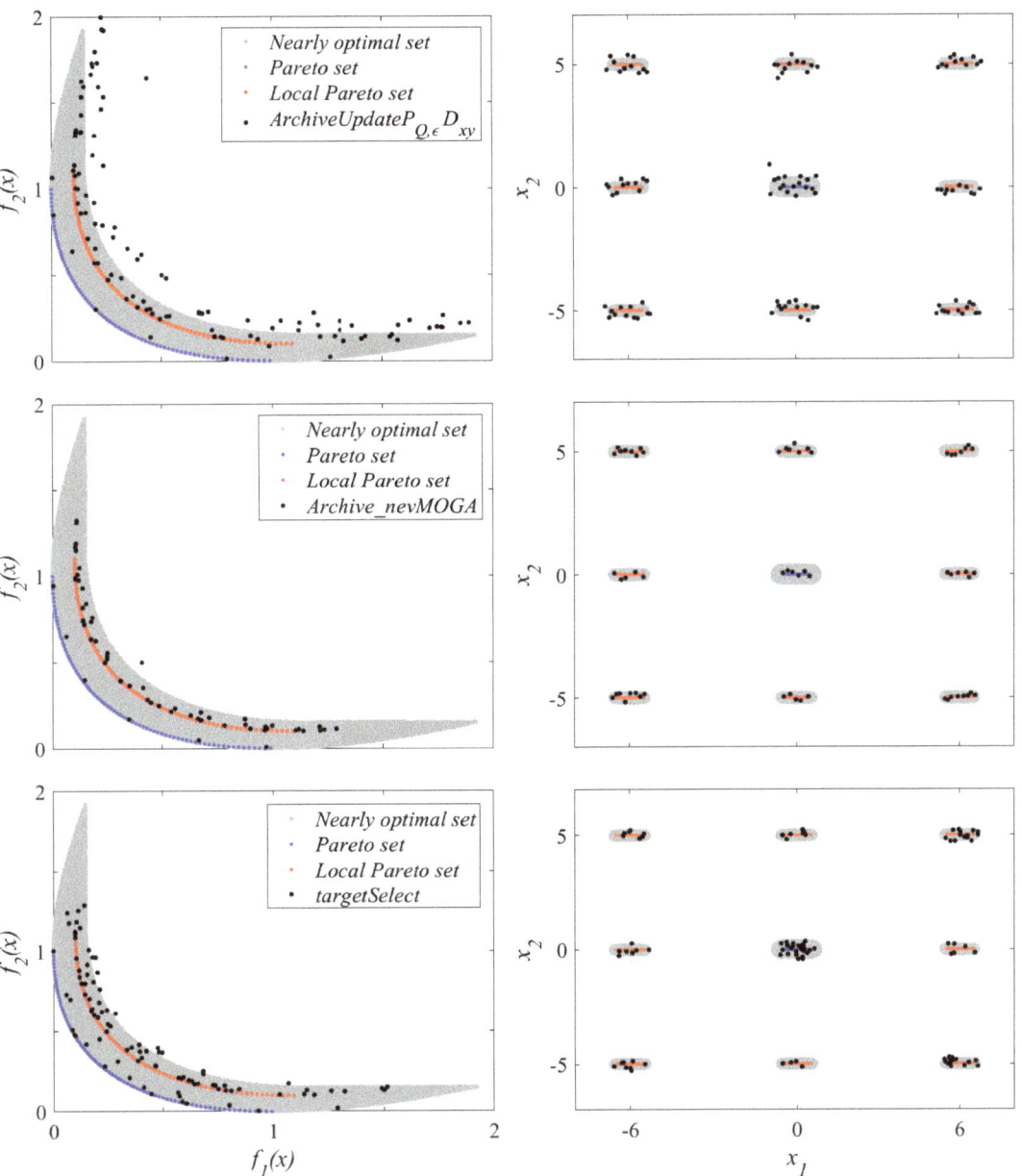

Figure 9. Median result of Archive A obtained by $ArchiveUpdateP_{Q,\epsilon}D_{xy}$, $Archive\_nevMOGA$ and *targetSelect* with a random dispersion on benchmark 1.

6.3. Benchmark 2 with Uniform Dispersion

The archive A, with the median result (in the decision space), for each archiver is shown in Figure 11. The archivers locate the neighborhoods where nearly optimal solutions are found. The archive of $Archive\_nevMOGA$ again obtains fewer solutions. Keep in mind that decreasing $\mu^t = 150$ for $targetSelect$ causes a considerable increase in the variability of the results obtained $\Delta_p(A, H)$ for the 25 tests. $targetSelect$ obtains a better approximation to the Pareto front due to the high value of the weight ω_o. However, $Archive\_nevMOGA$ obtains solutions closer to the local Pareto set. This is because $targetSelect$ seeks to achieve the greatest diversity in the decision space (through D_d) without taking into account whether these solutions are worse than a close solution (if they are not optimal).

Figure 12 shows the boxplot, of the 25 archives obtained for the archivers, for the indicator Δ_p in the decision and objective spaces and the archive size. Regarding the decision space, $Archive\_nevMOGA$ obtains a better approximation to the target set than its competitors. Regarding the objective space, $Archive\_nevMOGA$ and $targetSelect$ obtain a similar minimum value. However, $targetSelect$ obtains worse variability. Therefore, as occurred with benchmark 1, the archiver $Archive\_nevMOGA$ achieves a better approximation in both spaces and obtains a smaller number of solutions.

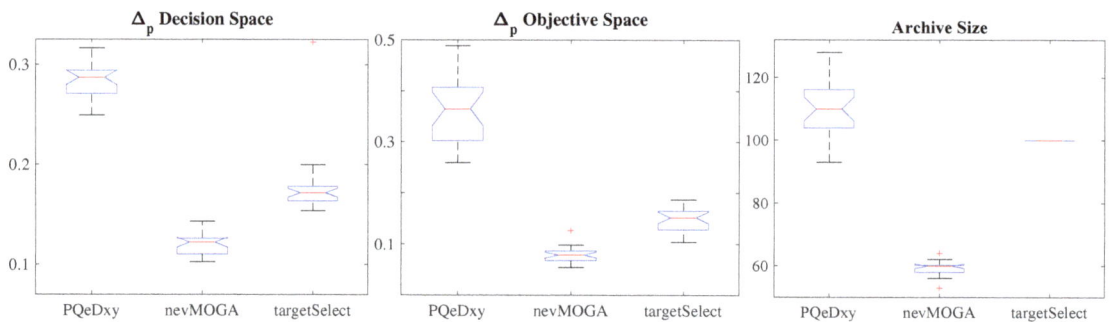

Figure 10. Boxplot of $\Delta_p(A, H)$ (decision space), $\Delta_p(f(A), f(H))$ (objective space) and archive size with a random dispersion on the benchmark 1.

6.4. Benchmark 2 with Random Dispersion

Figure 13 shows the archive A, which obtains a median result for $\Delta_p(A, H)$ for the archivers. Again, the archiver $Archive\_nevMOGA$ obtains significantly fewer solutions while also obtaining a better characterization of the target set H. Figure 14 shows the boxplot of the archivers. The archiver $Archive\_nevMOGA$ obtains a better value of Δ_p in both spaces. Regarding the objective space, $targetSelect$ obtains results similar to $Archive\_nevMOGA$ but worse variability. The archiver $Archive\_nevMOGA$ obtains fewer solutions, which simplifies the optimization and decision stages. Using the random search, $ArchiveUpdateP_{Q,\epsilon}D_{xy}$ and $Archive\_nevMOGA$ perform slightly worse than the uniform search. $targetSelect$ obtains slightly better results than the uniform search.

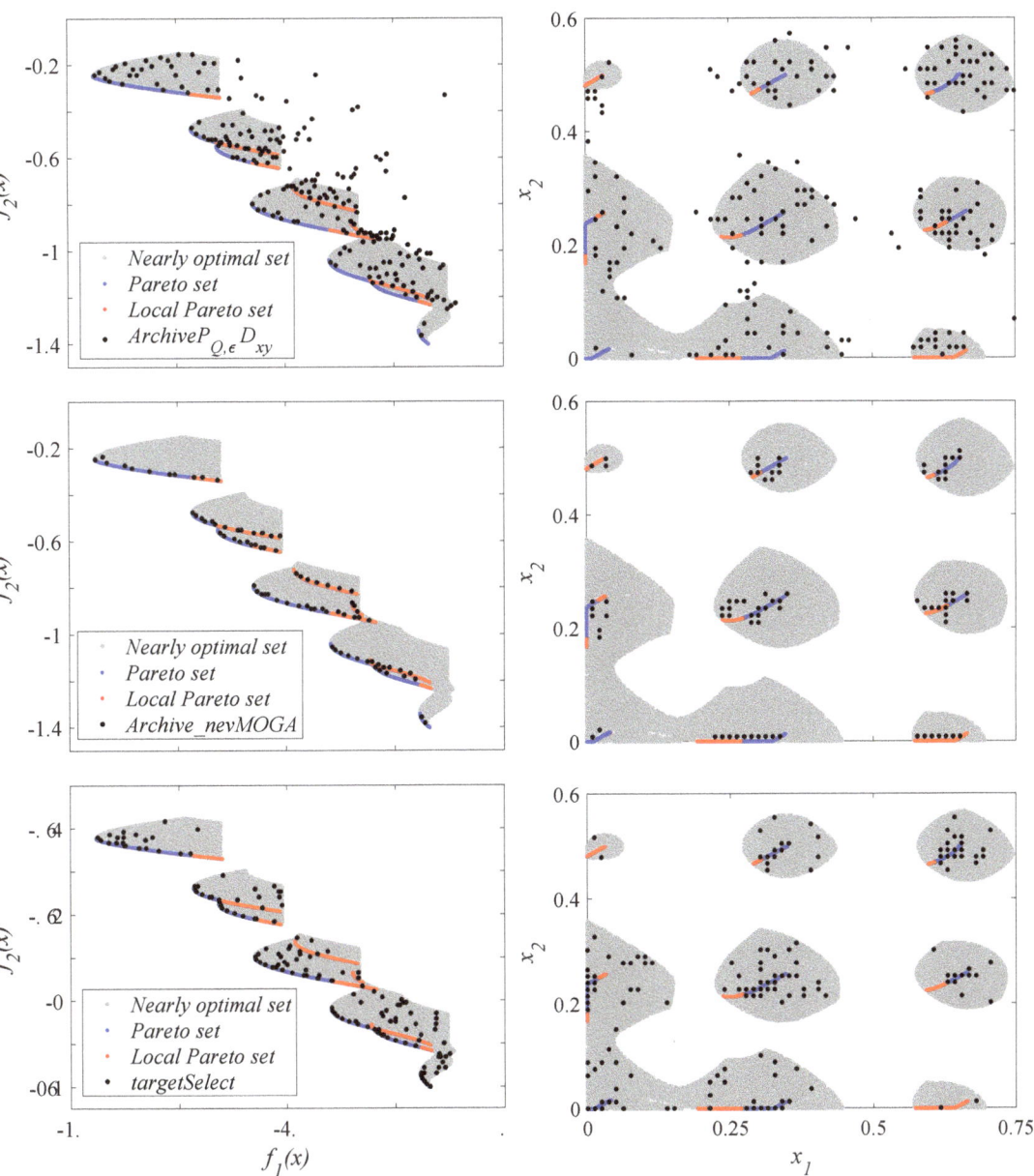

Figure 11. Median result of archive A obtained by $ArchiveUpdateP_{Q,\epsilon}D_{xy}$, $Archive\_nevMOGA$ and $targetSelect$ with a uniform dispersion on the benchmark 2.

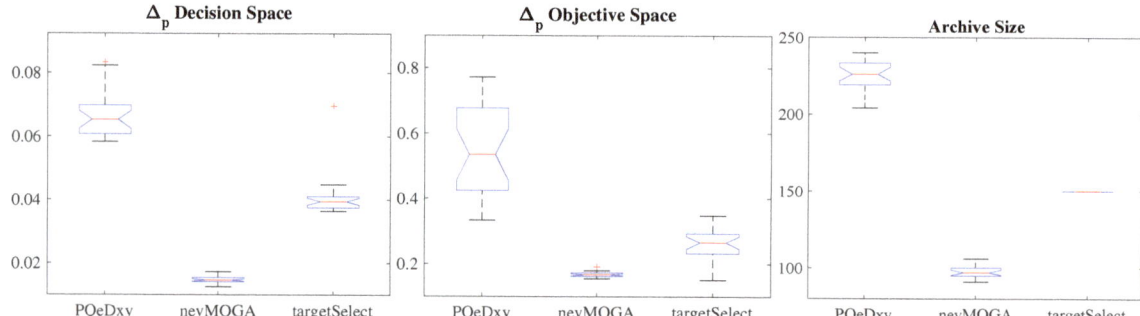

Figure 12. Boxplot of $\Delta_p(A, H)$ (decision space), $\Delta_p(f(A), f(H))$ (objective space) and archive size with a uniform dispersion on the benchmark 2.

6.5. Identification of a Thermal System

Figure 15 shows the final archive A obtained when using the three archivers inside a compared basic optimization algorithm. In this example, the obtained archives are compared by pairs to better observe the differences between them.

The three archivers obtain diversity in the decision space, and convergence in the Pareto front. The first thing that stands out is the large number of solutions (610 solutions) obtained by the archive $ArchiveUpdateP_{Q,\epsilon}D_{xy}$. This high number of solutions complicates the optimization and decision phases. For each iteration, the newly generated solutions must be compared with the solutions in the current file $A(t)$. Therefore, many solutions in the file $A(t)$ implies a higher computational cost. In addition, a high number of solutions makes the final decision of the DM more difficult. This large number of solutions can be reduced by increasing the parameters Δ_x and Δ_y. However, this increase also implies a worse discretization in both spaces. $ArchiveUpdateP_{Q,\epsilon}D_{xy}$ obtains a worse approximation to the Pareto front.

The results show more similarities with respect to the other two archivers. Archiver $Archive\_nevMOGA$ obtains 78 solutions. To compare under similar conditions, we set the size of the file obtained by $targetSelect$ to 78 solutions ($\mu^t = 78$). In this way, both archivers obtain the same number of solutions. The set of solutions found by both archivers are different. With respect to the Pareto front, both archivers achieve a good approximation in the range $f_1(x) = [0.3 \ 0.8]$. However, $Archive\_nevMOGA$ does not get solutions in the range $f_1(x) = [0.8 \ 0.9]$ of the Pareto front. Therefore, in this example, $targetselect$ gets a little more diversity in the Pareto front.

$targetSelect$ focuses on obtaining, in addition to a good convergence in the Pareto front, the greatest diversity in the decision space (using D_d, see Section 4.3). However, solutions that provide greater diversity may be worse than neighboring solutions. For example, Figure 15 shows the solution x^1 obtained by $Archive\_nevMOGA$. This solution has similar parameters to the $neighborhood_1$ solutions (see decision space). x^1 performs better ($f(x^1)$) than all the $neighborhood_1$ solutions obtained by $targetSelect$. Therefore, $Archive\_nevMOGA$ would eliminate all these solutions (dominated in its neighborhood by x^1). $targetSelect$ maintains them because they increase the diversity in the decision space. This happens repeatedly in this MOOP. For this reason, $targetSelect$ obtains nearly optimal solutions farther from the Pareto front than obtained by $Archive\_nevMOGA$. These solutions are in the contour/ends of the plane that form the optimal and nearly optimal solutions in the decision space, and therefore, they obtain a better diversity under the D_d indicator. $Archive\_nevMOGA$ could find solutions closer to the contour/ends of the plane formed in the decision space (as is the case with $targetSelect$) by reducing the parameter Δ_x, although this would imply obtaining a larger number of solutions. Therefore, depending on the needs or preferences of the DM, the use of one archiver or another may be more appropriate. This archiver can be embedded in most of the multi-objective algorithms available. *ArchiveUpdate$P_{Q,\epsilon}D_{xy}$*, *Archive_nevMOGA* and *targetSelect* archivers are currently built into the algorithms NϵSGA, *nevMOGA* and *DIOP* respectively.

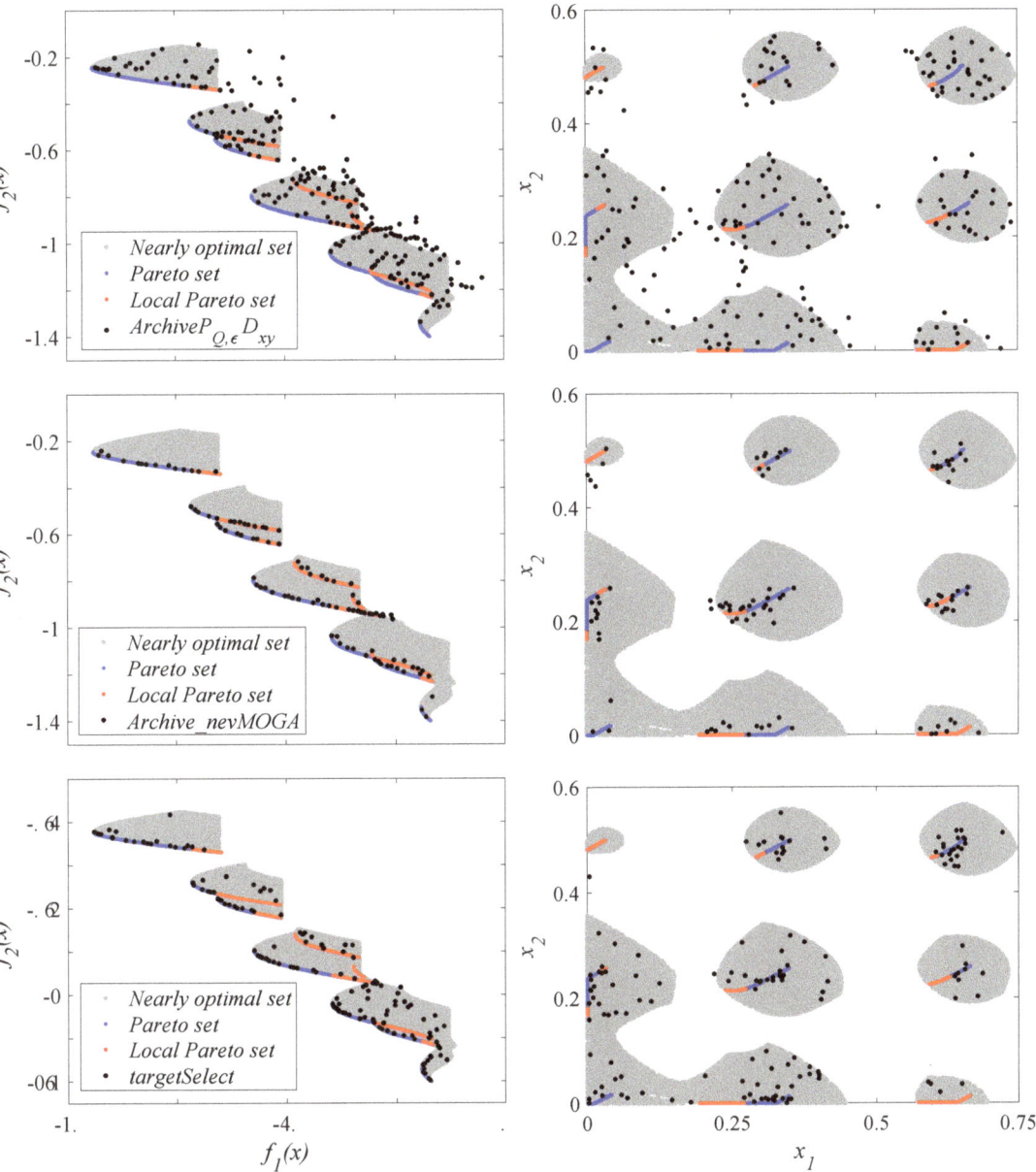

Figure 13. Median result of Archive A obtained by *ArchiveUpdate$P_{Q,\epsilon}D_{xy}$*, *Archive_nevMOGA* and *targetSelect* with a random dispersion on benchmark 2.

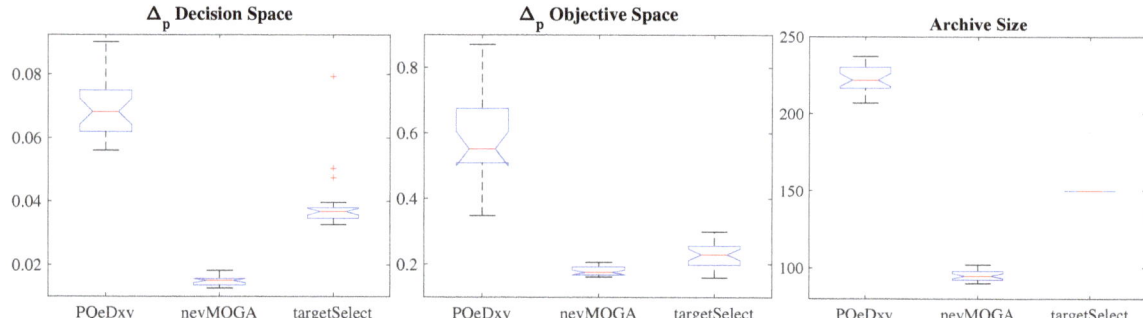

Figure 14. Boxplot of $\Delta_p(A, H)$ (decision space), $\Delta_p(f(A), f(H))$ (objective space) and archive size with a random dispersion on the benchmark 2.

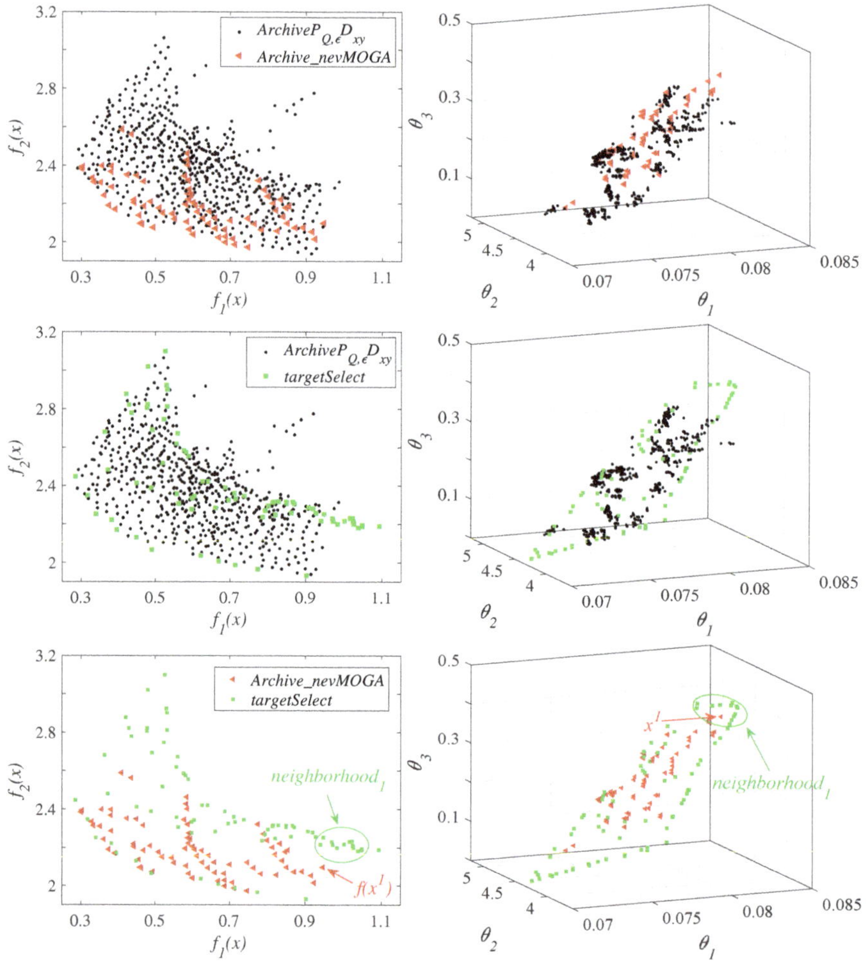

Figure 15. Archive A obtained with a generic algorithm and the archivers $ArchiveUpdateP_{Q,\epsilon}D_{xy}$, $Archive\_nevMOGA$ and $targetSelect$ for the identification of the thermal system.

7. Conclusions

In this paper, the characterization of nearly optimal solutions potentially useful in a MOOP has been addressed. In this type of problem, in practice, the DM may wish to obtain nearly optimal solutions, since they can play a relevant role in the decision-making stage. However, an adequate approximation to this set is necessary to avoid an excessive number of alternatives that could hinder the optimization and decision-making stages. Not all nearly optimal solutions provide the same useful information to the DM. To reduce the number of solutions to be analyzed, we consider potentially useful solutions (in addition to the optimals) that are close to the optimals in objective space—but which differ significantly in the decision space. To adequately characterize this set, it is necessary to discretize the nearly optimal solutions by analyzing the decision and objective spaces simultaneously.

This article compares different archiving strategies that perform this task: $ArchiveUpdateP_{Q,\epsilon}D_{xy}$, $Archive\_nevMOGA$ and $targetSelect$. The main objective of the archivers is to obtain potentially useful solutions. This analysis is of great help to designers of evolutionary algorithms who wish to obtain such solutions. In this way, designers will have more information to choose their archivers based on their preferences. $ArchiveUpdateP_{Q,\epsilon}D_{xy}$ and $Archive\_nevMOGA$ are two distance-based archivers. Both archivers simultaneously discard solutions that are similar in decision and objective spaces. However, $Archive\_nevMOGA$, in contrast to $ArchiveUpdateP_{Q,\epsilon}D_{xy}$, discards solutions dominated by a neighboring solution in the decision space. $targetSelect$ is an archive based on an indicator that measures the diversity in both spaces simultaneously. $targetSelect$, unlike the other archivers, can directly and arbitrarily specify the archive size. This can be an advantage. The archivers are evaluated using two benchmarks. They obtain a good approximation to the set of potentially useful solutions, characterizing the diversity existing in the set of nearly optimal solutions. As discussed in [19], the $ArchiveUpdateP_{Q,\epsilon}D_{xy}$ archiver is more practical than other archivers in the literature. However, this archiver, as demonstrated in this paper, obtains significantly more solutions than its competitors in this paper. This can make the optimization and decision phase more difficult. In addition, $Archive\_nevMOGA$ obtains a better approximation to the target set H under the averaged Hausdorff distance Δ_p. In addition, $Archive\_nevMOGA$ obtains a smaller number of solutions, which speeds up the optimization process and facilitates the decision-making stage. However, fewer solutions can also decrease diversity (which can lead to degraded global search capabilities).

Finally, the compared archivers are analyzed on a real engineering example. This real example is the identification of a thermal system. To carry out this analysis, a generic multi-objective optimization algorithm is used, in which it is possible to select different archivers. This enables a more realistic comparison of the impact of the archivers on the entire optimization process.

The three archivers obtain the existing diversity in the set of optimal and nearly optimal solutions. In this last example, we can see how the archive obtained by $ArchiveUpdateP_{Q,\epsilon}D_{xy}$ obtains a very high number of solutions, complicating the optimization and decision stages. $Archive\_nevMOGA$ and $targetSelect$ obtain the same number of solutions. Both archivers obtain an adequate Pareto front. However, $targetSelect$ gets more diversity on the Pareto front. The main difference between the two archivers is in the set of nearly optimal solutions. On the one hand, $Archive\_nevMOGA$ obtains solutions closer to the Pareto front, but significantly different in the decision space. On the other hand, $targetSelect$ obtains the solutions that provide the greatest diversity in the decision space, even though these solutions are farther away from the Pareto front, and therefore, offer significantly worse performance.

Finally, this analysis suggests two possible future lines of research: (1) design of new evolutionary algorithms, which characterize the nearly optimal solutions, using some of the archivers compared in this work and (2) design of new archivers that improve the current ones. For example, the clustering techniques could improve the archivers compared in this work. These techniques, not analyzed in this work, allow the location of new neighborhoods, allowing their exploration and evaluation. In this way, the optimization process could be improved.

Author Contributions: Conceptualization, A.P.; Methodology, A.P., X.B. and J.M.H.; Software developed, A.P.; Validation, A.P.; Writing—original draft preparation, A.P.; Writing—review and editing, X.B., J.M.H. and M.A.M.; Supervision, M.A.M.; Funding acquisition, X.B. and J.M.H. All authors have read and agreed to the published version of the manuscript.

Funding: This work was supported in part by the Ministerio de Ciencia, Innovación y Universidades (Spain) (grant number RTI2018-096904-B-I00), by the Generalitat Valenciana regional government through project AICO/2019/055 and by the Universitat Politècnica de València (grant number SP20200109).

Institutional Review Board Statement: Not applicable.

Informed Consent Statement: Not applicable.

Data Availability Statement: Not applicable.

Conflicts of Interest: The authors declare no conflict of interest.

References

1. Rhinehart, R.R. *Engineering Optimization: Applications, Methods and Analysis*; John Wiley & Sons: Hoboken, NJ, USA, 2018.
2. Rao, S.S. *Engineering Optimization: Theory and Practice*; John Wiley & Sons: Hoboken, NJ, USA, 2019.
3. Dai, C.; Lei, X. A multiobjective brain storm optimization algorithm based on decomposition. *Complexity* **2019**, *2019*. [CrossRef]
4. Miettinen, K. *Nonlinear Multiobjective Optimization*; Springer Science & Business Media: Berlin/Heidelberg, Germany, 2012; Volume 12.
5. Deb, K. *Multi-Objective Optimization Using Evolutionary Algorithms*; John Wiley & Sons: Hoboken, NJ, USA, 2001; Volume 16.
6. Reynoso-Meza, G.; Blasco, X.; Sanchis, J.; Herrero, J.M. *Controller Tuning with Evolutionary Multiobjective Optimization: A Holistic Multiobjective Optimization Design Procedure*; Springer: Berlin/Heidelberg, Germany, 2017.
7. Gunantara, N. A review of multi-objective optimization: Methods and its applications. *Cogent Eng.* **2018**, *5*, 1502242. [CrossRef]
8. Sanchis, J.; Martínez, M.A.; Blasco, X. Integrated multiobjective optimization and a priori preferences using genetic algorithms. *Inf. Sci.* **2008**, *178*, 931–951. [CrossRef]
9. Zitzler, E.; Deb, K.; Thiele, L. Comparison of multiobjective evolutionary algorithms: Empirical results. *Evol. Comput.* **2000**, *8*, 173–195. [CrossRef]
10. Ulrich, T.; Bader, J.; Thiele, L. Defining and optimizing indicator-based diversity measures in multiobjective search. In *International Conference on Parallel Problem Solving from Nature*; Springer: Berlin/Heidelberg, Germany, 2010; pp. 707–717.
11. Beyer, H.G.; Sendhoff, B. Robust optimization—A comprehensive survey. *Comput. Methods Appl. Mech. Eng.* **2007**, *196*, 3190–3218. [CrossRef]
12. Jin, Y.; Branke, J. Evolutionary optimization in uncertain environments—A survey. *IEEE Trans. Evol. Comput.* **2005**, *9*, 303–317. [CrossRef]
13. Deb, K.; Gupta, H. Introducing robustness in multi-objective optimization. *Evol. Comput.* **2006**, *14*, 463–494. [CrossRef]
14. Pajares, A.; Blasco, X.; Herrero, J.M.; Reynoso-Meza, G. A new point of view in multivariable controller tuning under multiobjetive optimization by considering nearly optimal solutions. *IEEE Access* **2019**, *7*, 66435–66452. [CrossRef]
15. Loridan, P. ε-solutions in vector minimization problems. *J. Optim. Theory Appl.* **1984**, *43*, 265–276. [CrossRef]
16. White, D.J. Epsilon efficiency. *J. Optim. Theory Appl.* **1986**, *49*, 319–337. [CrossRef]
17. Engau, A.; Wiecek, M.M. Generating ε-efficient solutions in multiobjective programming. *Eur. J. Oper. Res.* **2007**, *177*, 1566–1579. [CrossRef]
18. Hernández Castellanos, C.I.; Schütze, O.; Sun, J.Q.; Ober-Blöbaum, S. Non-Epsilon Dominated Evolutionary Algorithm for the Set of Approximate Solutions. *Math. Comput. Appl.* **2020**, *25*, 3. [CrossRef]
19. Schütze, O.; Hernandez, C.; Talbi, E.G.; Sun, J.Q.; Naranjani, Y.; Xiong, F.R. Archivers for the representation of the set of approximate solutions for MOPs. *J. Heuristics* **2019**, *25*, 71–105. [CrossRef]
20. Tanabe, R.; Ishibuchi, H. A review of evolutionary multimodal multiobjective optimization. *IEEE Trans. Evol. Comput.* **2019**, *24*, 193–200. [CrossRef]
21. Li, M.; Yao, X. An empirical investigation of the optimality and monotonicity properties of multiobjective archiving methods. In *International Conference on Evolutionary Multi-Criterion Optimization*; Springer: Berlin/Heidelberg, Germany, 2019; pp. 15–26.
22. Knowles, J.D.; Corne, D.W. Approximating the nondominated front using the Pareto archived evolution strategy. *Evol. Comput.* **2000**, *8*, 149–172. [CrossRef] [PubMed]
23. Laumanns, M.; Thiele, L.; Deb, K.; Zitzler, E. Combining convergence and diversity in evolutionary multiobjective optimization. *Evol. Comput.* **2002**, *10*, 263–282. [CrossRef]
24. Schütze, O.; Laumanns, M.; Coello, C.A.C.; Dellnitz, M.; Talbi, E.G. Convergence of stochastic search algorithms to finite size Pareto set approximations. *J. Glob. Optim.* **2008**, *41*, 559–577. [CrossRef]

25. Schütze, O.; Lara, A.; Coello, C.A.C.; Vasile, M. Computing approximate solutions of scalar optimization problems and applications in space mission design. In Proceedings of the IEEE Congress on Evolutionary Computation, Barcelona, Spain, 18–23 July 2010; pp. 1–8.
26. Zhou, X.; Long, J.; Xu, C.; Jia, G. An external archive-based constrained state transition algorithm for optimal power dispatch. *Complexity* **2019**, *2019*. [CrossRef]
27. Schütze, O.; Hernández, C. *Archiving Strategies for Evolutionary Multi-Objective Optimization Algorithms*; Springer: Berlin/Heidelberg, Germany, 2021.
28. Schütze, O.; Vasile, M.; Coello, C.A.C. Approximate solutions in space mission design. In *International Conference on Parallel Problem Solving from Nature*; Springer: Berlin/Heidelberg, Germany, 2008; pp. 805–814.
29. Pajares, A.; Blasco, X.; Herrero, J.M.; Reynoso-Meza, G. A multiobjective genetic algorithm for the localization of optimal and nearly optimal solutions which are potentially useful: nevMOGA. *Complexity* **2018**, *2018*. [CrossRef]
30. Schütze, O.; Vasile, M.; Coello, C.A.C. Computing the set of epsilon-efficient solutions in multiobjective space mission design. *J. Aerosp. Comput. Inf. Commun.* **2011**, *8*, 53–70. [CrossRef]
31. Pareto, V. *Manual of Political Economy*; Oxford University Press: Oxford, UK, 1971.
32. Schütze, O.; Coello, C.A.C.; Talbi, E.G. Approximating the ε-efficient set of an MOP with stochastic search algorithms. In *Mexican International Conference on Artificial Intelligence*; Springer: Berlin/Heidelberg, Germany, 2007; pp. 128–138.
33. Teytaud, O. On the hardness of offline multi-objective optimization. *Evol. Comput.* **2007**, *15*, 475–491. [CrossRef] [PubMed]
34. Schütze, O.; Esquivel, X.; Lara, A.; Coello, C.A.C. Using the averaged Hausdorff distance as a performance measure in evolutionary multiobjective optimization. *IEEE Trans. Evol. Comput.* **2012**, *16*, 504–522. [CrossRef]
35. Karimi, D.; Salcudean, S.E. Reducing the hausdorff distance in medical image segmentation with convolutional neural networks. *IEEE Trans. Med. Imaging* **2019**, *39*, 499–513. [CrossRef]
36. Van Veldhuizen, D.A.; Lamont, G.B. *Multiobjective Evolutionary Algorithm Research: A History and Analysis*; Technical Report; Air Force Institute of Technology: Dayton, OH, USA, 1998.
37. Zhou, A.; Jin, Y.; Zhang, Q.; Sendhoff, B.; Tsang, E. Combining model-based and genetics-based offspring generation for multi-objective optimization using a convergence criterion. In Proceedings of the 2006 IEEE International Conference on Evolutionary Computation, Vancouver, BC, Canada, 16–21 July 2006; pp. 892–899.
38. Mahbub, M.S.; Wagner, T.; Crema, L. Improving robustness of stopping multi-objective evolutionary algorithms by simultaneously monitoring objective and decision space. In Proceedings of the 2015 Annual Conference on Genetic and Evolutionary Computation, Lille, France, 10–14 July 2015; pp. 711–718.
39. Khare, V.; Yao, X.; Deb, K. Performance scaling of multi-objective evolutionary algorithms. In *International Conference on Evolutionary Multi-Criterion Optimization*; Springer: Berlin/Heidelberg, Germany, 2003; pp. 376–390.
40. Sebag, M.; Tarrisson, N.; Teytaud, O.; Lefevre, J.; Baillet, S. A Multi-Objective Multi-Modal Optimization Approach for Mining Stable Spatio-Temporal Patterns. In Proceedings of the 19th International Joint Conference on Artificial Intelligence, Edinburgh, UK, 30 July–5 August 2005; pp. 859–864.
41. Krmicek, V.; Sebag, M. Functional brain imaging with multi-objective multi-modal evolutionary optimization. In *Parallel Problem Solving from Nature-PPSN IX*; Springer: Berlin/Heidelberg, Germany, 2006; pp. 382–391.
42. Zechman, E.M.; Giacomoni, M.H.; Shafiee, M.E. An evolutionary algorithm approach to generate distinct sets of non-dominated solutions for wicked problems. *Eng. Appl. Artif. Intell.* **2013**, *26*, 1442–1457. [CrossRef]
43. Deb, K.; Agrawal, S.; Pratap, A.; Meyarivan, T. A fast elitist non-dominated sorting genetic algorithm for multi-objective optimization: NSGA-II. In *International Conference on Parallel Problem Solving from Nature*; Springer: Berlin/Heidelberg, Germany, 2000; pp. 849–858.
44. Hernández Castellanos, C.I. Set Oriented Methods for Multi-Objective Optimization. Ph.D. Thesis, Centro de Investigación y de Estudios Avanzados del Instituto Politécnico Nacional, Mexico City, Mexico, 2017.
45. Ulrich, T.; Bader, J.; Zitzler, E. Integrating decision space diversity into hypervolume-based multiobjective search. In Proceedings of the 12th Annual Conference on Genetic and Evolutionary Computation, Portland, OR, USA, 7–11 July 2010; pp. 455–462.
46. Solow, A.R.; Polasky, S. Measuring biological diversity. *Environ. Ecol. Stat.* **1994**, *1*, 95–103. [CrossRef]
47. Rudolph, G.; Naujoks, B.; Preuss, M. Capabilities of EMOA to detect and preserve equivalent Pareto subsets. In *International Conference on Evolutionary Multi-Criterion Optimization*; Springer: Berlin/Heidelberg, Germany, 2007; pp. 36–50.
48. Zhang, K.; Chen, M.; Xu, X.; Yen, G.G. Multi-objective evolution strategy for multimodal multi-objective optimization. *Appl. Soft Comput.* **2021**, *101*, 107004. [CrossRef]
49. Yue, C.; Qu, B.; Liang, J. A multiobjective particle swarm optimizer using ring topology for solving multimodal multiobjective problems. *IEEE Trans. Evol. Comput.* **2017**, *22*, 805–817. [CrossRef]
50. Zhang, X.; Liu, H.; Tu, L. A modified particle swarm optimization for multimodal multi-objective optimization. *Eng. Appl. Artif. Intell.* **2020**, *95*, 103905. [CrossRef]
51. Li, X.; Engelbrecht, A.; Epitropakis, M.G. *Benchmark Functions for CEC'2013 Special Session and Competition on Niching Methods for Multimodal Function Optimization*; RMIT University, Evolutionary Computation and Machine Learning Group: Melbourne, Australia, 2013.
52. Wang, H.; Jin, Y.; Doherty, J. Committee-based active learning for surrogate-assisted particle swarm optimization of expensive problems. *IEEE Trans. Cybern.* **2017**, *47*, 2664–2677. [CrossRef]

53. Ariyaratne, M.; Fernando, T.; Weerakoon, S. A Hybrid Algorithm to Solve Multi-model Optimization Problems Based on the Particle Swarm Optimization with a Modified Firefly Algorithm. In *Proceedings of the Future Technologies Conference*; Springer Berlin/Heidelberg, Germany, 2020; pp. 308–325.
54. Marek, M.; Kadlec, P.; Čapek, M. FOPS: A new framework for the optimization with variable number of dimensions. *Int. J. RF Microw. Comput. Aided Eng.* **2020**, *30*, e22335. [CrossRef]
55. Lienhard, I.; John, H. *A Heat Transfer Textbook*; Phlogiston Press: Cambridge, MA, USA, 2005.
56. Deb, K.; Agrawal, R.B. Simulated binary crossover for continuous search space. *Complex Syst.* **1995**, *9*, 115–148.

Article

A Simple Proposal for Including Designer Preferences in Multi-Objective Optimization Problems

Javier Blasco [1,*], Gilberto Reynoso-Meza [2], Enrique A. Sánchez-Pérez [3], Juan Vicente Sánchez-Pérez [4] and Natalia Jonard-Pérez [5]

1. Instituto Universitario de Automática e Informática Industrial, Universitat Politècnica de València, Camino de Vera s/n, 46022 Valencia, Spain
2. Programa de Pós-Graduação em Engenharia de Produção e Sistemas (PPGEPS), Pontifícia Universidade Católica do Paraná (PUCPR), Curitiba 80215-901, Brazil; g.reynosomeza@pucpr.br
3. Instituto Universitario de Matemática Pura y Aplicada (IUMPA), Universitat Politècnica de València, Camino de Vera s/n, 46022 Valencia, Spain; easancpe@mat.upv.es
4. Centro de Tecnologías Físicas: Acústica, Materiales y Astrofísica (CTF:AMA), Universitat Politècnica de València, Camino de Vera s/n, 46022 Valencia, Spain; jusanc@fis.upv.es
5. Departamento de Matemáticas, Facultad de Ciencias, Universidad Nacional Autónoma de México, Mexico City 04510, Mexico; nat@ciencias.unam.mx
* Correspondence: xblasco@isa.upv.es; Tel.: +34-963877007 (ext. 75713)

Citation: Blasco, X.; Reynoso-Meza, G.; Sánchez-Pérez, E.A.; Sánchez-Pérez, J.V.; Jonard-Pérez, N. A Simple Proposal for Including Designer Preferences in Multi-Objective Optimization Problems. *Mathematics* **2021**, *9*, 991. https://doi.org/10.3390/math9090991

Academic Editor: David Greiner

Received: 1 April 2021
Accepted: 26 April 2021
Published: 28 April 2021

Publisher's Note: MDPI stays neutral with regard to jurisdictional claims in published maps and institutional affiliations.

Copyright: © 2021 by the authors. Licensee MDPI, Basel, Switzerland. This article is an open access article distributed under the terms and conditions of the Creative Commons Attribution (CC BY) license (https://creativecommons.org/licenses/by/4.0/).

Abstract: Including designer preferences in every phase of the resolution of a multi-objective optimization problem is a fundamental issue to achieve a good quality in the final solution. To consider preferences, the proposal of this paper is based on the definition of what we call a preference basis that shows the preferred optimization directions in the objective space. Associated to this preference basis a new basis in the objective space—dominance basis—is computed. With this new basis the meaning of dominance is reinterpreted to include the designer's preferences. In this paper, we show the effect of changing the geometric properties of the underlying structure of the Euclidean objective space by including preferences. This way of incorporating preferences is very simple and can be used in two ways: by redefining the optimization problem and/or in the decision-making phase. The approach can be used with any multi-objective optimization algorithm. An advantage of including preferences in the optimization process is that the solutions obtained are focused on the region of interest to the designer and the number of solutions is reduced, which facilitates the interpretation and analysis of the results. The article shows an example of the use of the preference basis and its associated dominance basis in the reformulation of the optimization problem, as well as in the decision-making phase.

Keywords: multi-objective decision-making; Pareto front; multi-objective optimization; preference in multi-objective optimization

1. Introduction

In a design problem posed as a multi-objective optimization, there is not a single solution. Therefore, the designer—decision maker (DM in what follows)—must get, from the set of optimal solutions—Pareto set—, a suitable solution adjusted to a given set of preferences. It is accepted that the preferences of the designer might play a fundamental role in the resolution of this type of problems [1–4]. Using preference handling mechanisms in the optimization process has shown to be a valuable tool when facing multi-objective optimization problems [5]. These mechanisms also facilitate the decision-making process at the selection step, because the DM will focus its attention to the pertinent region of the Pareto front [6]. Nevertheless, this implies that it is necessary to have tools that allow to take into account the preferences in any of the phases of resolution of a design problem that is intended to be solved through multi-objective optimization. Thus, preferences of the DM affect the definition of the objectives, the optimization to reach an approximation to the

Pareto set, and/or the selection of the final solution. There are already some papers that explain how to bring some progress in all these steps but most of them are mainly focused to incorporating preferences in the optimization problem. For example, there are some works on how to build functions to optimize according preferences [7]. Mechanisms for *pertinency* have also been included in some multi-objective optimization algorithms [2,6,8–10] in which the preferences guide the approach to the front. Reference [11] proposes a new angle-based preference selection mechanism to be included in the optimization algorithm. Reference [12] investigates two methods to find simultaneously optimal (in the objective space) and practically desirable solutions (in the decision space). Reference [13] incorporates preference information into evolutionary multi-objective optimization in an interactive way, at each iteration, the decision maker has to include preference information (as points corresponding to aspiration levels for objectives). Reference [2] introduces a new preference relation based on a reference point approach that is used into an interactive optimization scheme that uses a multi-objective evolutionary algorithm (MOEA). Reference [14] integrates preferences in the optimization process by a nonuniform mapping of the objective space according to an aspiration level vector, and some other parameters (number of divisions, expected extend of the region of interest, etc.) supplied by the DM. A recent work [15] proposes a modification of a decomposition-based multi-objective evolutionary algorithm to obtain a denser set of solution closer to a reference point.

Most of the works found in the bibliography try to provide new developments applicable in the final decision phase or throughout the optimization via interactive mechanisms. Another large group of proposals tries to modify the optimization algorithm to incorporate preferences in the optimization process itself. The proposal described here is based on modifying the original problem to incorporate preferences without having to modify the optimization algorithm. That is there is no need to incorporate any additional mechanism or layer in the algorithm. In this paper, we show a very simple way to incorporate preferences in the objective definition phase. It allows the use of any current optimization algorithm without special requirements, being compatible with any advance made in the multi-objective optimization algorithm. Additionally, it is shown how to use this same methodology in the final decision phase helping in the visualization of the preferred solutions. This is applicable in case of problems in which the Pareto front has been obtained without taking into account the preferences, and it is necessary to incorporate them in the final decision stage.

The adaptation of the original problem to include preferences usually also depends on the problem itself. It involves the adaptation or design of functions to be optimized that reflect the preferences. A widely used way is to transform the multi-objective problem into a single-objective problem by means of scalarization [16]. A classic method widely used is the weighted sum of the objectives where it is necessary to adjust weighting factors that, in some way, incorporate the preferences of the designer. The idea of this work is somewhat similar; it consists of modifying the functions to be optimized but maintaining the multi-objective character, that is, without turning it into a single-objective function. Technically, we use what we call preference directions, that are introduced by choosing a special basis for the space of objectives $\mathbb{R}^n$. Broadly speaking, the elements of such a basis represent the directions in the space in which the DM feels that the optimization must be realized. Associated to these preference directions, a new basis, called dominance basis, is obtained. Their elements are defined as intersection of hyperplanes which are orthogonal to the preference directions. From the geometrical point of view, the cone generated by the dominance basis provides a representation of the dominance cone. Thus, the corresponding change of basis allows a reinterpretation of the classical dominance relationship that is used for defining the Pareto front, that represents the solution of a multi-objective optimization problem. The underlying idea is very simple and consists of the fact that the change of base produces a deformation of the objective space in a way that favors the preferences of the DM. The definition of these bases is beyond the scope of this paper, and it is problem dependent: it is assumed that the DM has established such preference directions using his own criteria. Nevertheless, for the aim of illustrating this notion, we will shown how to obtain it in some particular examples.

The paper is organized as follows. Section 2 introduces the main concepts involved in the proposal. In Section 3, we show how to use preference and dominance bases to reformulate the multi-objective problem and also to help in the decision-making step. Section 4 shows an application example, and the last Section 5 summarizes the main conclusion and future works.

2. Preference Directions and Dominance

The way proposed to introduce preferences in the geometric space is by means of the space basis of $\mathbb{R}^n$: $\mathcal{B}_p := \{v_i : i = 1, ..., n\}$ (preference basis). We will call the vectors of such basis *preference directions*. Our main idea is that these vectors model the preferred directions for minimization. In other words, each hyperplane that is orthogonal to a vector in this basis separates the points of the space into two sets: the points "below" the hyperplane dominate—with respect to this preference direction—the points "above" the hyperplane.

Thus, associated with $\mathcal{B}_p$, a new basis $\mathcal{B}_d := \{v_{di} : i = 1, ..., n\}$ of $\mathbb{R}^n$ can be defined by using a geometric procedure that will be explained later on (see Proposition 1). We will call it the dominance basis. The positive cone C_d^+ (dominance cone see Definition 1) generated by $\mathcal{B}_p$, and also associated with $\mathcal{B}_d$, allows to reinterpret dominance according to the new preference directions provided. Recall that the notion of dominance depends on the lattice order that is considered in the Euclidean space. The usual order in $\mathbb{R}^n$ is given by the "coordinates" ordering, when these coordinates correspond to the canonical basis in the Euclidean n-dimensional space. If we consider the coordinate order with respect to a *different* basis of $\mathbb{R}^n$—in our case, $\mathcal{B}_d$—, we obtain a different dominance relation among points. Our main contributions in the present paper are: (1) to show how to obtain the dominance basis from the preference basis; and (2) how to use it in the multi-objective optimization problem.

Figure 1 shows an example of the reinterpretation of dominance for a 2-objective space. The figure shows the canonical interpretation of dominance and how the dominance area changes when a different basis of preference directions is considered. Note that the effect of considering the dominance defined by the cone generated by $\mathcal{B}_d$ can be understood as a deformation of the area that is dominated by a given point. Remark that, if the canonical base is set as the preference basis, the dominance basis is also the canonical basis (dominance and preference basis are the same). In the next subsection, we will provide the equations to obtain the dominance basis $\mathcal{B}_d$ from the preference directions basis.

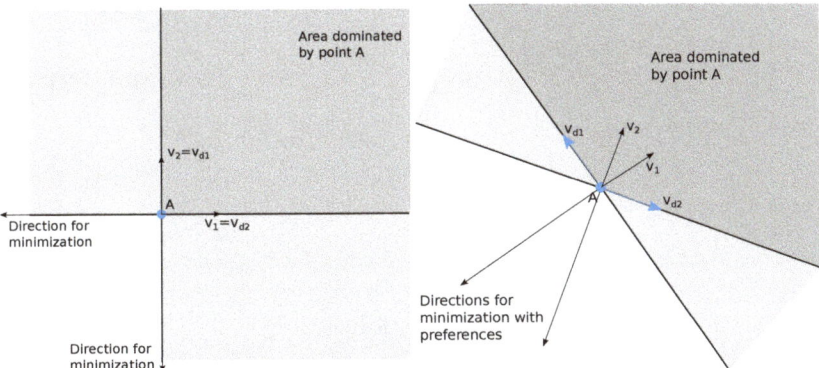

Figure 1. Graphical interpretation of canonical dominance and dominance with preferences. Dominated area by point A for different dominance basis (v_{d1}, v_{d2}). Dominance with the canonical basis (figure on the **left**) and dominance when preferences are included with a basis change (figure on the **right**).

Computation of Dominance Basis $\mathcal{B}_d$

Let $(\mathbb{R}^n, \|\cdot\|_2)$ be the n-dimensional Euclidean space. Consider a basis $\mathcal{B}_p := \{v_i : i = 1, ..., n\}$ norm one elements of $\mathbb{R}^n$ included in the positive cone of $\mathbb{R}^n$. Each of the elements of the basis defines what we have called a preference direction for the optimization. The optimization procedure that we present here proposes to consider the associated dominance cone as the set of all vectors that are over the hyperplane defined by each vector defining a preference direction (its projection on this vector must be positive). Therefore, we define the *positive semispace* $S_{v_i}^+$ associated with the preference direction $v_i, i = 1, ..., n$, as

$$S_{v_i}^+ := \{v \in \mathbb{R}^n : v \cdot v_i \geq 0\}.$$

Consequently, the dominance cone C_d^+ can be defined as follows.

Definition 1. *Let $\mathcal{B}_p := \{v_i : i = 1, ..., n\}$ be a basis of preference directions of $\mathbb{R}^n$. We define the **dominance cone** as the intersection of all the semispaces $S_{v_i}^+$, that is:*

$$C_d^+ := \bigcap_{i=1}^n S_{v_i}^+ = \{v \in \mathbb{R}^n : v \cdot v_i \geq 0, i = 1, ..., n\}.$$

The positive cone C_d^+ generated by $\mathcal{B}_p$ allows to reinterpret dominance according to the new preference directions: if A is a point in $\mathbb{R}^n$, $A + C_d^+$ gives the set of all the points that are dominated by A. In Figure 1, a 2D example is given, where the dark grey area shows the area dominated by point A, that is, $A + C_d^+$.

Definition 2. *Given a set of vectors $Z = \{z_1, ..., z_m\} \subset \mathbb{R}^n$, we define its **positive linear hull** as the set*

$$Pos(Z) := \{\sum_{i=1}^m \lambda_i z_i : \lambda_i \geq 0\}.$$

For the subsequent proposals described in Section 3.2, it is necessary to obtain the dominance basis $\mathcal{B}_d$ corresponding to the preference basis $\mathcal{B}_p$ defined by the DM. A way to obtain $\mathcal{B}_d$ is based on the equivalence of C_d^+ and $Pos(\mathcal{B}_d)$ (see Proposition 1). As we already mentioned, the dominance basis is defined by vectors that are orthogonal to vectors of the preference basis.

Proposition 1. *The dominance cone associated with a basis of preference directions $\mathcal{B}_p = \{v_i : i = 1, ..., n\}$ coincides with the positive linear hull $Pos(\mathcal{B}_d)$ of the set of vectors $\mathcal{B}_d := \{v_{di} : i = 1, ..., n\}$ defined as follows.*

For every $i = 1, ..., n$, $v_{di} \in \mathbb{R}^n$ is a (norm one) solution of the system

$$\begin{bmatrix} v_1 \cdot v_{di} = 0 \\ \vdots \\ v_{i-1} \cdot v_{di} = 0 \\ v_{i+1} \cdot v_{di} = 0 \\ \vdots \\ v_n \cdot v_{di} = 0 \end{bmatrix}$$

which also satisfies that $v_i \cdot v_{di} > 0$.

Proof. First, note that for each $i = 1, ..., n$, the linear system defined above always has a subspace of solutions of dimension 1, because it is made of $n - 1$ linear equations. So, we can choose a solution with norm equal to one, and also $v_i \cdot v_{di} > 0$ (if $v_i \cdot v_{di} = 0$ for all i, then $v_{di} = 0$, since $\mathcal{B}_p$ is a basis).

Let us prove first that $C_d^+ \subseteq Pos(\mathcal{B}_d)$. Take an element $v \in C_d^+$. By the definition of the set of preference directions—they are linearly independent—, we have that the set $\{v_{d1}, ..., v_{dn}\}$ defined as in the statement of the result, is a basis for $\mathbb{R}^n$. Then, v can be written as $v = \sum_{i=1}^n x_i v_{di}$, for real numbers $x_1, ..., x_n$. We have that

$$0 \leq v \cdot v_k = \sum_{i=1}^n x_i v_{di} \cdot v_k = x_k v_{dk} \cdot v_k$$

for all $k = 1, ..., n$. Since $v_{dk} \cdot v_k > 0$, we get that $x_k \geq 0$. This can be done for all $k = 1, ..., n$, and so we get that $C_d^+ \subseteq Pos(\mathcal{B}_d)$.

Conversely, take an element $v \in Pos(\mathcal{B}_d)$. Then, it can be written as $v = \sum_{i=1}^n x_i v_{di}$, where all the coordinates x_i are non-negative. Then,

$$v \cdot v_k = \sum_{i=1}^n x_i v_{di} \cdot v_k = x_k v_{dk} \cdot v_k.$$

But $x_k \geq 0$ and $v_{dk} \cdot v_k \geq 0$, and so $v \cdot v_k \geq 0$ for all k. Consequently, $v \in C_d^+$, and so $Pos(\mathcal{B}_d) \subseteq C_d^+$.

Therefore, $Pos(\mathcal{B}_d) = C_d^+$, and the result is proved. □

Based on Proposition 1, it is possible to give an algorithm that computes the vectors of the corresponding dominance basis $\mathcal{B}_d$. An example with MATLAB is shown in Figure 2.

```
function Md=dominanceCone(Mp)
% Mp matrix of preference directions (each column is a vector of the basis)
%       Mp=[v1' v2' ... vn']
% Md matrix that define dominance cone
%       Md=[vd1' vd2' ... vdn'] (each column is a vector of the basis)

n=size(Mp,1);
Md=zeros(n,n);

for i=1:n
    Maux=Mp;
    Maux(:,i)=[];    % Remove i column
    v=Mp(:,i)';      % Extract i column
    % Building system of linear equation
    for ii=1:n % to avoid det(A)=0
        A=[Maux'; zeros(1,n)];
        A(end,ii)=1;
        if det(A)~=0
            break;
        end
    end
    B=zeros(n,1);
    B(end)=1;
    % Solve system of linear equations Av = B for v
    vaux1=(A\B)';
    % Compute inverse vector (opposite direction)
    vaux2=-vaux1;
    % Checking convenient cone (v*vi>0)
    if (vaux1*v')>(vaux2*v')
        Md(:,i)=vaux1'/norm(vaux1);
    else
        Md(:,i)=vaux2'/norm(vaux2);
    end
end
```

Figure 2. MATLAB function based on proposition 1 that computes dominance basis from preference basis.

3. Using Preference Directions in Multi-Objective Optimization Problems

There are two direct uses of the dominance bases:

- Including preferences in the decision-making step.
- Including preferences by reformulating the optimization problem.

In both cases, the mathematical meaning underlying the concept of "optimization direction" refers to the definition of preference basis. From this basis, it is possible to obtain the dominance basis—as described in the previous section—which is the one that redefines the space of objectives, thus incorporating the preferences.

3.1. Including Preferences in the Decision Making Step

As a first application, we consider the case of a DM who is provided with a Pareto front and tries to select a solution based on his preferences. In this selection process, graphical tools are important to visually analyze the Pareto front. A standard way used is to color the different points of the Pareto front according to the designer's preferences [17].

To illustrate this use, a 2D example is shown below. Let us suppose that all possible solutions of a 2D multi-objective problem are placed on a disc of radius r = 1, and centered on the origin. Obviously, the perimeter of this set of solutions is formed by the circle of radius unit centered on the origin (see Figure 3a). If the optimization problem consists on minimizing both objectives, it is easy to locate the Pareto front, represented as a discrete approach formed by the points highlighted in black in Figure 3a. Once the Pareto front is obtained, the DM usually has to choose one of the solutions included in the front. Assuming that both objectives are equally relevant, it seems reasonable to select a solution close to the ideal point. The ideal point has the minimum possible value for each objective, which is highlighted with a diamond in Figure 3b. A commonly used tool in the decision phase is to color the points of the Pareto front according to their proximity to the designer's preferred area. In the case of Figure 3b, the points closest to the ideal point are colored dark blue, and the furthest points are colored red.

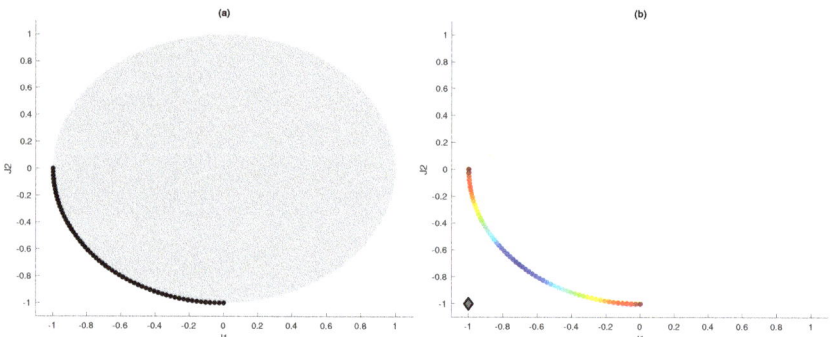

Figure 3. (**a**) Possible values of the objectives (grey disc) and approximation to Pareto front (black points). (**b**) Coloring Pareto front according to proximity to ideal point (diamond).

However, if the DM defines some preference directions, the analysis of the solutions of the Pareto front changes. If we suppose that $\mathcal{B}_p$ is defined by the DM, $\mathcal{B}_d$ can be easily obtained applying the algorithm provided by Proposition 1:

$$\mathcal{B}_p = \left\{ \begin{array}{ll} v_1 = & (0.3162, 0.9487) \\ v_2 = & (0,1) \end{array} \right\} \rightarrow \mathcal{B}_d = \left\{ \begin{array}{ll} v_{d1} = & (1,0) \\ v_{d2} = & (-0.9487, 0.3162) \end{array} \right\}. \quad (1)$$

With this new base, which incorporates the DM's preferences, the objective space is deformed by affecting the distances between points. With the associated dominance basis $\mathcal{B}_d$ and the matrix M_d, the new distances are easily calculated (see Appendix A). Figure 4a shows, in the canonical space, the Pareto front colored according to the proximity to the ideal point with the new $\mathcal{B}_d$ base. In the same figure, the vectors of the $\mathcal{B}_p$ preference base (in blue color) are shown, as well as the vectors corresponding to the $\mathcal{B}_d$ dominance base (in red color). In addition, to better understand how the relationship of distances to the ideal point has been changed, Figure 4b is shown. In this figure, the Pareto front and the ideal point in the space defined by $\mathcal{B}_d$ are represented.

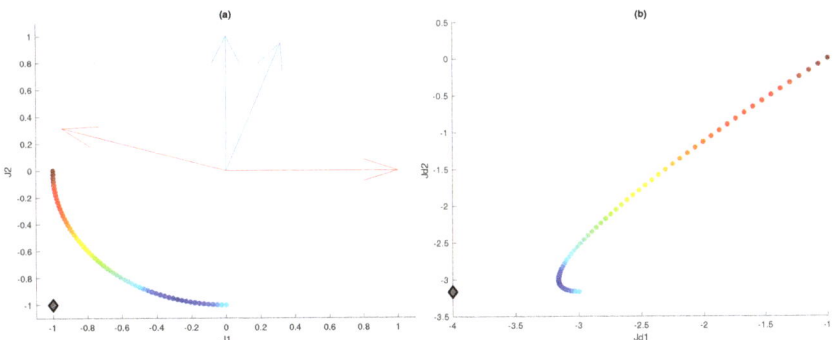

Figure 4. (**a**) Pareto front colored according to proximity to ideal point computed in the basis $\mathcal{B}_d$ and plotted in canonical basis. Vectors of preference directions in blue and vectors of dominance directions in red. (**b**) Same points but plotted in the basis $\mathcal{B}_d$.

As seen in the figure, this simple preference-based coloring allows the DM to select a subset of relevant solutions according to his preferences.

3.2. Including Preferences by Reformulating the Optimization Problem

This way of considering preferences—by selecting a new basis—is easy to apply with all the available optimization tools because it is just a change of coordinates. Defining a multi-objective problem as:

$$\min_{\theta} J(\theta) \qquad (2)$$

$$\text{subject to:} \quad \theta \in S \subset \mathbb{R}^m,$$

where $\theta = (\theta_1, \ldots, \theta_m) \in \mathbb{R}^m$ is the decision vector, $J(\theta) = (J_1(\theta), \ldots, J_n(\theta)) \in \mathbb{R}^n$ is objective vector, and S is the subspace that satisfies all the additional constraints of the problem.

We must define a new basis that collects the preference directions:

$$\mathcal{B}_p := \{v_i : i = 1, \ldots, n\}. \qquad (3)$$

The redefinition of the multi-objective optimization problem that include preferences is done using the associated dominance basis $\mathcal{B}_d := \{v_{di} : i = 1, \ldots, n\}$ and the matrix M_d (see Equation (A1), Appendix A):

$$\min_{\theta} J_d(\theta) \qquad (4)$$

$$J_d(\theta) = (M_d^{-1} J(\theta)^T)^T$$

$$\text{subject to:} \quad \theta \in S \subset \mathbb{R}^m.$$

To see the effect of the preference directions in the optimization problem, an example of a 2D problem (in the objective space) is shown. It is assumed that the space of possible objectives is on a disk of unit radius. Figure 5a shows the periphery of this set on a canonical basis, while Figure 5b shows the same circle but at the base $\mathcal{B}_d$. It is clearly shown how the circle is deformed and becomes an ellipse. The equivalent points have been colored identically in both figures to make them easier to locate.

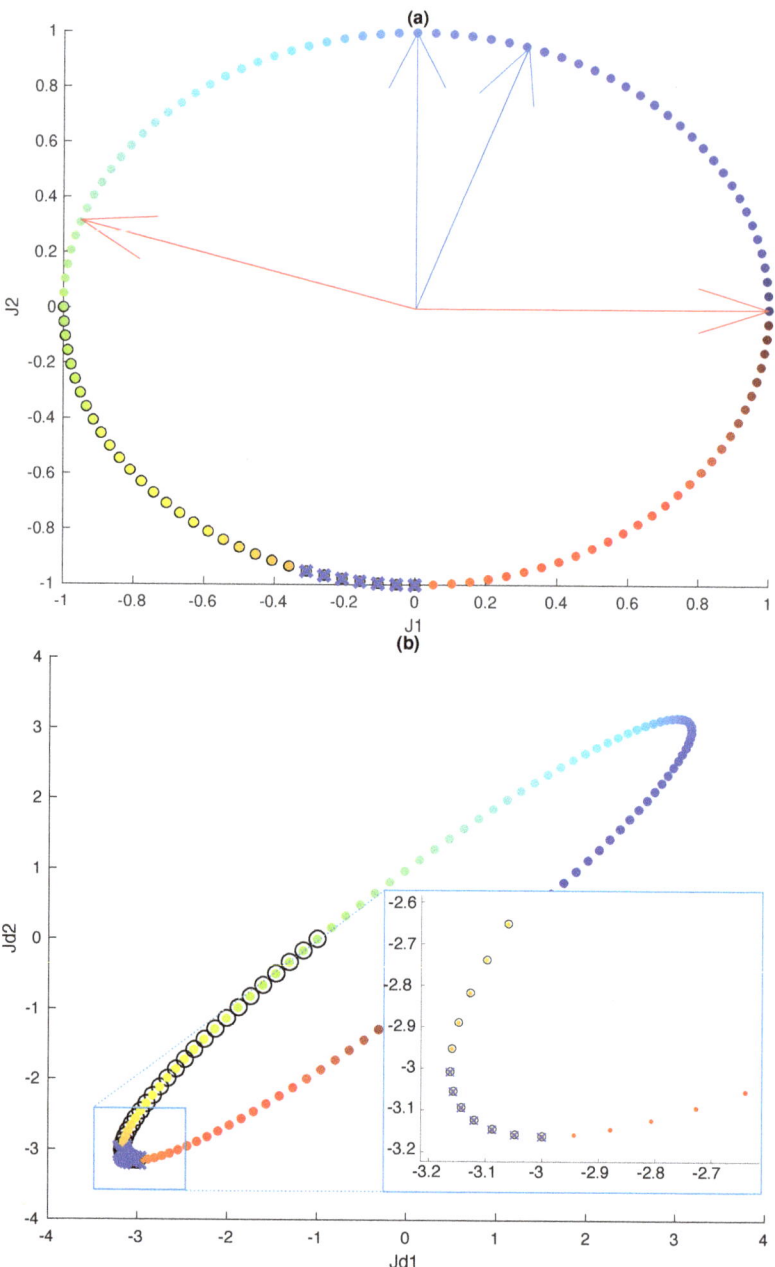

Figure 5. (**a**) Representation of the unit circle in the canonical basis, and the preference (red) and dominance (blue) vectors of the corresponding basis. Lines with 'o' and 'x' show Pareto front in canonical and $\mathcal{B}_d$ basis, respectively. (**b**) Same circle and Pareto fronts drawn in $\mathcal{B}_d$ basis.

Figure 5a also shows the vectors that correspond to the preferences base (in blue) and to the dominance base (in red). Note that Pareto front changes according to the base on which the points are represented. The points that correspond to an approximation of the front that would be obtained using the canonical base are highlighted with circles, and the points of the front that would be obtained using the dominance base $\mathcal{B}_d$ are marked with "x". It can be seen that the use of this basis allows the front to be focused on the area preferred by the DM, which has been previously defined by $\mathcal{B}_p$. As a additional comment, it can be observed that by focusing on the region of interest, in general fewer solutions are obtained. This facilitates the final decision stage, since the DM has to choose between fewer options but all oriented according to his preferences.

Therefore, the use of $\mathcal{B}_d$ directly in the formulation of the optimization problem provides the *pertinency* property without having to modify the optimization algorithm. This methodology can be used with any optimizer since it only involves a reformulation of the optimization problem.

A 3D example comparing canonical dominance and dominance with preferences is shown in Figure 6. For demonstration purposes, the set of possible objective values is a sphere, and the boundary of the sphere is represented in blue. The basis $\mathcal{B}_p$ defines the dominance cone $\mathcal{B}_d$. Remark that, if $\mathcal{B}_p$ is the canonical base, $\mathcal{B}_p = \mathcal{B}_d$.

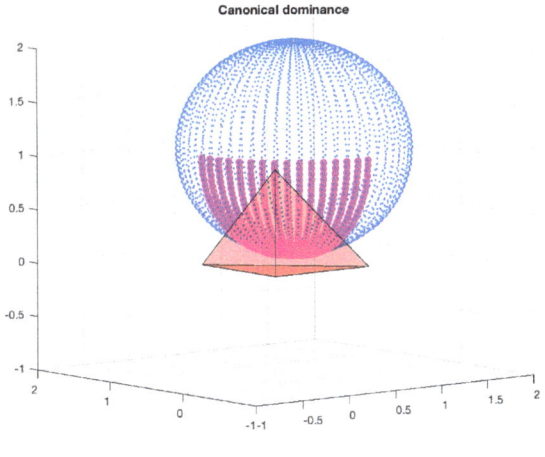

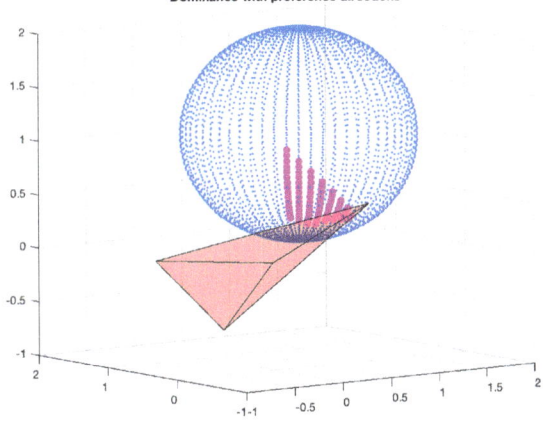

Figure 6. 3D Pareto front approximation (in magenta) with canonical dominance (**upper** figure) and dominance with preferences (**bottom** figure).

$$\mathcal{B}_p = \left\{ \begin{array}{ll} v_1 = & (1,0,0) \\ v_2 = & (0,1,0) \\ v_3 = & (0,0,1) \end{array} \right\} \rightarrow \mathcal{B}_d = \left\{ \begin{array}{ll} v_{d1} = & (1,0,0) \\ v_{d2} = & (0,1,0) \\ v_{d3} = & (0,0,1) \end{array} \right\}. \qquad (5)$$

The dominance cone defined by $\mathcal{B}_d$ is represented with a polyhedron with one of its vertex at the origin. Each of the edges starting at the origin represents a vector of $\mathcal{B}_d$. The polyhedron is colored in red. The resulting Pareto front has been highlighted in magenta

Figure 6 shows the resulting Pareto front for canonical dominance (upper figure) and Pareto front obtained with a base of preferences defined by $\mathcal{B}_p$ (bottom figure):

$$\mathcal{B}_p = \left\{ \begin{array}{ll} v_1 = & (0, 0.7071, 0.7071) \\ v_2 = & (0.5774, 0.5774, 0.5774) \\ v_3 = & (0.7071, 0.7071, 0) \end{array} \right\} \rightarrow$$
$$\mathcal{B}_d = \left\{ \begin{array}{ll} v_{d1} = & (-0.7071, 0.7071, 0) \\ v_{d2} = & (0.5774, -0.5774, 0.5774) \\ v_{d3} = & (0, 0.7071, -0.7071) \end{array} \right\}. \qquad (6)$$

The reader can notice how the selection of preference directions provides a narrower Pareto set when comparing with the canonical situation. Indeed, the preference direction basis $\mathcal{B}_p$ is strictly included in the canonical positive cone, what provides a wider dominance cone (see the picture on the bottom of Figure 6). This means that, in the new situation, each point in the Pareto front dominates more points of the original sphere of objective values, what gives a smaller set of dominating relevant points. To understand the deformation produced by the preferences, Figure 7 shows the same that Figure 6 (bottom), but drawn in $\mathcal{B}_p$ basis. The DM has to consider a smaller set of optimal points for his final election, all of them fitting with his original preference requirements.

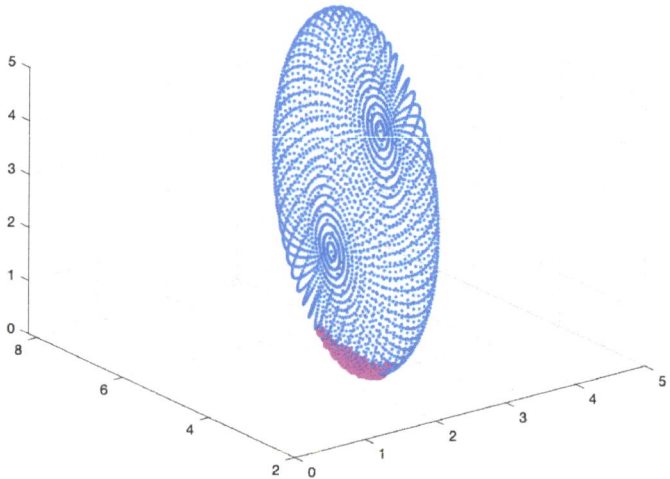

Figure 7. 3D deformed sphere (blue points) resulting from preferences and Pareto front approximation (in magenta) drawn in $\mathcal{B}_d$.

4. Application Example: The Daily Diet Design Problem

In this section, we will develop a complete example regarding the design of a healthy diet. The original problem can be found in Reference [18], where an alternative form of decision-making called TOPSIS is presented. The methodology tries to reach the compromise that the chosen alternative should have *the smallest distance from the positive ideal solution* and *the largest distance from the negative ideal solution*. Therefore, no preferences are proposed by the DM and cannot be compared with the proposal of the present work. In order to illustrate the methodology, we will propose some DM preferences.

The problem concerns the prescription of a diet for a patient who has some particular health characteristics. Suppose that the *decision variables* with their bound constraints are the following.

- Milk pints: $0.0 \leq x1 \leq 6.00$.
- Beef pounds: $0.0 \leq x2 \leq 1.00$.
- Eggs dozen: $0.0 \leq x3 \leq 0.25$.
- Bread ounces: $0.0 \leq x4 \leq 10.00$.
- Lettuce and salad ounces: $0.0 \leq x5 \leq 10.00$.
- Orange juice pints: $0.0 \leq x6 \leq 4.00$.

The problem has three *design objectives*, J1, J2 and J3:

$$\begin{aligned} J_1(x) &= 24x_1 + 27x_2 + 15x_4 + 1.1x_5 + 52x_6 \\ J_2(x) &= 10x_1 + 20x_2 + 120x_3 \\ J_3(x) &= 0.22x_1 + 2.2x_2 + 0.8x_3 + 0.1x_4 + 0.05x_5 + 0.26x_6 \end{aligned} \quad (7)$$

where $J_1(x)$ is the Carbohydrate intake [g], $J_2(x)$ is the Cholesterol intake [unit], and $J_3(x)$ is the Cost [\$].

There are also some constraints that affects the optimization calculations, regarding the amount of vitamin A [i.u] ($g_1(x)$), iron [mg] ($g_2(x)$), food energy [calories] ($g_3(x)$), and proteins [g] ($g_4(x)$).

$$\begin{aligned} g_1(x) &= 720x_1 + 107x_2 + 7080x_3 + 134x_5 + 1000x_6 \geq 5000 \\ g_2(x) &= 0.2x_1 + 10.1x_2 + 13.2x_3 + 0.75x_4 + 0.15x_5 + 1.2x_6 \geq 12.5 \\ g_3(x) &= 344x_1 + 1460x_2 + 1040x_3 + 75x_4 + 17.4x_5 + 240x_6 \geq 2500 \\ g_4(x) &= 18x_1 + 151x_2 + 78x_3 + 2.5x_4 + 0.2x_5 + 4.0x_6 \geq 63 \end{aligned} \quad (8)$$

The multi-objective problem to be solved can be written as follows:

$$\min_x J(x) \quad (9)$$

subject to:
$$J(x) = (J_1(x), J_2(x), J_3(x)) \in \mathbb{R}^3$$
$$\underline{x} \leq x \leq \overline{x}, \ x \in \mathbb{R}^6$$
$$g_i(x) \geq b_i, \ i = 1\ldots 4, \ g_i(x) \in \mathbb{R}, b_i \in \mathbb{R}.$$

Since it is a linear problem in both objectives and constraints, a classic optimization method obtains good results with a reasonable computational cost. Pareto front is obtained using a classic method, ϵ-constraint method, described in Reference [16]. Pareto front values are achieved by solving multiple mono-objective problems in which only one of the objectives is minimized, and the rest are considered as constraints. In order to obtain a suitable discretization of the front, it is necessary to vary the value of these constraints on the objectives in an adequate way. The steps are the following: first, the minimums of each independent objective are obtained. With these values we have the range of values where approximately the Pareto front is located. Subsequently, the increase to be applied to the restriction in each mono-objective problem is defined. In this particular case, the range of variation of two of the objectives (J_1 y J_3) has been divided into 50 parts, obtaining $\epsilon J1$ y $\epsilon J3$. Subsequently, multiple problems of a single objective are solved by adding constraints on the other objectives. Then, multiple problems ($50 \times 50 = 2500$) of a single objective are solved by adding constraints on the other objectives, that is, varying n from 1 to 50 and m from 1 to 50, and solving the following mono-objective problems:

$$\min_{x} J_2(x) \tag{10}$$

subject to:
$$\underline{x} \leq x \leq \overline{x},\ x \in \mathbb{R}^6$$
$$g_i(x) \geq b_i,\ i = 1\ldots 4,\ g_i(x) \in \mathbb{R}, b_i \in \mathbb{R}$$
$$J_1(x) \leq J1_{min} + \epsilon J1 \cdot n$$
$$J_1(x) \leq J3_{min} + \epsilon J3 \cdot m.$$

If a better approach to the front is required, the process can be repeated minimizing $J_1(x)$ or $J_3(x)$.

Since the problem is linear in both the target and the constraints, linear programming has been used. The approximation to the front obtained consists of 2049 points. Twenty-five hundred mono-objective problems have been executed, but some of them were not feasible; therefore, they did not provide a solution. Figure 8 shows a 3D representation of the Pareto front approximation obtained.

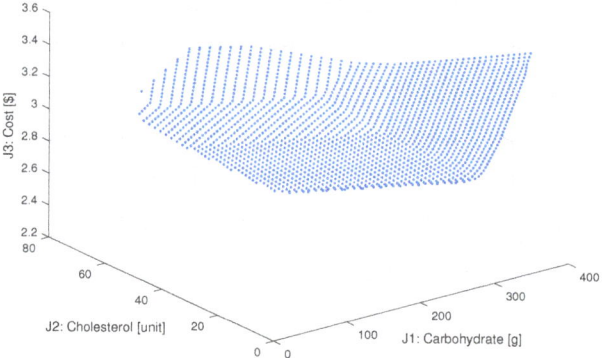

Figure 8. Pareto front approximation obtained for daily diet problem.

The DM must analyze this front and select a solution that fits his preferences. The first step, to have fewer points to choose from, is to reduce the number of points to 342 while maintaining a good representation of the front. This can be done reducing the discretization step.

Then, to help in the decision-making, the Pareto front is colored according to distance to the ideal point; see Figure 9 (upper). The ideal point corresponds to a possible solution obtained with the minimum values of each of the objectives.

This representation shows to the DM which are the points that present a more 'balanced' compromise between objectives. It should be noted that, in order to calculate the distance to the ideal point, the front has been scaled trying to avoid the distortions that can be caused by the different units and orders of magnitude of each of the objectives. Each objective is rescaled between 0 and 1. With these new values, the distances to the ideal point are calculated (that corresponds to the point $(0,0,0)$ when applying the scaling).

This 'balanced' solution is not always the preferred one, and the DM can decide on another order of preferences. It is very useful that these preferences will be incorporated in all phases of the MO problem resolution. In this case, a set of preferences defined by a vector base will be established.

Let us show now how to implement the preference directions by the DM, in this case, the doctor who treats the patient with special needs. Concretely, the doctor knows that the diet for the patient must satisfy preferably the following relations:

- The ratio among the Carbohydrate intake and the Cholesterol intake must be as near as possible to 2. This means that the preference equation $J1 = 2 \cdot J2$ must be taken into account. This condition supplies one of the preference directions: $v_1 = (2, 1, 0)$.
- Since there are a lot of bad quality Carbohydrates in the market, the doctor wants to promote the use of the correct ones.
  An indirect way to increase the use of good quality carbohydrates is through cost. It is estimated that the impact on the price of the diet of such carbohydrates is around 1\$ per 100 g of carbohydrates. Poor quality carbohydrates are significantly cheaper on the market. This allows to set a new direction of preference that is $J1/100 = J3$. This second conditions supplies another preference direction: $v_2 = (1/100, 0, 1)$.
- The third preference direction maintains the original objective of minimizing cost: $v_3 = (0, 0, 1)$.

Summing up all these assumptions and converting these vectors into unit vectors the preference basis $\mathcal{B}_p$ and its corresponding dominance basis are:

$$\mathcal{B}_p = \left\{ \begin{array}{l} v_1 = (0.8944, 0.4472, 0) \\ v_2 = (0.01, 0, 1) \\ v_3 = (0, 0, 1) \end{array} \right\} \rightarrow$$

$$\mathcal{B}_d = \left\{ \begin{array}{l} v_{d1} = (0, 1, 0) \\ v_{d2} = (0.4472, -0.8944, 0) \\ v_{d3} = (-0.4472, 0.8944, 0.0045) \end{array} \right\} \quad (11)$$

By changing the base, the calculation of the distances is modified. The example shows how the 'color' of the solutions closest to the ideal changes when these preferences are incorporated; see Figure 9 (bottom).

It can be seen that coloring, considering these new preferences (see Figure 9 (bottom)), orients the DM towards a different area of the front. Usually, these points would be more interesting for DM, and, consequently, he could choose more suitable solutions.

Table 1 shows the five best solutions extracted from each of the colored fronts. In particular, the points closest to the ideal point are shown. As you can see, the solutions obtained from coloring with preferences are closer to the preferences of the DM. The cost is lower, ratios among the Carbohydrate intake and the Cholesterol intake are nearest to two and ratios among Carbohydrate intake and Cost are nearest to 100. In this first case, the preference directions are used in the decision phase to assist the DM in making the final decision. This procedure does not reduce the number of solutions to be evaluated but enables a very simple way to visualize the results and select a subset of solutions of interest.

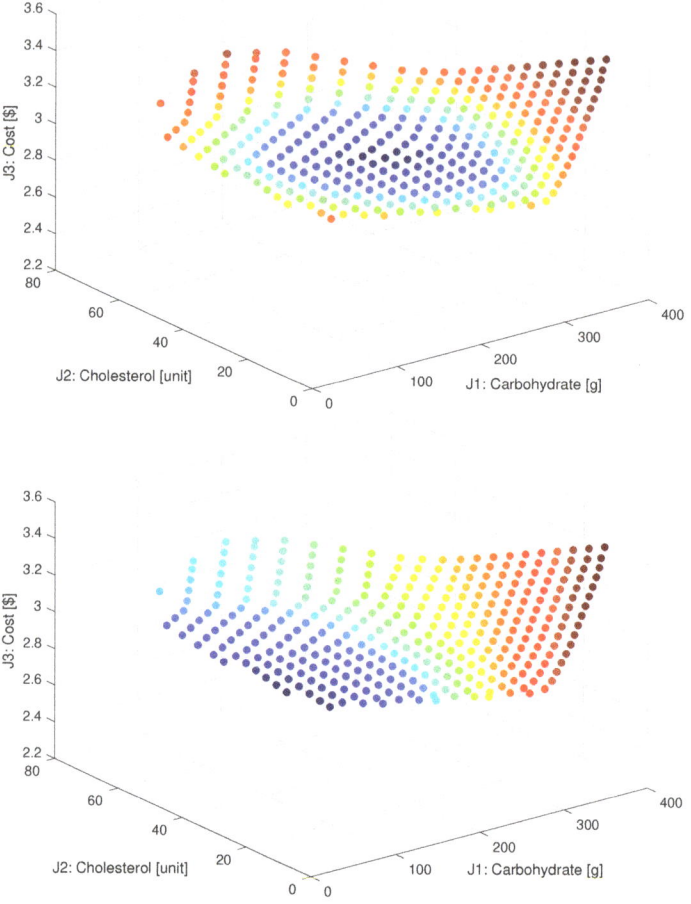

Figure 9. Pareto front obtained for daily diet problem. Upper figure shows the Pareto front colored according to the proximity to ideal point. Figure on the bottom shows the Pareto front colored according to the proximity to ideal point computed in basis $\mathcal{B}_d$. In both cases, dark blue color highlights the best solutions (dark red the worst).

Another alternative is to use the preference directions in the optimization process itself as proposed in Equation (4). The same multi-objective optimization method has been used to obtain the new front (constraint method). In this case, the range of variation of two of the objectives (J_{d1} and J_{d3}) has been divided into 10 parts and $10 \times 10 = 100$ mono-objective problems have been solved. The Pareto front achieved has 10 points (many of the mono-objective problems raised were not feasible and, therefore, did not provide a solution). Applying the preferences to modify the optimization problem has significantly reduced the computational cost and has focused the solution to the area of interest.

Table 1. *alt1*: The five nearest points to ideal point without considering preferences. *alt2*: The five nearest points to ideal point considering preferences.

Sol	J_1	J_2	J_3	x_1	x_2	x_3	x_4	x_5	x_6
alt1	236.87	39.51	2.85	1.78	0.69	0.07	0.69	0	3.17
	222.52	42.66	2.85	1.81	0.72	0.08	0.17	0	3.02
	236.87	35.97	2.91	1.60	0.74	0.04	0	0.24	3.43
	251.22	36.36	2.85	1.76	0.65	0.05	1.22	0	3.33
	222.52	39.79	2.91	1.74	0.74	0.06	0	1.13	3.07
alt2	236.87	69.26	2.36	3.52	0.21	0.25	7.40	0	0.68
	222.52	68.69	2.42	3.33	0.30	0.24	6.04	0	0.84
	281.67	67.83	2.24	3.78	0	0.25	10.00	0	0.79
	265.57	66.68	2.30	3.68	0.09	0.23	9.29	0	0.68
	251.22	66.11	2.36	3.50	0.18	0.23	7.92	0	0.84

Figure 10 shows the new front obtained compared to the previous one (Figure 8), the new front is colored in orange. There is a noticeable difference, since now the preferences have deformed the target space and have oriented the search towards a different zone adjusted to the DM's preferences. It should be noted that the number of points of the front obtained is 10, which is significantly lower than the number of points obtained in the original front without considering preferences (2049 points). Among these new values, the DM can select, for example, a more balanced solution. The points highlighted with black square in the figure show 3 examples of balanced solutions. The particular values of this 3 solutions are shown in Table 2.

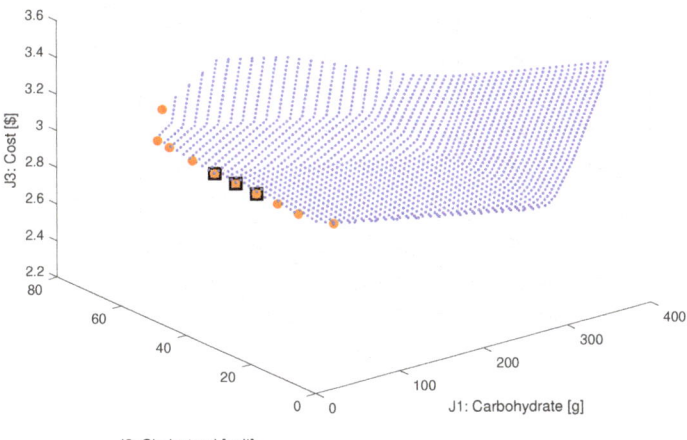

Figure 10. Pareto front obtained for daily diet problem. In orange, the Pareto front obtained including preferences into the optimization problem resolution. In blue, the original Pareto front obtained without preferences. Black square shows 3 balanced solutions.

Table 2. Balanced solutions obtained from the Pareto front with preferences.

J_1	J_2	J_3	x_1	x_2	x_3	x_4	x_5	x_6
151.65	70.34	2.65	2.66	0.69	0.25	0.31	0	1.24
175.46	70.13	2.57	2.91	0.55	0.25	2.31	0	1.08
199.27	69.91	2.49	3.15	0.42	0.25	4.31	0	0.92

5. Conclusions

This work shows a very easy alternative to introduce preferences both in the optimization problem formulation and/or in the final decision phases. The methodology is based on defining vectors that mark the directions of optimization and constitute what is called the preference base. It is shown and formally demonstrated how to obtain the dominance base from the preference one. In addition, the MATLAB code is provided for its calculation. This base is the one used to redefine the dominance relations, and it is the one in charge of 'deforming' the space to include the defined preferences. In Section 3, it is shown that the deformation of the space resulting from applying the dominance basis reconfigures the problem so that the points in the directions of preference dominate the rest of the solutions. This deformation is usable both to redefine the concept of dominance (this feature is usable by optimization algorithms) and to recalculate the distances that reorder the points in the space according to the preferences (this feature can be used in the decision phase to order solutions).

An important advantage of this methodology, when applied in the problem reformulation, is it can be used with any multi-objective optimization algorithm. In order to prove the use and usefulness of this methodology, a problem with three objectives has been raised. The results obtained show the simplicity of the use of this methodology and opens up a new range of possibilities for the inclusion of preferences. It has not been the objective of this work to show a general methodology for defining the directions of preferences; this aspect is particular to each problem and opens a very interesting research line that would allow to popularize the use of the methodology proposed in this article.

Author Contributions: Conceptualization, X.B. and E.A.S.-P.; Formal analysis, X.B., G.R.-M., E.A.S.-P. and N.J.-P.; Funding acquisition, X.B.; Methodology, X.B., G.R.-M. and E.A.S.-P.; Software, X.B.; Validation, G.R.-M., E.A.S.-P. and J.V.S.-P.; Writing—original draft, X.B. and E.A.S.-P.; Writing—review & editing, X.B., G.R.-M., E.A.S.-P., J.V.S.-P. and N.J.-P. All authors have read and agreed to the published version of the manuscript.

Funding: This work has been supported by the Ministerio de Ciencia, Innovación y Universidades, Spain, under Grant RTI2018-096904-B-I00.

Institutional Review Board Statement: Not applicable.

Informed Consent Statement: Not applicable.

Data Availability Statement: Not applicable.

Acknowledgments: *Conselho Nacional de Desenvolvimento Científico e Tecnológico*, CNPq-Brazil-Finance Code: 310079/2019-5-PQ2.

Conflicts of Interest: The authors declare no conflict of interest. The funders had no role in the design of the study; in the collection, analyses, or interpretation of data; in the writing of the manuscript, or in the decision to publish the results.

Abbreviations

The following abbreviations are used in this manuscript:

DM	Decision Maker
MOEA	Multi-Objective Evolutionary Algorithm
MOP	Multi-Objective Optimization Problem

Appendix A. Equations of the Change of Basis and Related Distance

Once the preference directions which define the dominating basis are established, we have to make a change of basis, which modifies also the equation of the distance in $\mathbb{R}^n$ that must be used. So, the equation for this new distance is provided just by considering the change of basis. Using elementary linear algebra, we get that changing coordinates from the canonical basis to another one—in this case, $\mathcal{B}_d$—is done by using the matrix M_d defined by the coordinates of the vectors $v_{d1} = (v_{d1}^1, ..., v_{d1}^n)$, $v_2..., v_n = (v_{dn}^1, ..., v_{dv}^n)$ as columns, that is,

$$M_d = \begin{bmatrix} v_{d1}^1 & \cdots & v_{dn}^1 \\ \vdots & \ddots & \vdots \\ v_{d1}^n & \cdots & v_{dn}^n \end{bmatrix}. \tag{A1}$$

Consider a vector $v = (x^1, ..., x^n) \in \mathbb{R}^n$, and write $(\alpha^1, ..., \alpha^n)$ for the coordinates of v with respect to the basis $\mathcal{B}_d$. Then,

$$M_d \cdot \begin{bmatrix} \alpha^1 \\ \vdots \\ \alpha^n \end{bmatrix} = \begin{bmatrix} x^1 \\ \vdots \\ x^n \end{bmatrix}, \tag{A2}$$

$$\begin{bmatrix} \alpha^1 \\ \vdots \\ \alpha^n \end{bmatrix} = M_d^{-1} \cdot \begin{bmatrix} x^1 \\ \vdots \\ x^n \end{bmatrix}, \tag{A3}$$

Thus, the definition of a new basis changes the associated norm, and this can also be used in the decision-making step for selecting/coloring points with the lower distance to a particular goal (see Section 3.2). Note that this distance takes into account the designer preferences. The definition of the norm with the new basis can be obtained as follows.

Let us find the equations that allow to write the scalar product $\cdot_{\mathcal{B}_d}$ in $\mathbb{R}^n$ that satisfies that the elements of $\mathcal{B}_d$ define an orthonormal basis of the space. For this, take the matrix M_d (see (A1)), if $w = (y^1, ..., y^n) \in \mathbb{R}^n$, the scalar product $v \cdot_{\mathcal{B}_d} w$ can be written as:

$$v \cdot_{\mathcal{B}_d} w = (x^1, ..., x^n) \cdot_{\mathcal{B}_d} \begin{bmatrix} y^1 \\ \vdots \\ y^n \end{bmatrix} = (x^1, ..., x^n) \cdot (M_d^{-1})^T \cdot (M_d^{-1}) \cdot \begin{bmatrix} y^1 \\ \vdots \\ y^n \end{bmatrix}. \tag{A4}$$

That is, the Gram matrix G associated with $\cdot_{\mathcal{B}_d}$ is given by $G = (M_d^{-1})^T \cdot (M_d^{-1})$. Therefore, the associated Euclidean norm can be computed as:

$$\|v\| := \sqrt{(x^1, ..., x^n)(M_d^{-1})^T \cdot (M_d^{-1}) \begin{bmatrix} x^1 \\ \vdots \\ x^n \end{bmatrix}} = \sqrt{\sum_{i=1}^n (\alpha^i)^2}, \tag{A5}$$

where $(\alpha^1, ..., \alpha^n)$ are the coordinates of v in $\mathcal{B}_p$.

Summing up, we have that the norm computed following the new dominance basis change the perception of distance according to the selected preference directions. Figure A1 shows all the points that are at a distance equal to 1, together with the preference directions and the dominance basis.

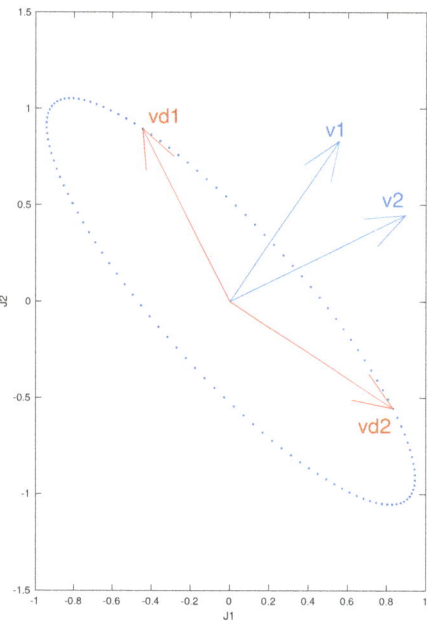

Figure A1. Ellipsoid representing the unit circle in the norm associated with the dominance basis. The preference directions fixed by the DM are shown (blue). The corresponding dominance basis is also shown (red).

References

1. Wang, H.; Olhofer, M.; Jin, Y. A mini-review on preference modeling and articulation in multi-objective optimization: Current status and challenges. *Complex Intell. Syst.* **2017**, *3*, 233–245. [CrossRef]
2. López-Jaimes, A.; Coello, C.A.C. Including preferences into a multiobjective evolutionary algorithm to deal with many-objective engineering optimization problems. *Inf. Sci.* **2014**, *277*, 1–20. [CrossRef]
3. Coello, C.A.C. Handling preferences in evolutionary multiobjective optimization: A survey. In Proceedings of the 2000 Congress on Evolutionary Computation, CEC00 (Cat. No.00TH8512), La Jolla, CA, USA, 16–19 July 2000; Volume 1, pp. 30–37. [CrossRef]
4. Bechikh, S.; Kessentini, M.; Said, L.B.; Ghédira, K. Preference Incorporation in Evolutionary Multiobjective Optimization. In *Advances in Computers*; Elsevier: Amsterdam, The Netherlands 2015; Chapter 4, pp. 141–207. [CrossRef]
5. Deb, K.; Sundar, J. Reference point based multi-objective optimization using evolutionary algorithms. In Proceedings of the 8th Annual Conference on Genetic and Evolutionary Computation—GECCO'06, Seattle, DC, USA, 8–12 July 2006; ACM Press: New York, NY, USA 2006. [CrossRef]
6. Reynoso-Meza, G.; Sanchis, J.; Blasco, X.; García-Nieto, S. Physical programming for preference driven evolutionary multi-objective optimization. *Appl. Soft Comput.* **2014**, *24*, 341–362. [CrossRef]
7. Messac, A. Physical programming—Effective optimization for computational design. *AIAA J.* **1996**, *34*, 149–158. [CrossRef]
8. Reynoso-Meza, G. Controller Tuning by Means of Evolutionary Multiobjective Optimization: A Holistic Multiobjective Optimization Design Procedure. Ph.D. Thesis, Universitat Politècnica de València, Camino de Vera s/n, Valencia, Spain, 2014 [CrossRef]
9. Sanchis, J.; Martínez, M.A.; Blasco, X. Integrated multiobjective optimization and a priori preferences using genetic algorithms *Inf. Sci.* **2008**, *178*, 931–951. [CrossRef]
10. Cruz-Reyes, L.; Fernandez, E.; Sanchez, P.; Coello Coello, C.A.; Gomez, C. Incorporation of implicit decision-maker preferences in multi-objective evolutionary optimization using a multi-criteria classification method. *Appl. Soft Comput.* **2017**, *50*, 48–57 [CrossRef]
11. Liu, R.; Li, J.; Feng, W.; Yu, X.; Jiao, L. A new angle-based preference selection mechanism for solving many-objective optimization problems. *Soft Comput.* **2018**, *22*, 6311–6327. [CrossRef]
12. Sagawa, M.; Kusuno, N.; Aguirre, H.; Tanaka, K.; Koishi, M. Evolutionary Multiobjective Optimization including Practically Desirable Solutions. *Adv. Oper. Res.* **2017**, *2017*, 1–16. [CrossRef]
13. Thiele, L.; Miettinen, K.; Korhonen, P.J.; Molina, J. A Preference-Based Evolutionary Algorithm for Multi-Objective Optimization *Evol. Comput.* **2009**, *17*, 411–436. [CrossRef] [PubMed]

4. Li, K.; Chen, R.; Min, G.; Yao, X. Integration of Preferences in Decomposition Multiobjective Optimization. *IEEE Trans. Cybern.* **2018**, *48*, 3359–3370. [CrossRef] [PubMed]
5. Zou, J.; He, Y.; Zheng, J.; Gong, D.; Yang, Q.; Fu, L.; Pei, T. Hierarchical preference algorithm based on decomposition multiobjective optimization. *Swarm Evol. Comput.* **2021**, *60*, 100771. [CrossRef]
6. Miettinen, K.M. *Nonlinear Multiobjective Optimization*; Kluwer Academic Publishers: New York, NY, USA, 1998. [CrossRef]
7. Blasco, X.; Herrero, J.M.; Sanchis, J.; Martínez, M. A new graphical visualization of n-dimensional Pareto front for decision-making in multiobjective optimization. *Inf. Sci.* **2008**, *178*, 3908–3924. [CrossRef]
8. Hwang, C.L.; Lai, Y.J.; Liu, T.Y. A new approach for multiple objective decision making. *Comput. Oper. Res.* **1993**, *20*, 889–899. [CrossRef]

Article

A Two-Stage Mono- and Multi-Objective Method for the Optimization of General UPS Parallel Manipulators

Alejandra Ríos [1], Eusebio E. Hernández [1,†] and S. Ivvan Valdez [2,*,†]

1. Instituto Politécnico Nacional, ESIME Ticomán, Mexico City 07738, Mexico; arioss1000@alumno.ipn.mx (A.R.); euhernandezm@ipn.mx (E.E.H.)
2. CONACYT, Centro de Investigación en Ciencias de Información Geoespacial, CENTROGEO A.C., Querétaro 76703, Mexico
* Correspondence: sergio.valdez@conacyt.mx
† These authors contributed equally to this work.

Abstract: This paper introduces a two-stage method based on bio-inspired algorithms for the design optimization of a class of general Stewart platforms. The first stage performs a mono-objective optimization in order to reach, with sufficient dexterity, a regular target workspace while minimizing the elements' lengths. For this optimization problem, we compare three bio-inspired algorithms: the Genetic Algorithm (GA), the Particle Swarm Optimization (PSO), and the Boltzman Univariate Marginal Distribution Algorithm (BUMDA). The second stage looks for the most suitable gains of a Proportional Integral Derivative (PID) control via the minimization of two conflicting objectives: one based on energy consumption and the tracking error of a target trajectory. To this effect, we compare two multi-objective algorithms: the Multiobjective Evolutionary Algorithm based on Decomposition (MOEA/D) and Non-dominated Sorting Genetic Algorithm-III (NSGA-III). The main contributions lie in the optimization model, the proposal of a two-stage optimization method, and the findings of the performance of different bio-inspired algorithms for each stage. Furthermore, we show optimized designs delivered by the proposed method and provide directions for the best-performing algorithms through performance metrics and statistical hypothesis tests.

Keywords: two-stage method; mono and multi-objective optimization; multi-objective optimization; optimal design; Gough–Stewart; parallel manipulator; performance metrics

1. Introduction

Over the last few decades, parallel robotic architectures have attracted considerable attention because they provide precise motion, high rigidity, and low inertia of moving parts under high loads. Compared to serial link manipulators, they do not have a long range of motion displacements, but they exhibit higher stiffness and better load capacity, as the external forces are distributed to several in-parallel kinematic chains [1]. In particular, the general 6 6-Degrees-Of-Freedom (DOF) Stewart platform is composed of closed kinematic chains sharing the same payload platform, a base, six extendable linear actuators, and two sets of six joints (see Figure 1). It was proposed first as the moving platform for aircraft simulators, but recently, it has been adopted in different domains, such as machine tools [2,3], vibration isolation devices [4], secondary positioning of radio telescopes [5,6], and non-destructive aeronautical inspection [7].

The design optimization of robotic manipulators deals with finding the best lengths, control gains, and other configuration parameters that minimize or maximize one or several objectives—for instance, minimizing the error or energy and maximizing dexterity or rigidity. Optimization algorithms from various families are employed for robotic design, and metaheuristics have shown impressive results in this engineering area. Working methods vary under a system's complexity and the number of objectives to optimize. According to Botello et al. [8,9], the optimization problems of robotic manipulators can be classified into three categories: The first is the static design problem, in which a homogeneous transformation or zero-order differential equation needs to be solved. Typical problems of this group are dexterity, stiffness, workspace volume, and manipulability maximization. The characteristic of this kind of problem is that the mathematical model does not depend on time. That is to say, the numerical simulation only computes indexes that are dependent on static positions. Usually, the output of this optimization is a set of lengths or sizes of the robot elements; this is called dimensional synthesis. In the state of the art, Panda et al. [10] proposed an evolutionary optimization algorithm based on the foraging behavior of *Escherichia Coli* bacteria for the optimization of the workspace volume of a three-revolute-type manipulator. The results were compared with the Genetic Algorithm (GA), Particle Swarm Optimization (PSO), and Differential Evolution (DE). Badescu and Mavroidis [11] proposed the optimization of the workspace of three-legged parallel platforms through a Monte Carlo method using three different indices: the workspace volume, the inverse of the condition number, and the global condition index. Lou et al. [12] addressed the workspace maximization of a Gough–Stewart platform by applying the Controlled Random Search (CRS) technique and measuring the dexterity index over the workspace. The solution of the inverse kinematics of serial robots also falls into this classification; in this subject, Ayyildiz and Cetinkaya applied the GA, PSO, Quantum Particle Swarm Optimization (QPSO), and Gravitational Search Algorithm (GSA) to the solution of the inverse kinematics of a four-DOF manipulator. Similarly, Dereli and Köker [13] addressed the inverse kinematics problem of a seven-DOF mechanism through the Firefly Algorithm (FA), PSO, and Artificial Bee Colony (ABC). Falconi et al. [14] proposed a solution to the inverse kinematics of generic serial robots in a more general way through the Behavioral-based Particle Swarm Optimization (B-PSO) algorithm. Considering that there is no need to solve differential equations, but rather algebraic expressions, this kind of problem usually demands the lowest computational resources compared to the next two categories. In addition, they usually do not require specialized robotics software.

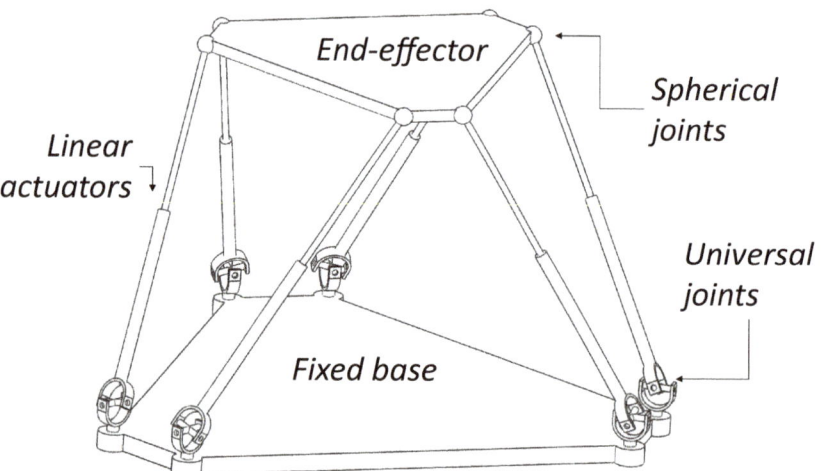

Figure 1. General scheme of the 6–6 degree-of-freedom (DOF) Stewart platform mechanism.

The second category is the kinematic design problem, which implies the solution of first-order ordinary differential equations, that is, the computation of velocities in a system. They are usually controlled by a Proportional Integral Derivative (PID) controller or some of its variations. This case is less common because it does not represent the complete physics of the problem, since neither the mass nor accelerations are involved in the equations. In this view, the kinematic design problem is useful when the involved forces are not considered. Hence, the most important and the main purpose is the control of velocities.

In the third category, the dynamic design problem, second-order differential equations are involved. Hatem et al. [15] developed a clear example of this category, in which trajectory planning subject to kinematic and dynamic constraints was approached. A Direct Search Optimization method was developed for determining the optimal task time and optimal joint torques of serial robots. Kuçuk [16] performed trajectory planning using the PSO for parallel manipulators. This optimization was divided into two stages: The first one was focused on the derivation of a minimum-time trajectory, and the second stage looked for the elimination of the jerk. Furthermore, more evolutionary algorithms have been implemented for the dynamic optimization of robotic design [17]. These works perform dynamic optimization, but they do not consider optimization of the dimensional structure as an initial step. This may cause the design to be inappropriate for the task.

In addition, the classification of the optimization tasks using metaheuristics takes another direction when the objectives are more than one. In this case, the optimization can be approached by dividing the process into the individual solution of each objective function. Most of the current approaches in robotics optimization address a separate analysis of the dimensional synthesis [18,19], accuracy, or dexterity maximization [20] and the control design [21], among other objectives. Nevertheless, there are a few proposals that consider an optimization framework that integrates several objectives into a single process [22–26]. For instance, a concurrent design has been tested through various design methods in delta parallel manipulators [8]. In this regard, a limitation for considering such a unified framework is that various objectives conflict. In the real-world design process, decision-makers select parameters through different design stages in an iterative process by using their knowledge and expertise to approach the best selection. This process requires several inputs, such as time, hardware, software, knowledge, and expertise. It outputs a design with different degrees of closeness to the best according to the set of decisions made. Thus, the aim of optimizing several objectives simultaneously is to support real-world designers, that is to say, it is not to replace the designer's final decision, but to provide a reduced set of close-to-optimal solutions by automatically discarding those solutions that, evidently, are not optimal. In this regard, evolutionary multi-objective algorithms naturally work on a set of candidate solutions that are evolved to approach a Pareto optimal set. The solutions in this set are incomparable in the sense that one cannot improve an objective without the detriment of another. From these solutions, the decision-maker must select the most convenient. Notice that the method does not substitute the designer's work, but reduces the dependence on experiments, time, and expertise. In addition, it foments the generation of high-performance designs.

In this context, research with multi-objective genetic algorithms for symmetric and unsymmetrical Stewart platforms has been applied for aeronautical purposes. Here, Cirillo et al. [27] intended to maximize the payload and minimize the forces used to experiment during positioning. Joumah and Albitar [28] approached the optimization of a six-DOF Revolute-Universal-Spherical (RUS) Stewart platform in two ways: The first stage solved three cost functions separately related to the workspace volume and the global conditioning index, as well as the stiffness index of the structure with a GA, while the second stage unified these indices into a single cost function with a GA and PSO. The output was a manipulator with optimal geometry and structure. Analogously to this work, Nabavi et al. [29] proposed a one-stage optimization for similar indices: the global workspace index, the global dexterity Index, and the global kinetic energy index. They found Stewart platform configurations with a Prismatic-Universal-Spheric (PUS) structure. In addition, they analyzed the obtained Pareto front and the interaction between the existing indices, elucidating that there exists an architecture that significantly lowers the maximum actuator's static and dynamic forces. Lara-Molina et al. [30] addressed the optimal design of parallel manipulators based on multi-objective optimization by considering as objective functions the global conditioning index, the global payload index, and the global gradient index. These indices were evaluated over a required workspace. This optimization method consisted of two stages: The first one, solved with a GA, comprised a simulation for the maximization of the global conditioning index, and the behavior of the rest of the indices was observed. The second was a simultaneous optimization of the three indices by applying a Multiobjective Evolutionary Algorithm (MOEA) based on the Control Elitist Non-dominated Sorting Genetic Algorithm (CENSGA) to find the Pareto front. The second stage derived an optimal geometric configuration for the parallel manipulator. According to the review of this research in the field of Stewart platform optimization and despite the application of many metaheuristic algorithms, multi-objective optimization for dynamic problems is still weak.

Contributions

We contemplate three possible cases in multi-objective optimization: The first occurs when minimizing one of the objectives, which leads to minimization of the other; this is called support. The second case is when the objectives can be independently optimized and minimizing one does not affect the other; mathematically this means that the different objective functions depend on sets of variables without intersection. The third case occurs when minimizing one objective, which leads to maximizing another; this case is called conflict.

In this proposal, we consider that dimensional synthesis for approaching a regular workspace with sufficient dexterity is independent of error and energy. This assumption is due to practical considerations. To optimize the geometry, we only use kinematics, which is cheap in the computational sense; in addition, a configuration with minimum lengths and high dexterity provokes less error and reduces the peaks of energy required in near-to-singular positions. For the sake of completeness, the problem could be approached as a three-objective problem, including the workspace fitting as an additional objective. Nevertheless, this will considerably increase the computational cost and it would increment the complexity of the software development in order include parametric design changes in a dynamic simulator. Optimizing the three objectives means that a different geometry is delivered for each trajectory, which is not the rule in practical applications. This dimensional synthesis is performed independently with three different trajectories given by parametric equations. In the second stage, we aim to minimize the error and energy; both objectives are in conflict and depend on the control gains. This optimization requires a dynamic simulation, which is the most computationally expensive part of the algorithm. In addition, different trajectories can be optimized for different control gains without modifying the geometry of the mechanism. This stage considers only the optimization of one of the three trajectories from the dimensional synthesis. To the best of our knowledge, there is not similar research on the state of the art that combines mono- and multi-objective optimization to define a methodology that integrates various design objectives. The existing research on optimizing the Stewart platform focuses separately on either the static or dynamic parameters. Moreover, many of the above-mentioned approaches for multi-objective optimization combine the many objectives into one single objective function using weighted sums. Hence, the designer obtains a single objective value instead of a Pareto front, which offers the designer a family of optimal solutions along the objectives. Consequently, the designer makes decisions about the importance of the objectives without the information of the affectation in the platform's performance. For this reason, this proposal considers treating the objectives in the second stage as separate issues so that the designer can decide on the most optimal parameters by analyzing only a reduced set of the best-performing solutions.

Our proposal considers the merging of both sequential and concurrent optimization methodologies as a design method for parallel robots considering both static and dynamic optimization. The static optimization ensures that the parallel robot is capable of reaching a desired dexterous workspace, and the dynamic optimization looks for high-performance operation conditions using the optimal control.

Finally, in light of the metaheuristics used, we contrast several mono- and multi-objective evolutionary algorithms to give directions about the best-performing optimizer for these kinds of applications.

2. Optimization Design Criteria

In this section, we briefly review the kinematics and dynamics of the general Universal Prismatic Spherical (UPS) Gough–Stewart platform to present the necessary concepts for defining the objective functions of the optimization problems.

2.1. Kinematics Computation

To define a reachable workspace, we need the computation of the inverse kinematics of the system in order to obtain the lengths effected by each linear actuator in terms of the position and orientation, as shown in Equation (1). The Jacobian expression in Equation (2) is derived from the inverse kinematics. It is used for defining the dexterity index.

Thus, the inverse kinematics can be derived from the schematic of Figure 2a. L_i represents the vector along the linear positioners and n_i represents its corresponding unit vector, acting in terms of the position x and orientation in the moving platform $[\psi, \theta, \phi]$, also called the end-effector. Vectors describing the positions of the spherical and universal joints are included to form closed kinematic chains. Thus, p_i and b_i denote the position of both spherical and universal joints, respectively, according to the inertial reference framework B, while p_i^P denotes the spherical joint with P as a reference framework. Moreover, the manipulator is assumed to be symmetric, and the locations of the universal and spherical joints are equally distant with a radius r_B in the base (Figure 2b) and r_P in the moving platform (Figure 2c), with equal aperture angles between pairs δ_B and δ_P. Thereby, the expression for the lengths L_i can be written as

$$l_i = |L_i| = |x + R_P^B p_i^P - b_i|, \tag{1}$$

and, after deriving it, we can obtain the Jacobian expression in the next form:

$$\dot{l} = J_1 \begin{bmatrix} \dot{x} \\ \omega \end{bmatrix} = \begin{bmatrix} n_1^T & (\frac{R_P^B}{r_P} p_1^P \times n_1)^T \\ \vdots & \vdots \\ n_6^T & (\frac{R_P^B}{r_P} p_6^P \times n_6)^T \end{bmatrix} \begin{bmatrix} \dot{x} \\ \omega \end{bmatrix}. \tag{2}$$

In Equation (1), R_P^B corresponds to the rotation matrix, which is derived in this work by following the $z - y - x$ sequence of axis rotation, while the angles ψ, θ, and ϕ denote the rotations made about the z, y, and x axis, most commonly known as yaw, pitch, and roll angles, respectively, and trigonometric functions are written in a short form: s denotes *sine* and c denotes *cosine*.

$$R_P^B = \begin{bmatrix} c\psi c\theta, & -s\psi c\theta + c\psi s\theta s\phi, & s\psi s\phi + c\psi s\theta c\phi, \\ s\psi c\theta, & c\psi c\theta + s\psi s\theta s\phi, & -c\psi s\phi + s\psi s\theta c\phi, \\ -s\theta, & c\theta s\phi, & c\theta c\phi, \end{bmatrix}. \tag{3}$$

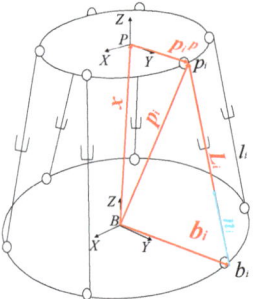
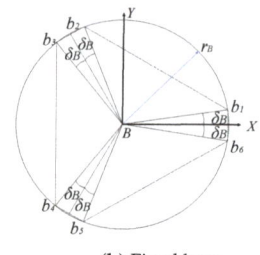

(a) Defining vectors for kinematic model

(b) Fixed base

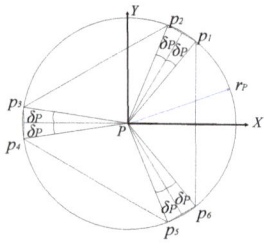

(c) End-effector

Figure 2. General Universal-Prismatic-Spherical (UPS) Gough-Stewart structure. Stage 1 looks for the optimal $r_B, r_P, \delta_B, \delta_P$, and maximum stroke l_{max} to cover a dexterous workspace. Stage 2 uses the resulting geometry to provide an optimal dynamic control.

2.2. Regular Workspace Computation Considering Sufficient Dexterity

The optimization algorithms generate candidate length configurations corresponding to the defining sizes of the parallel robot. Each length configuration is called a candidate solution (an individual in the GA context). In order to evaluate them, we must quantify whether they are capable of reaching any point in a set of trajectories with sufficient dexterity. We measure the dexterity of the system with the inverse of the global condition number in Equation (4) using the maximum and minimum singular values of the dimensionally homogeneous Jacobian matrix shown in Equation (2). The dexterity index decreases its value when the robot is close to a singular non-desirable position and increases otherwise. Thus, an acceptable workspace should be restricted to the permitted limits of this value over its complete volume according to the designer's decision.

$$\kappa^{-1} = \frac{\sigma_{min}(J)}{\sigma_{max}(J)}. \tag{4}$$

For this purpose, consider a set of tasks given as target trajectories. They are performed inside a bounding box built by the maximum and minimum coordinates of the target trajectories (see Figure 3) within a limited period. Hence, the platform must be designed to reach, with sufficient dexterity, any point inside such a bounding box.

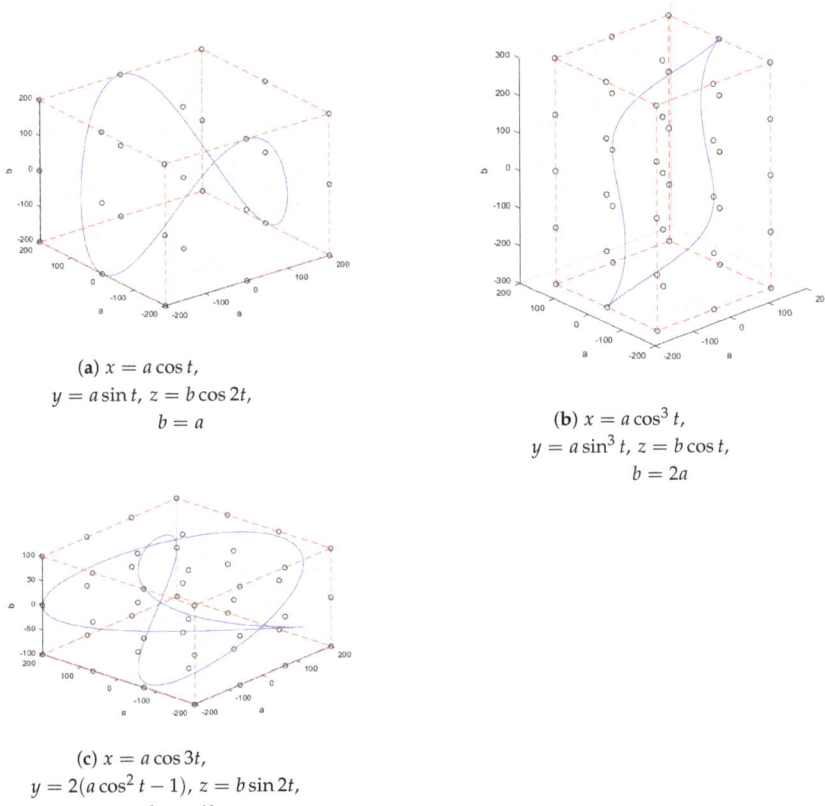

Figure 3. Target trajectories, named hereafter (**a**–**c**), labelled correspondingly, within prismatic bounding boxes, and points of the discretized workspace.

Here, we introduce the procedure for quantifying the volume of the bounding box reached by a candidate platform. To ensure that the target trajectories lie in allowable dexterity boundaries (in terms of the condition number), the bounding boxes are discretized in points that are spatially distributed along the surfaces and interior, and then, the inverse condition number is evaluated at each of these points. For the case of the cube-shaped trajectory shown in Figure 3a, which is called Trajectory **a** hereafter, points are uniformly allocated to ensure their position along the edges, corners, and intermediate positions. This distribution can be seen as the positioning of three equal planes that are parallel to the xy plane and composed of nine points along three different equidistant heights in the z axis, that is to say, a total of 27 points. In a similar way, the trajectory of Figure 3b, called Trajectory b, is discretized by placing the nine-point planes at five equidistant heights of the z axis, resulting in a total of 45 distributed points. For the trajectory of Figure 3c, Trajectory c, it is considered that the bounding box height measures half of the sides' lengths. Here, sixteen points distributed in planes parallel to xy were placed at three different equidistant heights of z, giving a total of 48 points.

For every candidate solution, namely, the lengths of a Stewart platform, the inverse condition number is evaluated at all of the n discretized points, in which the orientation is maintained constant and parallel to the platform base. If the manipulator is not capable of reaching the total of the bounding box points within the allowable dexterity, and with the stroke limitation of each linear actuator, the candidate solution is not considered suitable.

Mono-Objective Optimization for Dimensional Synthesis

The candidate solutions are generated from a vector of decision variables, which mainly denote the geometrical characteristics of the Stewart platform, $d = [r_B, r_P, \delta_B, \delta_P, l_{max}, c_m]$. Here, l_{max} denotes the maximum stroke of the linear actuators, and a last term c_m, is added as a decision variable, indicating the position along the z axis of the center of mass of the bounding box. This way, not only a dimensional synthesis of the robot is made, but also a decision about the best place to locate the trajectory according to this candidate solution.

The objective function in Equation (5) comprises two terms. For the n discretized points, the first term evaluates the competence of the Stewart platform of: (1) reaching the position x_w with the proposed maximum stroke l_{max} and (2) fulfilling the dexterity desired for this point. Each time that a point fulfilling stroke and dexterity conditions is reached, I takes a unit value, and zero otherwise. If all points in the workspace are adequately reached, the sum is equal to n. The second term looks for the minimization of the radius of the base, the radius of the end-effector, and the maximum stroke of the linear actuators. Hence, the optimization parameters of the second term are defined as $d_m = [r_B/r_B^{max}, r_P/r_P^{max}, l_{max}/l_{max}^{max}]$. In this vector, the variables to minimize are normalized with respect to their upper boundaries in such a manner that, in the second term of the objective function, the sum does not take a value higher than the unit, and the smaller this value is, the better the dimensional synthesis. Notice that all feasible geometries that reach the whole workspace report an objective value greater than $n - 1$, and the best—that with the minimum lengths—is the maximum value.

$$f(d) = \sum_{w=1}^{n} I(x_w) - \frac{1}{3}\sum |d_m|, \quad (5)$$

$$\text{where } I(x_w) = \begin{cases} 1 & \text{if } \kappa^{-1}(x_w) \geq 0.2 \text{ and } l_i < l_{max} \\ 0 & \text{otherwise} \end{cases},$$

for the x_w positions of the $w = 1,\ldots n$ discretized points and for the $i = 1,\ldots 6$ linear actuators. The search limits of the decision variables are given in terms of the bounding boxes' sizes, as shown in Table 1.

Table 1. Upper and lower boundaries of the decision variables for the dimensional synthesis.

Trajectory a	Trajectory b	Trajectory c
$0.1a < r_B < 3a$	$0.1b < r_B < 3b$	$0.1a < r_B < 3a$
$0.3a < r_P < 3a$	$0.4b < r_P < 3b$	$0.4a < r_P < 4a$
$5° < \delta_B < 30°$	$5° < \delta_B < 30°$	$5° < \delta_B < 30°$
$7° < \delta_P < 30°$	$7° < \delta_P < 30°$	$7° < \delta_P < 30°$
$0.1a < l_{max} < 3a$	$0.1b < l_{max} < 3b$	$0.1a < l_{max} < 4a$
$b < c_m < 5b$	$b < c_m < 5b$	$b < c_m < 5b$

2.3. Computation of the Dynamics

The dynamics of the Stewart platform are solved through the simulation software Simwise 4D [31]. First, the Stewart platform is constructed according to the best geometry obtained in the first stage with the mono-objective optimization using a CAD software; then, the assembly is exported to Simwise 4D for the simulation of the dynamics, with the linear actuators controlled from MATLAB Simulink. This simulation is used to measure the dynamic properties during the tracking of trajectories along with the time steps.

Error and Energy Multi-Objective Optimization

The optimization algorithms look for the minimal accumulated error and energy consumption during the tracking of a trajectory. To this effect, a PID control (see Figure 4) is applied on the linear actuators, which is limited in stroke by the l_{max} resulting from the first stage. Our decision variables are the PID gains for controlling the Stewart platform, which are $d = [k_p, k_i, k_d]$. Similarly to the first stage, decision variables are limited with lower and upper boundaries so that $[0,0,0] \leq d \leq [2,20,2]$. Notice that the force of the actuators is intrinsically limited by the control gains; hence, they are intrinsically bounded by the search limits. Joint constraints are not considered in this simulation. The PID control law appears in Equation (6).

$$k_p + k_i \frac{1}{s} + k_d \frac{1}{1+\frac{1}{s}}. \qquad (6)$$

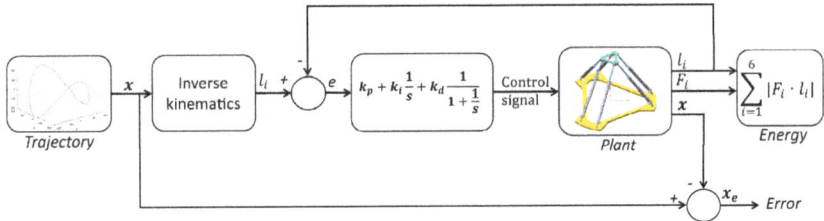

Figure 4. Proportional Integral Derivative (PID) control scheme of the lengths. It is used for the error and energy optimization.

The objective function of this second stage is shown in Equation (7). In the first term, it is the integral over the time of the absolute error of the trajectory. The second term is the integral of the absolute multiplication of force times displacement; this term measures a function related to the accumulated energy through the simulation time. Nevertheless, the first lengths have a larger weight than the last because l_i is the total displacement at time t. This objective function helps us to emphasize the energy expended during the initial steps rather than the rest of the trajectory, considering that the initial steps are those with the greatest energy consumption and greatest energy,

$$f(d) = \begin{bmatrix} \int_0^t |e| dt \\ \int_0^t \sum |F_i \cdot l_i| dt \end{bmatrix}, \text{ for the } i = 1,...,6 \text{ linear actuators.} \qquad (7)$$

3. Optimization Design Process: Implementation and Results

The integral optimization proposal is defined in two stages as follows:

1. Dimensional synthesis of the platform: Using a mono-objective optimization algorithm, we determine the lengths of a platform with minimum dimensions that, with a dexterity measure equal to or greater than 0.2, reaches the points inside a regular workspace.
2. Error and energy optimization: Using the lengths of the previous stage and a multi-objective evolutionary algorithm, we approximate the Pareto set, that is, we obtain a set of arrays of gains of the PID controller with the best performance in both objectives.

For Stage 1, we compare three bio-inspired algorithms, and for Stage 2, we compare two multi-objective algorithms from the state of the art. This methodology is briefly explained in Figure 5.

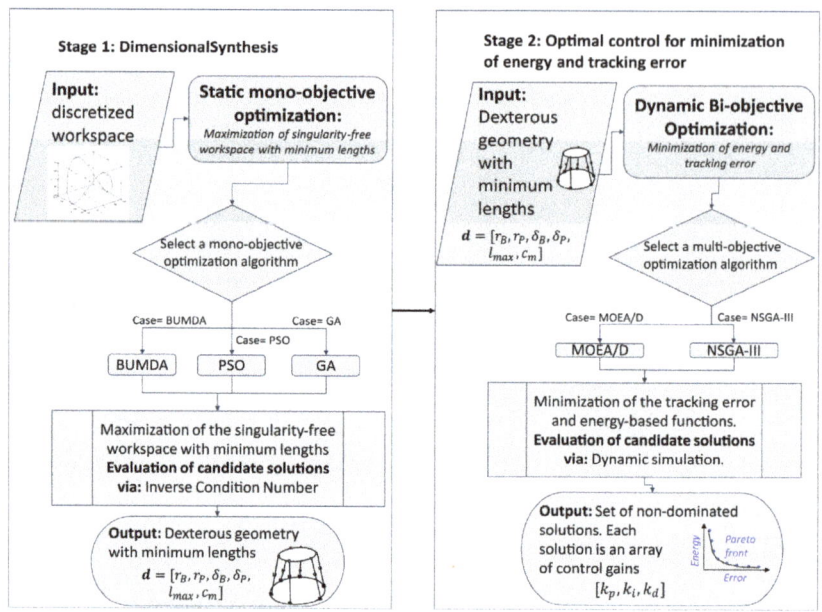

Figure 5. Flowchart of the general optimization process. Stage 1, on the left, finds the geometrical configuration, and Stage 2, on the right, finds the optimal control parameters. Notice that both stages are connected.

3.1. Mono-Objective Optimization Algorithms

For the workspace optimization, a comparison among three different population-based algorithms was made: the Boltzmann Univariate Marginal Distribution Algorithm (BUMDA) [32], which uses a probabilistic method to sample the optimum values, Particle Swarm Optimization, inspired by flocks of birds or swarming insects [33–35], and a Genetic Algorithm based on the research performed by Goldberg [36] and Conn et al. [37,38]. The Particle Swarm Optimization and Genetic Algorithm were taken from the MATLAB routines `particleswarm` [39] and `ga` [40].

The PSO used a weight of the neighborhood best and global best equal to 1.49, and the GA used a scattered crossover, that is to say, a binary random array was generated; the positions with 1 were inherited from parent 1; otherwise, they were inherited from parent 2, with a crossover fraction and Gaussian mutation of 0.8.

For the sake of a fair comparison of the mono-objective optimization algorithms, we executed the three algorithms with an equal population size of 50 and 200 maximum iterations, that is, approximately 10,000 maximum function evaluations; the algorithms also stopped if the search suffered stagnation during the last generations. The stagnation of the objective value was measured as follows:

- For the BUMDA, the standard deviation of the objective function values of the population was less than 10^{-8}.
- For the PSO, the relative change in the best objective function value over the last 20 iterations was less than 10^{-6}, or the maximum number of iterations is met. The relative change was computed with the formula $|f_{best}^{20} - f_{best} / \max(1, |f_{best}|)|$, where f_{best}^k is the best function value in the k-th iteration.
- For the GA, the average relative change in the best objective function value over the last 50 generations was less than or equal to 10^{-6}, or the maximum number of iterations is reached. The average relative change was computed as $|f_{best}^1 - f_{best}^{50}| / (50 \cdot \max(1, |f_{best}^{50}|))$, where f_{best}^k is the best function value in the k-th iteration.

Nevertheless, these settings could favor the BUMDA, which converges faster than the other algorithms, considering that the best results for Trajectory a were [26.561535, 26.560189, 26.502771], for Trajectory b, they were [44.645261, 44.714242, 44.620334], and for Trajectory c, they were [47.713794, 45.584584, 39.524893] for the BUMDA, PSO, and GA, respectively. In consequence, we changed the population sizes to the recommended values [39,40] of 60 and 200 for the PSO and GA, respectively. For both cases, the number of iterations was unlimited. Using these last settings, we obtained the results in Table 2. Detailed information about the algorithms' results can be found by following the "Data Availability" link.

3.2. Multi-Objective Optimization Algorithms

The implementation of two algorithms of different natures [41,42], whose performance has been been recently compared, MOEA/D and NSGA-III [43,44], are executed to find the optimal gains of a PID control given three defined trajectories to track with the end-effector while maintaining a constant orientation. In this second stage, the bi-objective optimization focuses on minimizing the error in the trajectory, as well as the energy consumed. The MOEA/D implementation solves the optimization by decomposing the multi-objective problem into multiple single problems; meanwhile, the NSGA-III proceeds with the Pareto dominance for fitness assignment.

For the sake of completeness, the optimization algorithms were executed with the following parameters: MOEA/D used a weighted crossover as follows: $\alpha \cdot p1 + (1-\alpha) \cdot p2$, where $\cdot$ is a position by position vector multiplication, and α is an array of the same size as the parents, randomly picked from $[-\gamma, 1+\gamma]$ and $\gamma = 0.5$ without mutation, as presented in [41]. NSGA-III used a crossover percentage of 0.5 and mutation percentage of 0.02; it used the same crossover as the MOEA/D, but two children were generated in a single operation from two parents by inverting the positions of the parents. The children mutated a single position using Gaussian mutation with $\sigma = 0.1(Var_{max} - Var_{min})$.

3.3. Mono-Objective Optimization Results

Optimization algorithms use stochastic operators; hence, the results may differ one from another. With the intention of obtaining representative results from each algorithm, we executed them 15 times each and performed a statistical analysis to evaluate the performance of the algorithms in solving these problems. In Table 2, we show the best value of the objective function f resulting from this mono-objective stage for each trajectory. These results show the decision variables, that is, the best geometrical configuration of the Stewart platform yielding the best value of the objective function. The last column shows the number of function evaluations required by the algorithm to reach the resulting objective function value. These results are used for the construction of the geometrical models for the tracking of their related trajectories, as shown in Figure 6.

Table 2. Mono-objective optimization results: the optimized parameters and best objective function values from 15 independent executions of each algorithm.

Traj.	Alg.	$r_{B(mm)}$	$r_{P(mm)}$	$\delta_{B(rad)}$	$\delta_{P(rad)}$	$l_{max(mm)}$	$c_{m(mm)}$	f	n_{eval}
a	BUMDA	748.45	160.21	0.21	0.23	669.82	1099.29	**26.56154**	589
	PSO	815.85	120.02	0.12	0.12	647.26	999.99	26.56024	17,100
	GA	876.03	120.03	0.26	0.13	664.66	999.99	26.53869	50,400
b	BUMDA	927.03	151.24	0.19	0.23	837.33	1316.65	44.64526	442
	PSO	592.71	240	0.09	0.12	710.25	1262.62	**44.71426**	10,080
	GA	691.70	260.54	0.21	0.12	736.93	1296.01	44.68719	20,600
c	BUMDA	495.40	322.40	0.18	0.27	391.22	791.50	**47.71379**	2843
	PSO	1113.84	160.00	0.11	0.12	578.97	500	47.53664	8100
	GA	988.34	160.04	0.12	0.13	521.48	499.88	45.58347	52,800

We compared our Stewart platform's design from Stage 1 with a generic methodology proposed in [45]. In this design method, the author proposed a set of decision variables to define the dimensions of a Stewart platform manipulator; then, he used a simulator to verify that it complied with the desired workspace. If the proposed dimensions do not fit the workspace, then the designer should modify them until a satisfactory solution is found. We applied this methodology to find a configuration for comparing and an objective function value for Trajectory a within the same lower and upper boundaries. The best value found after many iterations in the decision variables was $f(d) = 23.5278$ for the configuration $d = [800, 300, 0.14, 0.40, 600, 1000]$, while the best values found by the metaheuristics were 26.56152, 26.56024, and 26.53869 by the BUMDA, PSO, and GA, respectively. Despite the many attempts, we could not find a greater value of the objective function with the methodology in [45]. In addition, since f did not meet $f >= n - 1$, the desired workspace was not completely fulfilled. In comparison, our methodology found an objective function value greater than $n - 1$ and also optimizes it in seconds, while the contrasting method took hours.

(a) Model for Trajectory a

(b) Model for Trajectory b

(c) Model for Trajectory c

Figure 6. Results for dimensional synthesis based on target trajectories.

Despite the best results obtained with our optimization, we performed statistical analyses among the performances of the algorithms in this stage. This way, we could demonstrate whether one algorithm was better than the others in each trajectory. Table 3 shows a statistical comparison using a non-parametric test, the Wilcoxon rank sum test, and a t-test. These tests evaluated the null hypothesis—that the mean objective values from two algorithms are equal at the 5% significance level. The samples consist in the objective function values f obtained from 15 executions of each algorithm. The rejection of the null hypothesis shows that the two samples compared are significantly different. We compared the algorithms with the best results for each trajectory against the others.

Table 3. *p*-values and acceptance/rejection of the null hypothesis of the Wilcoxon rank sum test and the *t*-test. × indicates that the left algorithm is better than the right one, while √ indicates that there is not a significant difference between them.

Traj.	Algorithm	Wilcoxon		*t*-Test	
		p	×/√	*p*	×/√
a	BUMDA vs. PSO	0.000576	×	0.000642	×
	BUMDA vs. GA	0.198456	√	0.268204	√
b	PSO vs. BUMDA	0.00000	×	0.000003	×
	PSO vs. GA	0.000001	×	0.000003	×
c	BUMDA vs. PSO	0.861813	√	0.213323	√
	BUMDA vs. GA	0.000007	×	0.000003	×

3.4. Multi-Objective Optimization Results

This section shows the results obtained from the multi-objective optimization over Trajectory *a*. The second stage of this approach yields a set of Pareto solutions that help the designer to decide on the best gains for the PID controllers of the lengths of the actuators, since each individual in the Pareto front is a PID gain combination. The outputs of the optimization algorithms are highly influenced by the population and number of generations used for the executions. We executed the MOEA/D and NSGA-III 15 times with the configurations of [50, 200] and [100, 100] in population size and generations, respectively (60 total executions). The Pareto sets resulting from all of the executions were combined to yield the reference Pareto front, shown in Figure 7. This Pareto front only considered the non-dominated solutions, shown in red dots, while the dominated (blue crosses) were discarded. Then, each algorithm with the two sets of parameters was compared to the reference Pareto front using performance metrics to find which configuration was the best for the solution of the multi-objective problem. The results from each algorithm separately are depicted in Figure 8. The first performance metric was the Approximated Hypervolume (HV), which, in practical terms, reveals the extension and covering of the Pareto front. The other performance metrics were the Generational Distance (GD) and the Positive Inverse Generational Distance (IGD+); this last metric is a unique weak Pareto-compliant binary metric. These metrics expose how close the current Pareto set is in comparison to the reference one. The lower the value of the GD, the better the proximity to the ideal set, while the IGD+ acts inversely. Table 4 shows the metric results obtained from 15 executions of each algorithm configuration. On average, the MOEA/D performed the best, although these average values did not reveal whether the difference between the compared algorithms was statistically significant. In this regard, Table 5 shows the results of the Wilcoxon test and the *t*-test to statistically compare the mean of the IGD+ of the 15 independent executions. In the discussion section, we argue that this metric is the best suited for determining the best multi-objective solution from an engineering point of view. As a final result of this stage, we highlighted three results in the Pareto front in Figure 7, corresponding to three different regions of the objective function's space, that is to say, in the middle and extremes of both objectives. Then, we simulated these optimal results to visually analyze the tracking error; the simulations and their associated PID gains are shown in Figure 9.

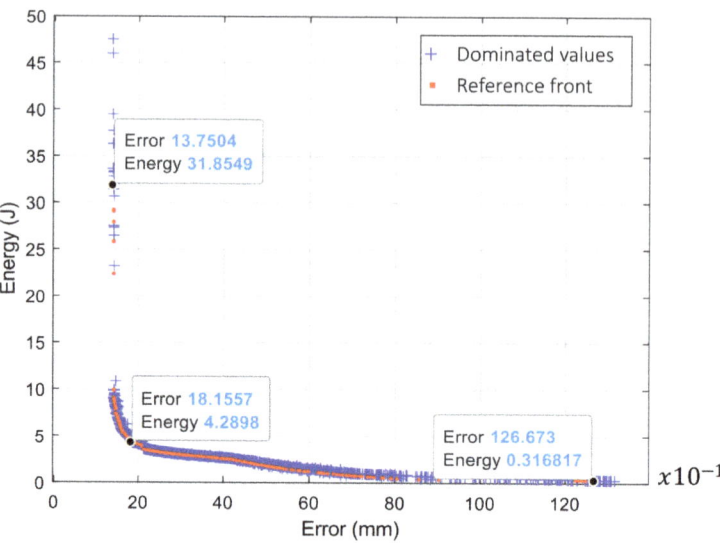

Figure 7. The non-dominated Pareto front versus the dominated error–energy values of both the MOEA/D and NSGA-III executions; labeled points are the configurations in Figure 9.

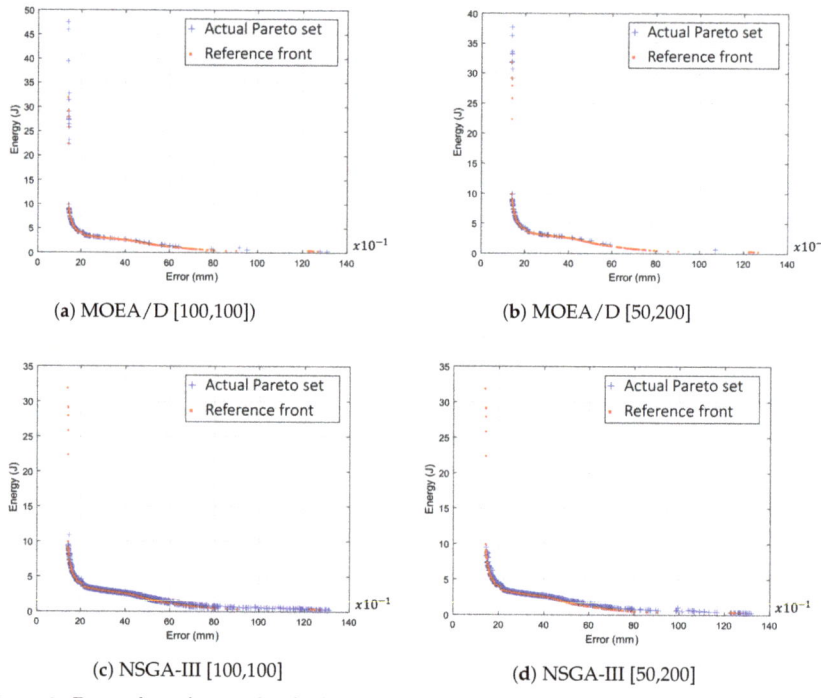

Figure 8. Pareto front from individual configurations of MOEA/D and NSGA-III from 15 independent executions. (**a**) MOEA/D with population size of 100 and 100 generations, (**b**) MOEA/D with population size of 50 and 200 generations, (**c**) NAGS-III with population size of 100 and 100 generations, and (**d**) NSGA-III with population size of 50 and 200 generations.

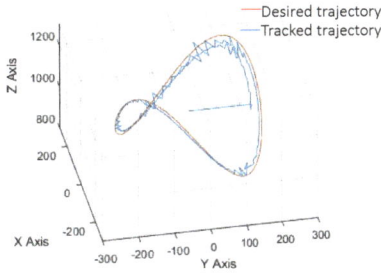
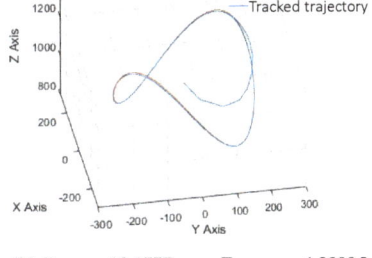

(a) Error= 13.7504 mm, Energy= 31.8549 J; ($k_p = 1.9965$, $k_i = 20$, $k_d = 1.7312$)

(b) Error= 18.1557 mm, Energy= 4.2898 J; ($k_p = 0.7099$, $k_i = 19.7958$, $k_d = 0.0011$)

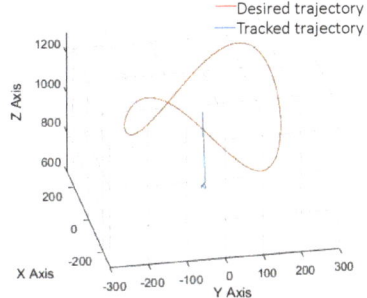

(c) Error= 126.6734 mm, Energy= 0.3168 J; ($k_p = 0.0272$, $k_i = 0$, $k_d = 0.0122$)

Figure 9. Three different simulations from the extremes and middle of the obtained Pareto front of Figure 7 and associated PID gains. (**a**) Low error but high energy consumption; (**b**) an error–energy balance; (**c**) denotes low energy but high error.

Table 4. Averages of performance metrics over the multi-objective results. Best values in bold.

Algorithm [nPop,nGen]	Approx. Hypervolume (Higher Is Better)	Generational Distance (Lower Is Better)	Positive Inverse Generational Distance (Higher Is Better)
MOEA/D [100,100]	**5345.740519**	1.823100	0.299273
MOEA/D [50,200]	3297.389387	**1.158101**	**0.348441**
NSGA-III [100,100]	1102.987694	1.623127	0.049558
NSGA-III [50,200]	927.873283	2.064423	0.181521

Table 5. p-values and acceptance/rejection of the Wilcoxon test and t-test for the multi-objective results of the Positive Inverse Generational Distance (IGD+) for 15 independent executions. Algorithms with statistical differences are in bold. × indicates that the left algorithm is better than the right one, while √ indicates that there is not a significant difference between them.

Algorithm Comparison [nPop,nGen]	Wilcoxon p	×/√	t-Test p	×/√
MOEA/D ([100,100] vs. [50,200])	1.000000	√	0.632716	√
MOEA/D [100,100] vs. NSGA-III [100,100]	**0.000003**	×	**0.000246**	×
MOEA/D [100,100] vs. NSGA-III [50,200]	**0.020191**	×	0.191011	√
MOEA/D [50,200] vs. NSGA-III [100,100]	**0.000003**	×	**0.000004**	×
MOEA/D [50,200] vs. NSGA-III [50,200]	**0.018054**	×	0.231068	√
NSGA-III ([100,100] vs. [50,200])	**0.000057**	×	**0.006621**	×

4. Discussion

4.1. Mono-Objective Optimization

The results in Table 2 show that the BUMDA delivers the best results for two of the three cases, and the PSO delivers the best result for the other; then, according to Table 3 for the BUMDA for Trajectory *a*, the tests reject the null hypothesis when comparing the BUMDA to the PSO and accept it when comparing BUMDA with the GA. For the second trajectory, the PSO performs the best, and for the third, the BUMDA is better than the GA but it is not significantly better than the PSO.

In general, the GA is not the best for any case, and the best algorithm for each trajectory is significantly better than it. Hence, the adequate algorithm selection is between the PSO and the BUMDA. Both algorithms were executed with their default parameters; in this sense, it is possible that the PSO improves its performance if the best execution parameters are used; nevertheless, this requires additional research and computational time for each design problem. On the other hand, the BUMDA requires a single execution parameter, the population size, which is not required to be tuned, according to the recommended settings of 10 times the number of optimization variables [8,32]. Thus, we recommend the use of the BUMDA by default, and to use the PSO if the solution with the BUMDA is not satisfactory or if the trajectory is highly complex, although using the PSO could require one to look for optimal execution parameters.

4.2. Multi-Objective Optimization

The results of the hypothesis tests in Table 5 show that the MOEA/D performed better than the NSGA-III for all cases; nevertheless, there are no statistical differences when using the MOEA/D with different execution parameters. According to the observed results, the largest population size increases the coverage of the Pareto front, while the shortest benefited the convergence. Usually, in the research field of multi-objective optimization, researchers aim to find Pareto front approximations with the maximum extension or coverage of the solution space. Nevertheless, although the results in the Pareto front represent a minimization in one or both objectives, as observed in Figure 7, the extreme solutions that indicate the full extension of the Pareto front are not the most interesting from an engineering point of view, considering that they have the greatest pay-off in the objective functions. That is to say, one extreme of the Pareto front shows a solution that must substantially increase the energy consumption in order to reduce the error by a small quantity, while the second extreme increases the energy for reducing the error with a shorter pay-off than the other. However, even in this case, the solution is not of interest because the energy is too high in contrast with the solutions in the middle, and even though the error is lower than the other solutions, the error signal oscillates due to the large value of the proportional gain, as noticed in Figure 9. Hence, the most useful solutions from an engineering point of view are those with the most balanced pay-off, namely, those in the middle of the Pareto front. For this reason, in the statistical comparison shown in Table 5, we preferred to use the IGD+ metric to emphasize the proximity of the Pareto fronts to the reference instead of using the HV to prioritize the optimality of the solutions rather than the extension of the Pareto front. Once we identified that the middle of the Pareto front was our interest area, comparing the HV was incorrect, since it does not represent the measure of our target. Thus, according to Table 5, we preferred the set of parameters that benefited the convergence, that is, MOEA/D with a population size of 50 and 200 generations.

Table 4 shows the HV of the two algorithms and parameters that were compared. Usually, the performance of multi-objective evolutionary algorithms is measured according to three features: spreading, distribution, and convergence. The first is the extension of the solutions of the approximation of the Pareto front, the second is the representativeness of the solutions—usually, a finite set of equally distributed solutions is desirable to represent a possibly infinite set of optimal solutions—and convergence is measured as the distance to the real Pareto front. The HV measures the first two features; the hypervolume in 2D, such as in this case, is an area measured from the zenith point (the minimum value for each objective) to the points in the Pareto front. Notice that clustered points result in intersecting areas, and the more spread the points are, the larger the covered area is. In this regard, in Table 4, the NSGA-III delivers hypervolumes of approximately one-fifth of the greatest delivered by the MOEA/D, which means that the MOEA/D delivers an approximation to the Pareto front with a larger extension and better-distributed solutions than that delivered by the NSGA-III. In other words, the area generated by the NSGA-III solutions is one-fifth of that delivered by the MOEA/D.

Moreover, the simulation results in Figure 9 confirm that the desired path tracking belongs to the solutions in the middle of the Pareto front, where both error and energy have intermediate values (Figure 9b). In this case, the controller has the greatest gains for the proportional and integrative terms and the lowest for the derivative. For the case in Figure 9a, we see that many oscillations are made around the desired path, and a very abrupt initial step is made because of the high energy expended. This might cause the platform's elements to experiment with high values of stress, and may also be derived from a failure of the linear actuators, which operate under many limitations, such as the payload and velocities. On the other hand, when the energy is kept at the minimal values, the tracked trajectory is not even close to the expected one (Figure 9c), thus highly increasing the error value. As a final point, notice that even the three presented solutions are not dominated; they are not applicable from an engineering point of view. However, this information is a priori unknown, and the desired solutions can only be selected after the multi-objective optimization.

5. Conclusions

In this work, we introduced a method comprised of two stages for the optimization of the structure and control of a Stewart platform by considering a task of trajectory tracking. The purpose of the two stages is to reduce the computational cost by means of partitioning the three-objective problem into mono-objective and bi-objective sub-problems.

In a first stage, we search for the dimensional synthesis of the structure, reaching a bounding box of the trajectory that fulfills a dexterity index to circumvent singular positions. This process is applied to three different trajectories by means of three optimization algorithms: the BUMDA, PSO, and GA. The BUMDA performs the best most of the time, with a significant difference for Trajectories a and c; meanwhile, for Trajectory b, the best algorithm is the PSO. Hence, we suggest using the BUMDA for dimensional synthesis because of its low computational cost and its use of a single execution parameter. Then, for the second stage, we use the geometry generated by the BUMDA for Trajectory *a* to model the geometry of the Stewart platform for the second stage.

In the second stage, we look for the best set of gains of a PID controller using the simulation software Simwise 4D for the dynamics of the robot. The MOEA/D and NSGA-III are used for the optimization of the energy and accumulated error. There exists a conflict between the error and energy; in other words, when the error decreases, the energy increases, and vice versa. As a result of this optimization, we obtain a Pareto front approximation and recall that it is unknown until the multi-objective algorithms compute it. Hence, a posteriori, we detect that the most desirable values for the energy and error are those that are most balanced, that is to say, those in the middle of the Pareto front. The associated PID gains for these solutions reveal that the controller acts with a very small derivative gain, with the highest values in the PI terms. In contrast, for the solutions with the smallest values of error, the energy is highly increased, resulting in non-desired oscillations around the trajectory and a very abrupt initial step that might cause the linear positioners to fail. On the contrary, when the energy is the lowest, the error is the highest, making the trajectory highly distant from the desired path.

Since the initial population is selected randomly, the results from algorithms vary in each execution. We executed each algorithm 15 times in the first and second stages to obtain the best results. The samples obtained from the 15 executions were statistically compared through the Wilcoxon rank sum test and *t*-tests to show which algorithm performed the best for each optimization problem. For the first stage, it was the BUMDA in most of the cases, while in the second stage, the samples for the statistical comparison were made with the IGD+ metric because the interest was in measuring how close the Pareto front approximation was in comparison to the reference front, considering that the best results are located in the middle of the Pareto front, and the solutions in the extremes are not desirable in engineering. Based on this perspective, we observed that the MOEA/D performed better than the NSGA-III in the IGD+ metric, a fact that was further validated by the statistical tests. In this same regard, the population size and iterations were varied for the purpose of obtaining the best execution parameters; they were a population size of 50 and 200 iterations.

For the sake of completeness, three metrics were used for the numerical evaluation of the performance of each algorithm: the Approximate Hypervolume, the Generational Distance, and the Positive Inverse Generational Distance. The MOEA/D performed the best in most cases.

In essence, the main contributions of this work are the proposal of the objective functions and the two-stage method. Furthermore, the implementation of a simulation of the dynamics contributes to decision-makers in selecting practically feasible solutions from a set of optimal solutions. Finally, we contributed with the application of metrics and statistical tests to support the evaluation of the optimization algorithms' performance and to provide directions for the adequate metaheuristics for this problem.

Author Contributions: Conceptualization, E.E.H. and S.I.V.; methodology, E.E.H., S.I.V., and A.R. software, A.R.; validation, S.I.V. and A.R.; investigation, E.E.H., S.I.V., and A.R.; resources, E.E.H.; writing—original draft preparation, S.I.V. and A.R.; writing—review and editing, E.E.H., S.I.V. and A.R.; visualization, A.R.; supervision, E.E.H. and S.I.V.; project administration, E.E.H.; funding acquisition, E.E.H. All authors have read and agreed to the published version of the manuscript.

Funding: The authors are grateful to SIP-Instituto Politécnico Nacional for supporting part of this work through the grants SIP-2109 and SIP-20201545. S. Ivvan Valdez is supported by Cátedras-CONACYT grant 7795.

Institutional Review Board Statement: Not applicable.

Informed Consent Statement: Excluded.

Data Availability Statement: Data from the experiments performed in this article are available via https://drive.google.com/drive/folders/1tJzGleSpmQ_8Xjn35QDO7NBbDfTbHnij?usp=sharing (accessed on 1 March 2021).

Conflicts of Interest: The authors declare no conflict of interest.

References

1. Sun, T.; Lian, B. Stiffness and mass optimization of parallel kinematic machine. *Mech. Mach. Theory* **2018**, *120*, 73–88. [CrossRef]
2. Pedrammehr, S.; Mahboubkhah, M.; Khani, N. A study on vibration of Stewart platform-based machine tool table. *Int. J. Adv. Manufac. Technol.* **2013**, *65*, 991–1007. [CrossRef]
3. Pugazhenthi, S.; Nagarajan, T.; Singaperumal, M. Optimal trajectory planning for a hexapod machine tool during contour machining. *Proc. Inst. Mech. Eng. Part C J. Mech. Eng. Sci.* **2002**, *216*, 1247–1256. [CrossRef]
4. Geng, Z.; Haynes, L.S. Six-degree-of-freedom active vibration isolation using a stewart platform mechanism. *J. Robot. Syst.* **1993**, *10*, 725–744. [CrossRef]
5. Kazezkhan, G.; Xiang, B.; Wang, N.; Yusup, A. Dynamic modeling of the Stewart platform for the NanShan Radio Telescope. *Adv. Mech. Eng.* **2020**, *12*, 1–10. [CrossRef]
6. Keshtkar, S.; Hernandez, E.; Oropeza, A.; Poznyak, A. Orientation of radio-telescope secondary mirror via adaptive sliding mode control. *Neurocomputing* **2017**, *233*, 43–51. [CrossRef]
7. Velasco, J.; Calvo, I.; Barambones, O.; Venegas, P.; Napole, C. Experimental Validation of a Sliding Mode Control for a Stewart Platform Used in Aerospace Inspection Applications. *Mathematics* **2020**, *8*, 2051. [CrossRef]
8. Botello-Aceves, S.; Valdez, S.I.; Becerra, H.M.; Hernandez, E. Evaluating concurrent design approaches for a Delta parallel manipulator. *Robotica* **2018**, *36*, 697–714. [CrossRef]
9. Valdez, S.I.; Botello-Aceves, S.; Becerra, H.M.; Hernández, E.E. Comparison Between a Concurrent and a Sequential Optimization Methodology for Serial Manipulators Using Metaheuristics. *IEEE Trans. Ind. Inform.* **2018**, *14*, 3155–3165. [CrossRef]

20. Panda, S.; Mishra, D.; Biswal, B. Revolute manipulator workspace optimization: A comparative study. *Appl. Soft Comput.* **2013**, *13*, 899–910. [CrossRef]
21. Badescu, M.; Mavroidis, C. Workspace Optimization of 3-Legged UPU and UPS Parallel Platforms with Joint Constraints. *J. Mech. Des.* **2004**, *126*, 291–300. [CrossRef]
22. Lou, Y.; Liu, G.; Chen, N.; Li, Z. Optimal design of parallel manipulators for maximum effective regular workspace. In Proceedings of the 2005 IEEE/RSJ International Conference on Intelligent Robots and Systems, Edmonton, AB, Canada, 2–6 August 2005; pp. 795–800. [CrossRef]
23. Dereli, S.; Köker, R. A meta-heuristic proposal for inverse kinematics solution of 7-DOF serial robotic manipulator: Quantum behaved particle swarm algorithm. *Artif. Intell. Rev.* **2016**, *53*, 949–964. [CrossRef]
24. Falconi, R.; Grandi, R.; Melchiorri, C. Inverse Kinematics of Serial Manipulators in Cluttered Environments using a new Paradigm of Particle Swarm Optimization. *IFAC Proc. Vol.* **2014**, *47*, 8475–8480. [CrossRef]
25. Al-Dois, H.; Jha, A.K.; Mishra, R.B. Task-based design optimization of serial robot manipulators. *Eng. Optim.* **2013**, *45*, 647–658. [CrossRef]
26. Kuçuk, S. Optimal trajectory generation algorithm for serial and parallel manipulators. *Robot. Comput. Integrat. Manufac.* **2017**, *48*, 219–232. [CrossRef]
27. Ravichandran, R.; Heppler, G.; Wang, D. Task-based optimal manipulator/controller design using evolutionary algorithms. *Proc. Dynam. Control Syst. Struc. Space* **2004**, 1–10.
28. Boudreau, R.; Gosselin, C.M. The Synthesis of Planar Parallel Manipulators with a Genetic Algorithm. *J. Mech. Des.* **1999**, *121*, 533–537. [CrossRef]
29. Patel, S.; Sobh, T. Task based synthesis of serial manipulators. *J. Adv. Res.* **2015**, *6*, 479–492. [CrossRef] [PubMed]
30. Lou, Y.; Zhang, Y.; Huang, R.; Chen, X.; Li, Z. Optimization Algorithms for Kinematically Optimal Design of Parallel Manipulators. *IEEE Trans. Autom. Sci. Eng.* **2014**, *11*, 574–584. [CrossRef]
31. Soltanpour, M.R.; Khooban, M.H. A particle swarm optimization approach for fuzzy sliding mode control for tracking the robot manipulator. *Nonlinear Dyn.* **2013**, *74*, 467–478. [CrossRef]
32. Zhang, X.; Nelson, C.A. Multiple-Criteria Kinematic Optimization for the Design of Spherical Serial Mechanisms Using Genetic Algorithms. *J. Mech. Des.* **2011**, *133*, 011005. [CrossRef]
33. Miller, K. Optimal Design and Modeling of Spatial Parallel Manipulators. *Int. J. Robot. Res.* **2004**, *23*, 127–140. [CrossRef]
34. Yang, C.; Li, Q.; Chen, Q. Multi-objective optimization of parallel manipulators using a game algorithm. *Appl. Math. Modell.* **2019**, *74*, 217–243. [CrossRef]
35. Hultmann Ayala, H.V.; dos Santos Coelho, L. Tuning of PID controller based on a multiobjective genetic algorithm applied to a robotic manipulator. *Expert Syst. Appl.* **2012**, *39*, 8968–8974. [CrossRef]
36. Zhang, D.; Gao, Z. Forward kinematics, performance analysis, and multi-objective optimization of a bio-inspired parallel manipulator. *Robot. Comput. Integrat. Manufac.* **2012**, *28*, 484–492. [CrossRef]
37. Cirillo, A.; Cirillo, P.; De Maria, G.; Marino, A.; Natale, C.; Pirozzi, S. Optimal custom design of both symmetric and unsymmetrical hexapod robots for aeronautics applications. *Robot. Comput. Integrat. Manufac.* **2017**, *44*, 1–16. [CrossRef]
38. Joumah, A.A.; Albitar, C. Design Optimization of 6-RUS Parallel Manipulator Using Hybrid Algorithm. *Mod. Educ. Comput. Sci. Press* **2018**, *10*, 83–95. [CrossRef]
39. Nabavi, S.N.; Shariatee, M.; Enferadi, J.; Akbarzadeh, A. Parametric design and multi-objective optimization of a general 6-PUS parallel manipulator. *Mech. Mach. Theory* **2020**, *152*, 103913. [CrossRef]
40. Lara-Molina, F.A.; Rosário, J.M.; Dumur, D. Multi-Objective Design of Parallel Manipulator Using Global Indices. *Benthnam Open* **2010**, *4*, 37–47. [CrossRef]
41. SimWise 4D, 2020. Available online: https://www.design-simulation.com/SimWise4d/ (accessed on 1 March 2021).
42. Valdez, S.I.; Hernández, A.; Botello, S. A Boltzmann based estimation of distribution algorithm. *Inform. Sci.* **2013**, *236*, 126–137. [CrossRef]
43. Kennedy, J.; Eberhart, R. Particle swarm optimization. In Proceedings of the ICNN'95—International Conference on Neural Networks, Perth, WA, Australia, 27 November–1 December 1995; Volume 4; pp. 1942–1948. [CrossRef]
44. Mezura-Montes, E.; Coello Coello, C.A. Constraint-handling in nature-inspired numerical optimization: Past, present and future. *Swarm Evolution. Comput.* **2011**, 173–194. [CrossRef]
45. Pedersen, M.E. *Good Parameters for Particle Swarm Optimization*; Hvass Laboratories: Luxembourg, 2010.
46. Goldberg, D.E. *Genetic Algorithms in Search, Optimization and Machine Learning*, 1st ed.; Addison-Wesley Longman Publishing Co. Inc.: Boston, MA, USA, 1989.
47. Conn, A.R.; Gould, N.I.M.; Toint, P.L. A Globally Convergent Augmented Lagrangian Algorithm for Optimization with General Constraints and Simple Bounds. *SIAM J. Num. Anal.* **1991**, *28*, 545–572. [CrossRef]
48. Conn, A.R.; Gould, N.I.M.; Toint, P.L. A Globally Convergent Augmented Lagrangian Barrier Algorithm for Optimization with General Inequality Constraints and Simple Bounds. *Math. Comput.* **1997**, *66*, 261–288. [CrossRef]
49. MATLAB, Help Particleswarm, Mathworks, 2020. Available online: https://www.mathworks.com/help/gads/particleswarm.html (accessed on 1 March 2021).
50. MATLAB, Help ga, Mathworks, 2020. Available online: https://www.mathworks.com/help/gads/ga.html (accessed on 1 March 2021).

41. Yarpiz. Multi-Objective Evolutionary Algorithm based on Decomposition (MOEA/D), 2016. Available online: https://yarpiz.com/456/ypea126-nsga3 (accessed on 1 March 2021).
42. Yarpiz. Implementation of Non-Dominated Sorting Genetic Algorithm III in MATLAB, 2015. Available online: https://yarpiz.com/95/ypea124-moead (accessed on 1 March 2021).
43. Deb, K.; Jain, H. An Evolutionary Many-Objective Optimization Algorithm Using Reference-Point-Based Nondominated Sorting Approach, Part I: Solving Problems With Box Constraints. *IEEE Trans. Evolution. Comput.* **2014**, *18*, 577–601. [CrossRef]
44. Li, H.; Deb, K.; Zhang, Q.; Suganthan, P.; Chen, L. Comparison between MOEA/D and NSGA-III on a set of novel many and multi-objective benchmark problems with challenging difficulties. *Swarm Evolution. Comput.* **2019**, *46*, 104–117. [CrossRef]
45. Gong, Y. Design Analysis of a Stewart Platform for Vehicle Emulator Systems. Master's Thesis, Department of Mechanical Engineering, Massachusetts Institute of Technology, Cambridge, MA, USA, 1992.

Assessment of Optimization Methods for Aeroacoustic Prediction of Trailing-Edge Interaction Noise in Axisymmetric Jets

Sarah Stirrat*, Mohammed Z. Afsar* and Edmondo Minisci

Department of Mechanical and Aerospace Engineering, James Weir Building, 75 Montrose Street, Glasgow G1 1XJ, UK; edmondo.minisci@strath.ac.uk
* Correspondence: sarah.stirrat@strath.ac.uk (S.S.); mohammed.afsar@strath.ac.uk (M.Z.A.)

Abstract: Our concern in this paper is in the fine-tuning of the arbitrary parameters within the upstream turbulence structure for the acoustic spectrum of a rapid-distortion theory (RDT)-based model of trailing-edge noise. RDT models are based on an appropriate asymptotic limit of the Linearized Euler Equations and apply when the interaction time of the turbulence with the surface edge discontinuity is small compared to the eddy turnover time. When an arbitrary transversely sheared jet mean flow convects a finite region of nonhomogeneous turbulence, the acoustic spectrum of the pressure field scattered by the trailing-edge depends on (among other things) the upstream turbulence via the Fourier transform of the correlation function, R_{22} (where subscript 2 refers to a co-ordinate surface normal to the plate). We show that the length and time scale parameters that govern the spatial and temporal de-correlation of R_{22} can be found using formal optimization methods to avoid any uncertainty in their selection by hand-tuning. We assess various optimization methods that are broadly categorized into an 'evolutionary' and 'non-evolutionary' paradigm. That is, we optimize the acoustic spectrum using the Multi-Start algorithm, Particle Swarm Optimization and the Multi-Population Adaptive Inflationary Differential Evolution Algorithm. The optimization is based upon different objective functions for the acoustic spectrum and/or turbulence structure. We show that this approach, while resulting in the total modest increase in computation time (on average 2 h), gives excellent prediction over most frequencies (within 2–4 dB) where the trailing-edge noise associated amplification in sound exists.

Keywords: aeroacoustics; trailing-edge noise; global optimization; evolutionary algorithms

1. Introduction

The advent of the jet engine in the mid-twentieth century [1] brought with it the intrusion caused by aircraft noise to the communities living near airports. In its entirety, however, the aircraft noise problem is an incredibly complex one [2]. For example, Dobrzynski [3] shows that the radiated sound on approach for both short and long stage length aircraft is evenly split between the engine and the airframe. Engine associated noise is generated from the internal moving surfaces within the engine components as well as from the exhaust gas emanating at the nozzle exit. The breakdown of the latter results in both jet noise and jet-surface interaction noise components when the turbulent air interacts with the airframe, wing edges and other external surfaces.

The trailing-edge component is a particularly dangerous noise source owing to the large increase in low frequency sound when the observation point is above or below the plate surface and vertical location (h) of the trailing-edge is of the order of jet diameter, $h \sim D$. Experiments on edge noise began in the 1970s by Olsen & Boldman [4] (discussed below). In the early 1980's, Wang [5] showed that the presence of an external surface increased the noise measured on the same side as the jet flow in comparison to the isolated jet. This amplification of sound due to the interaction with the trailing-edge is mainly at low frequencies up to the peak Strouhal number (i.e., the normalized angular frequency, $St = fD/U_j$ based on jet exit velocity, U_j and diameter, D); typically this is at $St \sim 0.1$. It also dominates at larger observation angles θ to the jet axis [6,7]. Bridges' recent experiments [8] confirm the work done in the 1970s by Olsen & Boldman [4] and Wang's result by showing that the amplification in sound perpendicular to the jet axis (i.e., $\theta = 90°$) is typically of the order of 10 dB for a high speed jet at an acoustic Mach number based on the speed of sound at infinity, $Ma = U/c_\infty$, of $Ma = 0.9$. As (θ) reduces, the jet noise contribution increases until, at shallow angles (e.g., $\theta = 30°$), the latter jet noise dominates the total noise radiation signature at almost all measured frequencies; typically this covers Strouhal numbers, $St = [0.01, 2.0]$. The Bridges' [8] and Bridges et al. [9] datasets also covered the parameter range of acoustic Mach number, Ma, edge location with respect to the nozzle lower lip line and the nozzle shape itself. We briefly summarize these trends now (1). The amplification in sound is greater at lower Ma, e.g., at $Ma = 0.5$ cf. 0.9; this is consistent with the 'dipole' directionality of the edge noise source. (2). As the vertical standoff distance, h/D, is increased the edge noise reduces in magnitude. At the limiting condition where $h \gg D$, the amplification in sound due to the edge vanishes and the total sound owes itself to the jet noise alone. The streamwise location also has an important impact on the magnitude of low frequency noise amplification. Bridges' results show that the edge must be placed in the vicinity of where the jet potential core terminates for the amplification to reach its greatest magnitude. (3). The round jet appears to result in greater noise amplification than the high-aspect ratio rectangular (i.e., planar) jet flow [9].

From an analytical and numerical standpoint, a lot of work has been done on this problem [10–15]. The discontinuity in the solid surface boundary condition can be treated formally using the Wiener-Hopf technique for a flat plate that is doubly infinite in the spanwise direction and lies parallel to the level curves of the streamwise mean flow. See Figure 1 for a depiction of this problem in the $(y_1 - y_2)$ plane. The so-called 'gust solution' then acts as the input to an inhomogeneous boundary value problem in which the scattered pressure field is determined at the output. Goldstein et al. [10–12] used the method of matched asymptotic expansions at the low frequency limit to construct the gust-induced boundary condition and the homogeneous solutions to the Rayleigh equation that enter in the solution to the Wiener-Hopf problem for the acoustic field scattered by the edge. This solution (Equations (6.26) and (6.27) in [12]) was analytically continued to high frequencies, and (Equations (6.28)–(6.30)) show that the mean square scattered pressure depends on the upstream structure of the two-point time-delayed turbulence correlation R_{22} where the subscript 2 denotes the co-ordinate plane normal the plate surface. Just as in acoustic analogy models of jet noise [16,17], the correlation function is modeled by comparing to a wide bank of experimental data (see, e.g., ref. [18]) on jet flow turbulence.

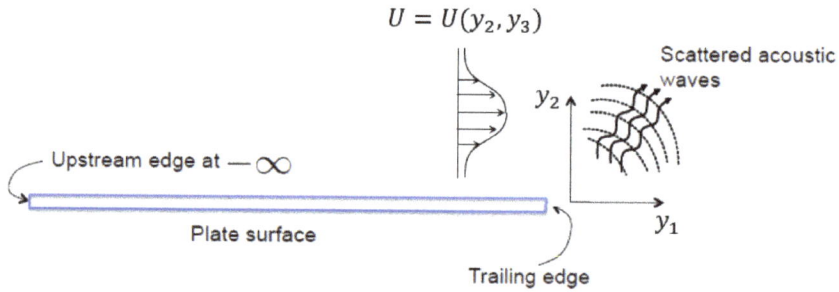

Figure 1. Canonical trailing-edge noise problem seen in the $(y_1 - y_2)$-plane.

All models are formulated with arbitrary 'tuning' parameters to quantify the degree of spatial and temporal de-correlation as well as the permanence of a finite anti-correlation region in spatial and temporal separation [14]. Previous modeling approaches tuned these scales by hand to obtain good agreement with the acoustic data. In this paper we show that this form of empiricism can be avoided entirely by using an appropriate numerical optimization routine to determine the parameters for an objective function that seeks to minimize the difference between the functional form of the turbulence model and turbulence data as well as minimizing the difference between acoustic predictions and acoustic data.

A (single-objective) optimization problem can be described in terms of minimizing the objective function $J(x, \psi)$, where $x_i (i = 1, n)$ are the n design variables which are modified to find the optimum, and ψ are the state parameters which describe the system [19]. The objective function may also be subject to (in general) a total (m, p) of (inequality/equality) constraints that take the form:

$$\begin{aligned} g_j(x) \leq 0 & \quad j = 1, m \\ h_k(x) = 0 & \quad k = 1, p \end{aligned} \quad (1)$$

Additionally, the design variables may be bounded ($x_{i,\text{LOWER}} \leq x_i \leq x_{i,\text{UPPER}}, i = 1, n$), which are known as side constraints.

Optimization algorithms can find multiple solutions to this problem which are known as local optima. Hence, the algorithms can be split into two categories: local optimization methods (which find the local minimum for the starting conditions) and global optimization methods (which aim to find the global minimum of the search space). The majority of local methods use gradient information to find the optimum and there are several methods to do this [20–23]. However if the problem has multiple local minima, these gradient methods will converge to the closest minimum not necessarily the global. Each minimum has a basin of attraction, where a design point initialised within the basin converges to that local minimum. Hence, local optimization strongly depends on the location of the initial design point. Global optimization methods aim to find the global optimum, however, it should be noted that it cannot be guaranteed that the global optimum will be found, only that it would be if the algorithm could run indefinitely. Global optimization can be split into three types of algorithm: Multi-Start algorithms, evolutionary algorithms and deterministic algorithms. Multi-Start algorithms perform local optimizations at different starting locations, and choose the best local optimum to be the global optimum. Evolutionary algorithms are stochastic and heuristic, they advance a population of design parameters through the search space to find the global optimum. Deterministic algorithms typically require manipulation of the objective function and are designed to solve specific classes of problem [24], an overview of these is described in [25], and will not be further discussed here.

Optimization methods have been used for this kind of problem in aeroacoustics before, for example the Multipoint Approximation Method (MAM) [26–28]. This method was designed for computationally expensive and noisy objective functions, so it uses trust regions and a series of approximations to the objective function. It is similar to the Multi-Start algorithm in that it uses multiple starting locations, however it differs by approximating the objective function. In this paper, the objective function is not computationally expensive, therefore we investigate the Multi-Start method instead.

The rest of the paper is organized as follows. In Section 2 we summarize the general theory to determine the acoustic field scattered by the trailing-edge and show the final formula that we use for the subsequent optimization experiments that we perform in order to fine tune the modeling of the correlation function R_{22}. Section 3 then reviews the various types of optimization that can be used for problems of this type paying particular attention to evolutionary algorithms. Popular evolutionary algorithms include the Genetic Algorithm (GA) [29] which was inspired by Darwin's principle of survival of the fittest, Particle Swarm Optimization (PSO) [30] which is based on a social model and differential evolution (DE) [31].

Specifically, we discuss the advantages and disadvantages of three methods (the non-evolutionary Multi-Start [32–34], and the evolutionary algorithms: Particle Swarm Optimization (PSO) [30] and Multi-Population Adaptive Inflationary Differential Evolution

Algorithm (MP-AIDEA) [35] which is an extension to the original differential evolution algorithm) for a problem of this type and the results that are obtained for the parameters under different objective functions for the turbulence and/or final acoustic predictions (i.e., when comparison is made to turbulence and/or acoustic data). Finally in Section 4 we conclude by discussing the applicability of using such optimization approaches in acoustic modeling problems.

2. Summary of the Mathematical Modeling of Trailing-Edge Noise and Defining the Objective Functions

Rapid-distortion theory (RDT) analyzes the changes in turbulent flows by using linearized equations. It is, therefore, ideally suited to analyze the rapid changes that occur when a turbulent flow interacts with a discontinuity at the boundary of a solid surface embedded in the flow. It applies whenever the turbulence intensity is small and the length (or time) scale over which the changes take place is short compared to the length (or time) scale over which the turbulent eddies evolve. These assumptions imply, among other things, that the resulting flow is inviscid and non-heat conducting and is, therefore, governed by the Linearized Euler Equations, i.e., the Euler equations linearized about an arbitrary, usually steady, solution (the base flow) to the nonlinear equations. Goldstein et al. [11] showed that the upstream boundary conditions can be imposed infinitely far upstream in a region where the flow is undisturbed by the interaction.

Goldstein et al. [12] used rapid distortion theory to determine the trailing-edge noise spectrum above the flat plate (given the Fourier transform of the mean square scattered pressure $[p'^s(x,t)]^2$), denoted by $I(x,\omega)$ for the axisymmetric jet $U(y_T)$ interacting with the flat plate as depicted in Figure 1, and is given by (Equations (6.26) and (6.27) in their paper):

$$I(x,\omega) \to \left(\frac{k_\infty}{4\pi |\bar{x}|}\right)^2 \int_{-\infty}^{0} \int_{-\infty}^{0} D(u,\tilde{u};\theta)\bar{S}(u,\tilde{u};\omega)\,du\,d\tilde{u}, \tag{2}$$

where $D(u,\tilde{u};\theta)$ is the round jet directivity factor determined by application of the Wiener-Hopf technique (i.e., Equation (13) in Afsar et al. [13]), $y_T = (y_2, y_3)$ are transverse coordinates and,

$$\bar{S}(u,\tilde{u};\omega) = (\rho_\infty c_\infty^2)^2 \int_{-\pi}^{\pi} \int_{-\pi}^{\pi} S(u,\tilde{u}|v,\tilde{v};\omega) \left|\frac{dz}{dW}\right|^2 \left|\frac{d\tilde{z}}{d\tilde{W}}\right|^2 dv\,d\tilde{v}. \tag{3}$$

The function $S(u,\tilde{u}|v,\tilde{v};\omega)$ derived in [12] is

$$S(u,\tilde{u}|v,\tilde{v};\omega) = \left[\frac{dU/du}{U^2(u)}\frac{dU/d\tilde{u}}{U^2(\tilde{u})}\nabla u\tilde{\nabla}\tilde{u}\omega^2\right] F(u,\tilde{u},v,\tilde{v}) \tag{4}$$

This result shows, among other things, that $S(u,\tilde{u}|v,\tilde{v};\omega)$ is directly proportional to the Fourier transform of the streamwise-independent turbulence statistical quantity, $\hat{R}_{22}(\tilde{u}-u,\tilde{v}-v;\omega)$ which represents the Fourier transform of the two-point time delayed correlation function of the transverse velocity normal to the plate surface at $u=0$ (see Afsar et al. [13]).

Our starting point is, however, a spectral function $F(u,\tilde{u},v,\tilde{v})$ which is slightly more general than that used in [12] (Equation (4.22) in their paper). Since correlation functions of the type R_{22} will anti-correlate (i.e., become negative) for some values of its arguments [13], including such effects means that a mathematical model for R_{22} must have additional algebraic behaviour, for example given by the a_1 term in the following formula

$$R_{22}(\tau) \sim (1 - a_1\tau)e^{-\tau}, \tag{5}$$

where τ is the time-delay between the two space-time points being correlated far upstream of the interaction region near the trailing edge as required by the theory in [11]. Hence, the spectral function, $F(u, \tilde{u}, v, \tilde{v})$, from Equation (4) is given by the following formula:

$$F(u,\tilde{u},v,\tilde{v}) = l_2^4 A(u,\tilde{u}) \left[(1-a_1) \frac{\tau_0 f}{\pi\sqrt{1+\tilde{\omega}^2}} K_1(f\sqrt{1+\tilde{\omega}^2}) + \frac{a_1 \tau_0 f^2 \tilde{\omega}^2}{\pi(1+\tilde{\omega}^2)} \left[\frac{1}{2}(K_0(f\sqrt{1+\tilde{\omega}^2}) + K_2(f\sqrt{1+\tilde{\omega}^2})) + \frac{K_1(f\sqrt{1+\tilde{\omega}^2})}{f\sqrt{1+\tilde{\omega}^2}} \right] \right], \quad (6)$$

where $K_0, K_1,$ and K_2 are the modified Bessel functions of the second kind of order 0, 1, and 2 respectively, $A(u, \tilde{u})$ is the amplitude function and $f = |(\tilde{u} - u)/l_2 + (\tilde{v} - v)/l_3|$ controls the (u, v) de-correlation via the parameters (l_2, l_3).

The objective functions are then defined as the mean squared error between the acoustic/turbulence models (Equations (2) and (5)) and the relevant experimental data [36,37] is given by the following optimization norms:

$$J_A(x,\psi) = \frac{1}{N_A} \sum_{i=1}^{N_A} (I(x,\omega_i,\psi) - E_A(x,\omega_i,\psi))^2$$

$$J_R(x,\psi) = \frac{1}{N_R} \sum_{i=1}^{N_R} (R_{22}(x,\tau_i,\psi) - E_R(x,\tau_i,\psi))^2, \quad (7)$$

where: N_A, N_R is the number of experimental data points we are optimizing against for acoustics and R_{22} respectively, $I(x, \omega_i, \psi)$ is the acoustic spectrum result using the model for the ith frequency ω_i, and $E_A(x, \omega_i, \psi)$ is the corresponding experimental data. Likewise $R_{22}(x, \tau_i, \psi)$ is the result using our turbulence model at the ith time delay τ_i, and E_R is the corresponding experimental data.

The vector of state parameters (ψ) is the minimum set of parameters which describe the system and how it responds to input [19]. Our problem can be thought of as an "input/output system" where an input turbulence spectrum interacts with the streamwise discontinuity at the trailing-edge and produces noise. The sound radiation will depend on the acoustic Mach number of the jet (Ma), and the location where the noise measurements take place (far field angle, θ, measured with respect to the jet axis and azimuthal angle, ϕ). On the other hand the turbulence correlation function, R_{22}, is independent of these parameters, and instead depends on the location where the turbulence is measured $(y_1/D, r/D)$, where y_1/D is the streamwise location from the nozzle exit normalized by the nozzle diameter and r/D is the radial location from the jet centerline ($r^2 = y_2^2 + y_3^2$). The experimental set-up can be found in several papers [11,12,37].

The $\mathcal{O}(1)$ parameters (a_1, l_2, l_3, τ_0) in the spectral function $F(u, \tilde{u}, v, \tilde{v})$ are selected in order to find the optimum acoustic spectrum predictions across acoustic Mach number (Ma) and far-field angle θ, whilst maintaining a physically admissible turbulence structure. In the following sections we discuss the different optimization methods that can be used to achieve this. Since it was found that there were multiple local minima, only global optimization routines were considered. The location of the trailing edge $(x_{TE}/D, y/D)$ in all of the numerical tests investigated below is taken to be the same as Goldstein et al. [12] (see their Figure 4). In the following sections of this paper, the acoustic spectrum $I(x, \omega)$ is calculated in the form of the power spectral density of the far-field pressure fluctuation versus Strouhal number, which is defined in this case as $St = fh/U$, and is presented in the usual dB scale where $PSD = 10log(4\pi I(x,\omega)U/hp_{ref}^2)$ (relative to $p_{ref} = 20$ µPa). We show results only at the observation point of $\theta = 90°$ where the trailing edge noise is largest [36] to illustrate the benefit of optimization.

3. Evolutionary Versus Non-Evolutionary Optimization Methods

The non-evolutionary Multi-Start method is the most straightforward global optimization routine. It follows on from local optimization in that it simply performs a local optimization algorithm at several different starting points within the design space [32–34]. The local optima can then be compared to find the global optimum. As the number of starting points increase, the probability of finding the global optimum also increase. Often a Design of Experiments (DOE) is performed prior to Multi-Start to initialise design points within known basins.

Evolutionary algorithms are specifically designed to work on black box problems, i.e., they do not need direct access to the inner workings of the objective function nor do they need gradient information. Consequently, they can be used for non-smooth functions and it is not required that the programmer knows anything about the structure of the objective function. Evolutionary algorithms are known to be robust and have a good chance of finding the global optimum since they advance a fixed population of design variables through the search space. However, they are computationally expensive and require the tuning of parameters to solve each problem [24]. There are several types of evolutionary algorithm, two of the most popular are Particle Swarm Optimization (PSO) and Differential Evolution (DE).

Particle Swarm Optimization (PSO) [30] was developed from a social model. Each particle utilizes not only its own past experience to find an optimum but also that of the group at large. It involves initializing the population and a velocity vector for each particle. The velocity vector is then updated by including information from the particles past and from the group. More recently there have been modifications to PSO to better handle optimization problems with constraints [38]. There are three parameters which need to be tuned for the specific optimization problem. These are, the inertia parameter w, and the trust parameters c_1, c_2. The choices of parameters are very important and some recommendations are given in [39]. This paper uses the default parameters given in Matlab which adapts the inertia weight within bounds ($0.1 \leq w \leq 1.1$) and uses the trust parameters ($c_1 = c_2 = 1.49$).

Differential Evolution (DE) [31] also initializes a population of design points and then utilizes information from these points to mutate and find the next generation of design points. There are several different methods of differential evolution but the 'classic' method (DE/rand/1/bin) mutates the design parameters through the equation: $x_i^{'g+1} = x_{r_1}^g + F(x_{r_2}^g - x_{r_3}^g)$ where g is the current generation, i is the individual in the population, and ($r_1, r_2, r_3 \neq i$) are random parents in the population. The parameter F is the differential weight and controls the amplification of the differential, it typically lies within the interval 0.4–1 [40]. The mutation is demonstrated in Figure 2 for two dimensions.

Following mutation, crossover is used to increase the diversity of the population, after which the parent and child designs are compared and the best is selected to be the design point for the next generation. The second parameter of differential evolution is the crossover ratio: $0 < CR < 1$, this and the differential weight, F, need to be selected by the programmer.

There are several variations of the evolutionary algorithms which make them more complex and robust. An extension of the differential evolution algorithm, which we use in this paper, is the Multi-Population Adaptive Inflationary Differential Evolution Algorithm (MP-AIDEA) [35]. This uses multiple populations and combines basic differential evolution with monotonic basin hopping (MBH) to reduce the risk of converging to a minimum which is not global, it also adapts the optimization parameters autonomously. It is a further advancement on the inflationary differential evolution algorithm (IDEA) which only uses a single population and requires the parameters to be chosen by the programmer [41].

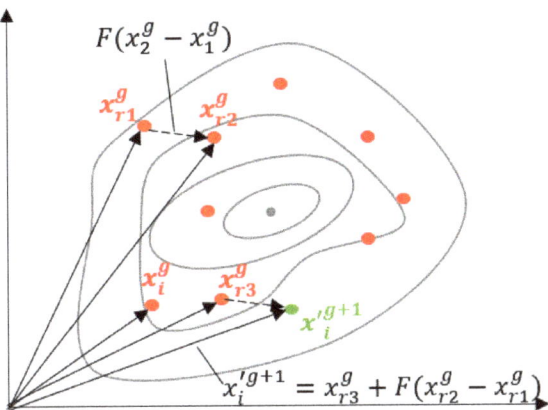

Figure 2. Example of Differential Evolution mutation in 2D (DE/rand/1/bin) (Adapted by permission from licensor: Springer Nature, Journal of Global Optimization, 1997 [31] https://www.springer.com/journal/10898 accessed on 18 February 21).

When a minimum has been found the monotonic basin hopping (MBH) [42] method generates a new point within the neighbourhood of this minimum (where the neighbourhood is defined as 2Δ). A local search is performed from this point and if the minimum found is better than the previous one it is chosen and a new point generated in its neighbourhood, and so on. If no better points are found for $n_{samples}$ then a restart can be performed.

IDEA uses MBH when the population contracts within a radius defined as the contraction limit (a parameter to be defined), when the population reaches this limit it is unlikely to be able to escape and search elsewhere in the design space, hence the need for a restart. Instead of using a local search within MBH it uses differential evolution. MP-AIDEA adapted IDEA to adjust the main parameters (crossover probability CR, differential weight F, local restart bubble δ_{local}, and the number of local restarts n_{LR}) autonomously. This makes the algorithm easier to apply to different problems. For full details of the algorithm refer to [35]. To adapt the values of δ_{local} and n_{LR} the restart of the population needs to be evaluated, therefore, multiple populations are used and evolved in parallel. The parameter n_{LR} is removed in this algorithm and a procedure to decide whether a local or global restart should be run is implemented instead. Figure 3 demonstrates the algorithm.

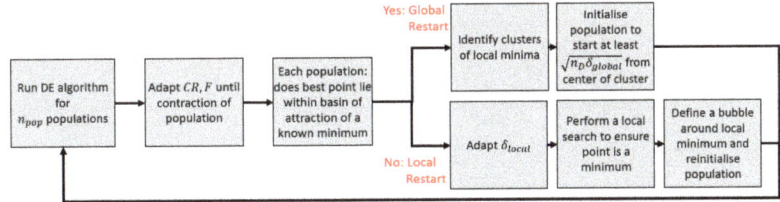

Figure 3. Description of the Multi-Population Adaptive Inflationary Differential Evolution Algorithm (MP-AIDEA).

4. Possible Routes to Minimizing the Objective Function in Equation (7)

There are various approaches to determine the parameters in the spectral function $F(u, \tilde{u}, v, \tilde{v})$. One way is by hand as in [12], but here we use the following methods:

Method 1: Optimize the acoustic model to find the 4 parameters.

Method 2: Optimize the R_{22} model to find a_1 and hand-tune the other 3 parameters for acoustic predictions.

Method 3: Optimize the R_{22} model to find a_1, and optimize the acoustic model to find the other 3 parameters.

Table 1 sets out the optimization problem which is to be solved for each method where the objective functions J_A, J_R were defined in Equation (7).

Table 1. Optimization problem statement for each method (hyphens indicate that no optimization was carried out).

	Method 1		Method 2		Method 3	
	Acoustics	R_{22}	Acoustics	R_{22}	Acoustics	R_{22}
Objective function	$J_A(x, \psi) = 0$	-	-	$J_R(x, \psi) = 0$	$J_A(x, \psi) = 0$	$J_R(x, \psi) = 0$
State parameters (ψ)	Ma, θ, ϕ	-	-	$x/D, y/D$	Ma, θ, ϕ	$x/D, y/D$
Design parameters (x)	a_1, l_2, l_3, τ_0	-	-	a_1	l_2, l_3, τ_0	a_1
Constraints	g_1	-	-	g_2	g_3	g_2

There are no equality or inequality constraints for this problem, only side constraints which were chosen to be:

$$g_1 : 0 < a_1 < 1,\ 0 < l_2 < 5,\ 1 < l_3 < 10,\ 1 < \tau_0 < 10$$
$$g_2 : 0 < a_1 < 1 \qquad (8)$$
$$g_3 : 0 < l_2 < 5,\ 1 < l_3 < 10,\ 1 < \tau_0 < 10$$

The acoustic spectrum results using these methods will be compared to experimental results from [37] for three acoustic Mach numbers $Ma = 0.5, 0.7, 0.9$ above the plate ($\phi = 90$), and at the far field angle ($\theta = 90$) where jet surface interaction is greatest.

We also compare the R_{22} model using the values found for a_1 against experimental data from Bridges [36] at the end of the potential core on the shear layer ($y_1/D = 6, r/D = 0.5$).

To reduce the time taken in optimization, for each acoustic Mach number 30 points were chosen from the experimental acoustic data (see Figure 4). The acoustic model was then run for each of these points to calculate the objective function.

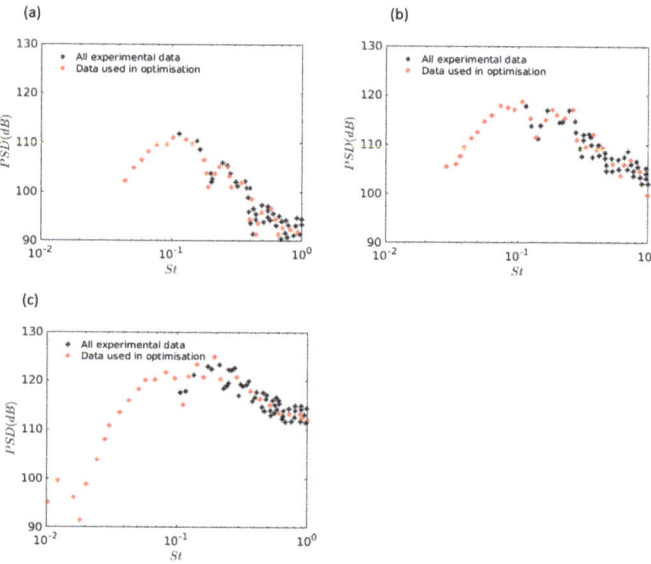

Figure 4. Points chosen from experimental data to calculate the objective function (**a**) $Ma = 0.5$ (**b**) $Ma = 0.7$ (**c**) $Ma = 0.9$.

We compare the results from three optimization routines: The Multi-Population Adaptive Inflationary Differential Evolution Algorithm (MP-AIDEA), Particle Swarm Optimization (PSO), and Multi-Start. The Multi-Start and PSO optimizations were done using the in-built Matlab routines. Multi-Start was carried out using 500 starting points.

5. Results

Since the optimization routines are stochastic, they were ran several times for $Ma = 0.9, \theta = 90$ to see if they consistently converged. Figure 5 shows the variance of parameters found using Particle Swarm Optimization. Likewise, Figures 6 and 7 show the variance of parameters for MP-AIDEA and Multi-Start respectively. The range of objective function values corresponding to these parameters obtained from each routine are shown in Figure 8.

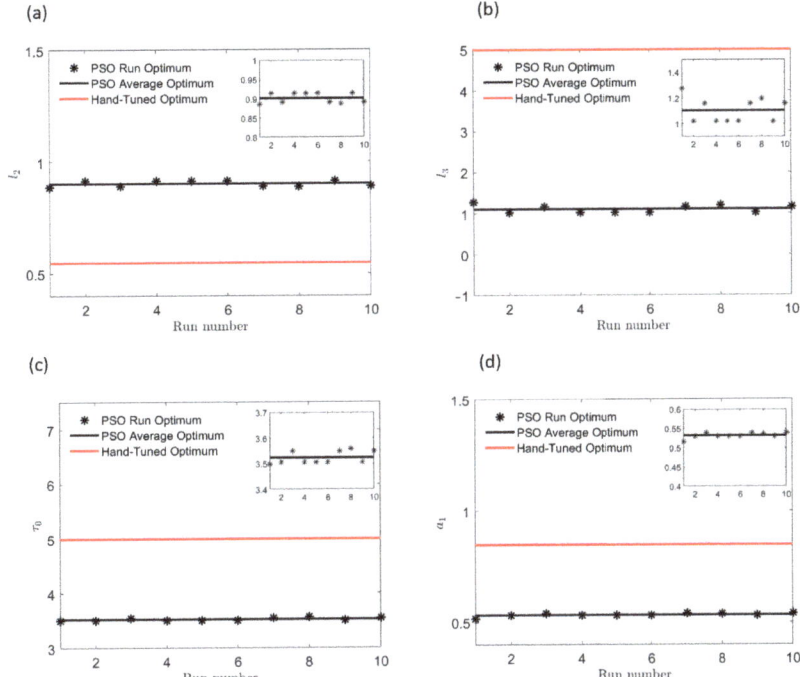

Figure 5. Parameter variance across 10 runs of PSO (**a**) l_2 (**b**) l_3 (**c**) τ_0 (**d**) a_1.

Tables 2–4 compare the parameters found through each method and optimization routine, the resulting objective function value ($fval$) and the time taken for the optimization routine to run.

Table 2. Ma = 0.5, θ = 90: Comparison of parameters found from the optimization methods.

	Ma = 0.5, $\theta = 90$						
	Method 1			Method 2	Method 3		
	MP-AIDEA	Multi-Start	PSO	Hand-Tuned	MP-AIDEA	Multi-Start	PSO
a_1	0.01	0.10	0.00	0.85	0.85	0.85	0.85
l_2	2.52	2.18	2.57	1.30	1.51	1.30	1.49
l_3	1.01	1.01	1.00	5.00	1.00	2.50	1.00
τ_0	2.61	2.04	2.69	5.00	3.68	3.52	3.54
fval	14.29	15.47	14.24	20.75	15.68	17.90	15.66
Time (s)	9538	6044	7766	-	7098	4700	7507

Table 3. Ma = 0.7, $\theta = 90$: Comparison of parameters found from the optimization methods.

	Ma = 0.7, $\theta = 90$						
	Method 1			Method 2	Method 3		
	MP-AIDEA	Multi-Start	PSO	Hand-Tuned	MP-AIDEA	Multi-Start	PSO
a_1	0.59	0.56	0.60	0.85	0.85	0.85	0.85
l_2	1.39	1.14	1.37	0.90	1.05	0.89	1.06
l_3	1.02	3.22	1.00	5.00	1.00	2.80	1.00
τ_0	3.71	3.73	3.64	5.00	3.75	4.28	3.75
fval	4.91	5.15	4.18	5.70	4.27	5.12	4.26
Time (s)	9596	5841	8944	-	7183	4610	7264

Table 4. Ma = 0.9, $\theta = 90$: Comparison of parameters found from the optimization methods.

	Ma = 0.9, $\theta = 90$						
	Method 1			Method 2	Method 3		
	MP-AIDEA	Multi-Start	PSO	Hand-Tuned	MP-AIDEA	Multi-Start	PSO
a_1	0.52	0.56	0.54	0.85	0.85	0.85	0.85
l_2	0.89	0.82	0.89	0.55	0.68	0.57	0.69
l_3	1.22	2.37	1.16	5.00	1.12	9.19	1.07
τ_0	3.54	4.35	3.55	5.00	4.08	5.48	4.08
fval	9.21	9.73	9.20	11.36	9.97	10.48	9.96
Time (s)	9501	5684	14,921	-	7075	4450	7393

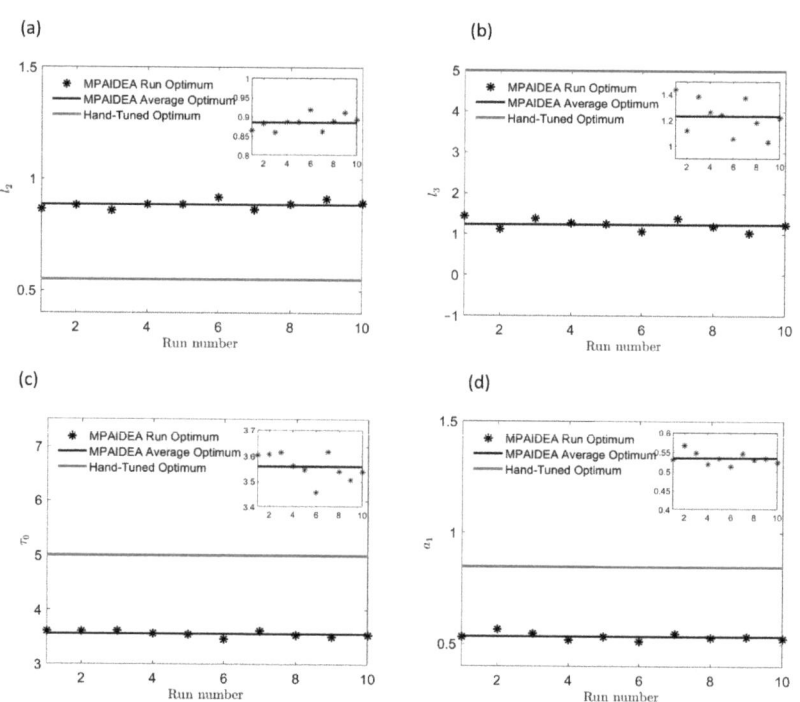

Figure 6. Parameter variance across 10 runs of MP-AIDEA (**a**) l_2 (**b**) l_3 (**c**) τ_0 (**d**) a_1.

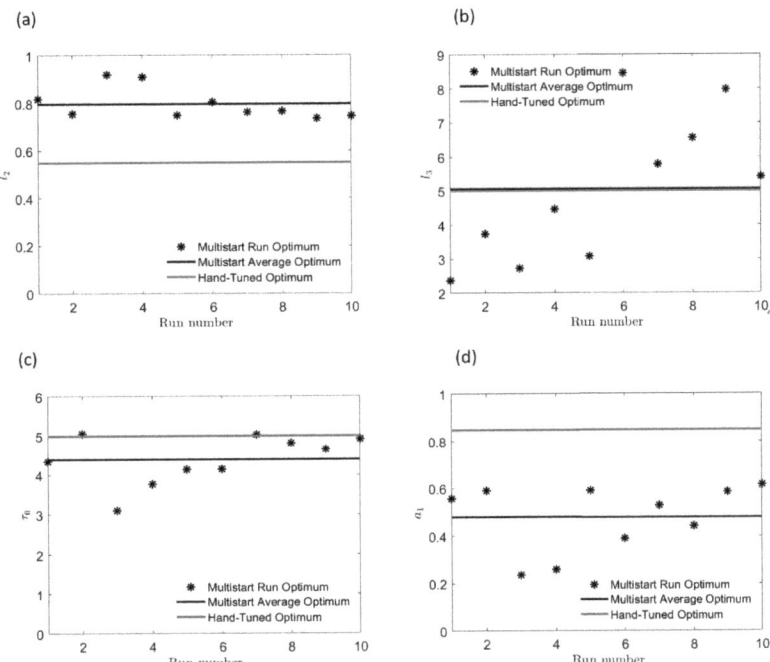

Figure 7. Parameter variance across 10 runs of Multi-Start (**a**) l_2 (**b**) l_3 (**c**) τ_0 (**d**) a_1.

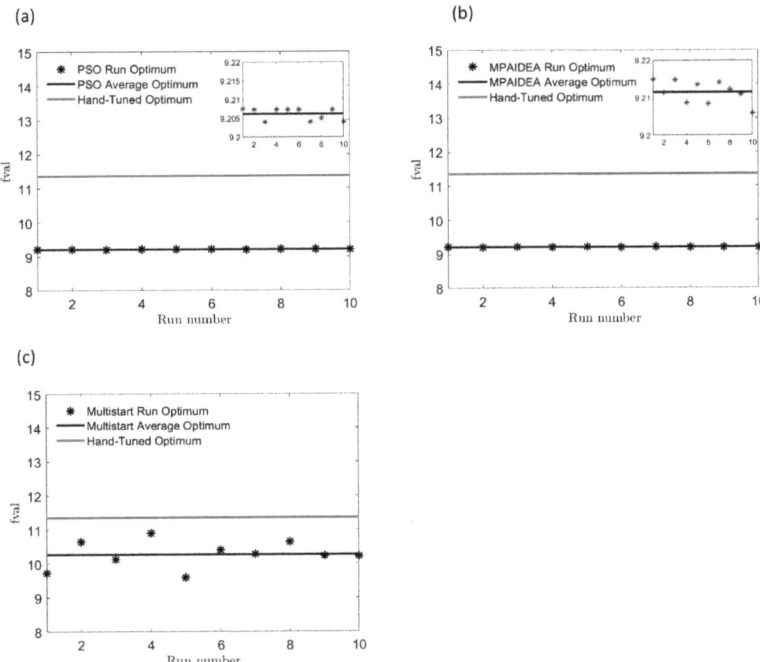

Figure 8. Objective function value range across 10 runs (**a**) PSO (**b**) MP-AIDEA (**c**) Multi-Start.

Figure 9 compares the turbulence correlation function R_{22} for methods 1, 2, and 3 Since the model for R_{22} only depends on the parameter a_1, methods 2 and 3 give identical results. In Method 1, a_1 varies for each optimization routine.

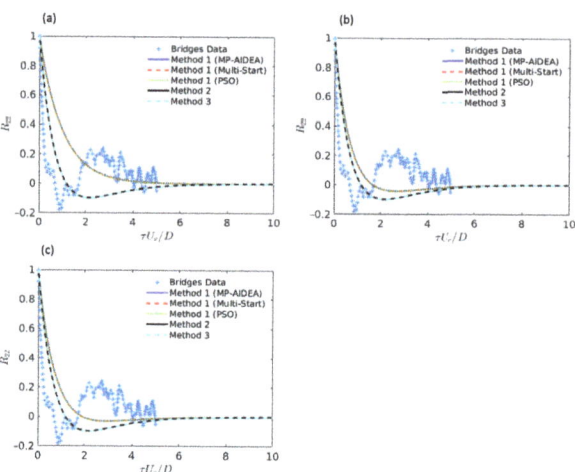

Figure 9. Comparison of R_{22} for different optimization methods and routines (**a**) $Ma = 0.5$ (**b**) $Ma = 0.7$ (**c**) $Ma = 0.9$.

Figure 10 compares the acoustic spectrum for method 1 using each optimization routine. Similarly, Figure 11 compares the routines for method 3. In Figures 12–14 we compare the acoustic spectrum for methods 1, 2, and 3 using Particle Swarm Optimization, MP-AIDEA, and Multi-Start respectively.

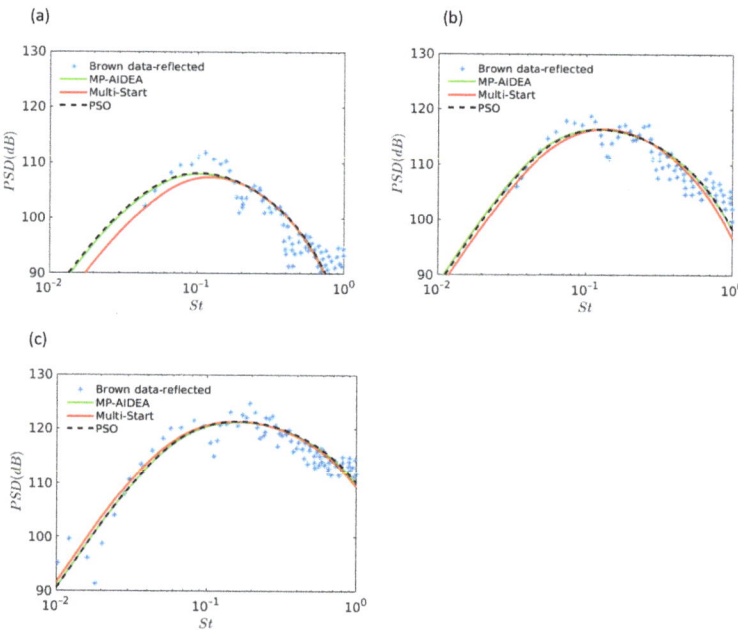

Figure 10. Comparison of acoustic predictions using different optimization routines for method 1 (**a**) $Ma = 0.5$ (**b**) $Ma = 0.7$ (**c**) $Ma = 0.9$.

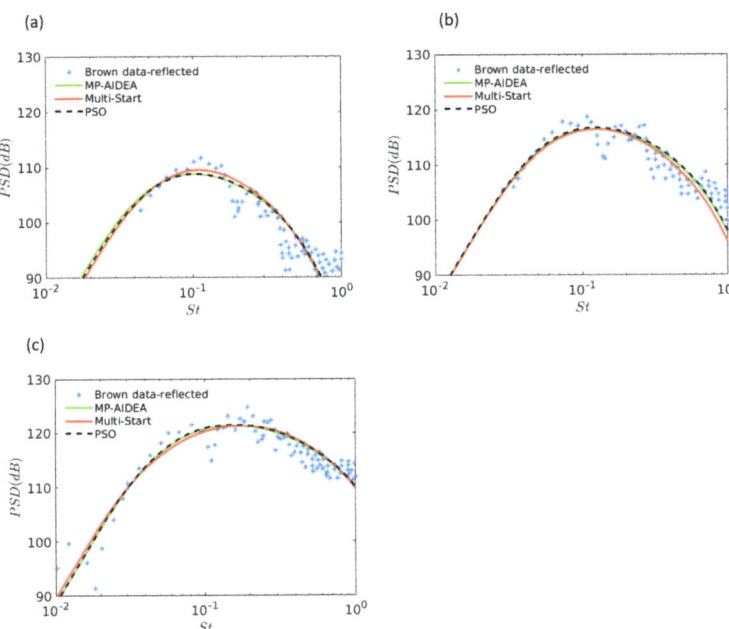

Figure 11. Comparison of acoustic predictions using different optimization routines for method 3 (**a**) $Ma = 0.5$ (**b**) $Ma = 0.7$ (**c**) $Ma = 0.9$.

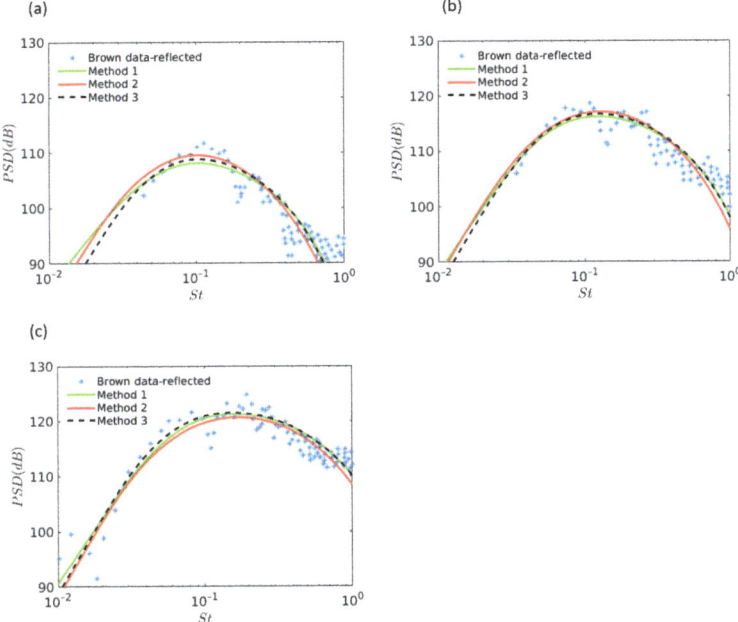

Figure 12. Comparison of the acoustic predictions for methods 1, 2 and 3 using Particle Swarm Optimization (**a**) $Ma = 0.5$ (**b**) $Ma = 0.7$ (**c**) $Ma = 0.9$.

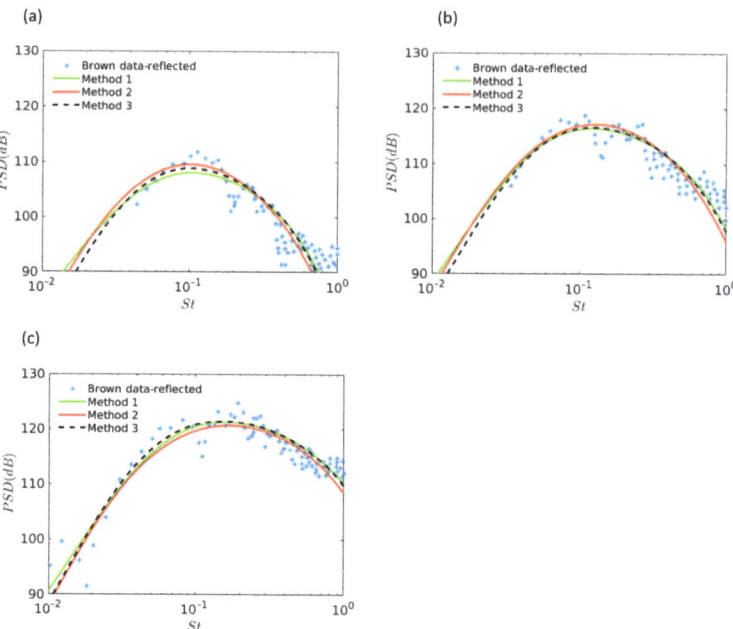

Figure 13. Comparison of the acoustic predictions for methods 1, 2 and 3 using MP-AIDEA (**a**) $Ma = 0.5$ (**b**) $Ma = 0.7$ (**c**) $Ma = 0.9$.

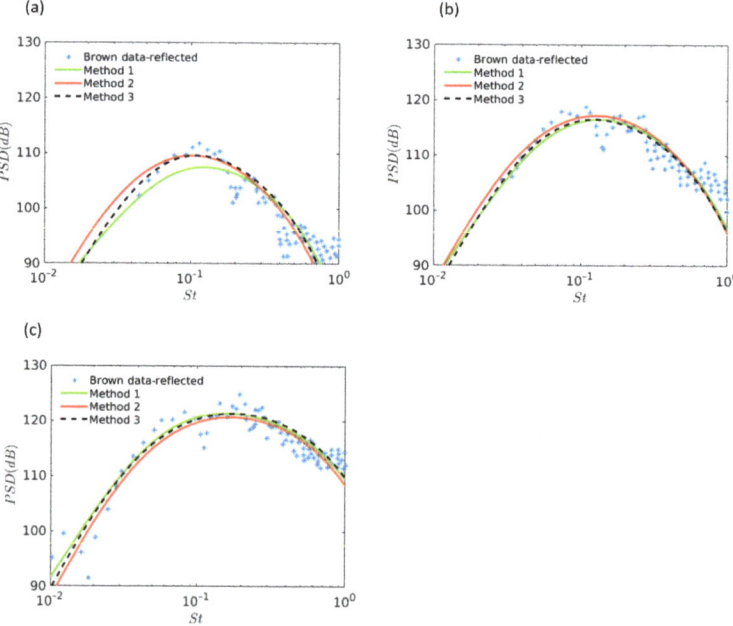

Figure 14. Comparison of the acoustic predictions for methods 1, 2 and 3 using Multi-Start (**a**) $Ma = 0.5$ (**b**) $Ma = 0.7$ (**c**) $Ma = 0.9$.

6. Discussion

Method 1 was used to find the four parameters through optimization of our acoustic model against the experimental data in [12]. Figure 10 shows that the predictions are particularly good for $Ma = 0.7$, and $Ma = 0.9$ for all optimization routines. For $Ma = 0.5$ Multi-Start gives a poorer prediction due to the change in l_3 which affects low frequency roll-off. From this figure, it is shown that MP-AIDEA and PSO give more or less the same acoustic predictions for all acoustic Mach numbers.

However, the turbulence correlation function R_{22} that we use (Equation (5)) is a function of a_1, and the values for a_1 that were found from this method for each optimization routine do not give a good representation of R_{22}, as shown in Figure 9. The initial decay is too slow and there is little to no anti-correlation region, this is particularly the case for $Ma = 0.5$. Note, that we've allowed $\tau U_c / D \rightarrow 10$ to show where the model goes to zero, there is no turbulence data at these locations most likely due to measurement difficulties.

In methods 2 and 3, we optimized the R_{22} model separately against experimental data from Bridges ($y_1/D = 6, r/D = 0.5$) [36]. This means that the anti-correlation region is represented and the initial decay is steeper, as shown in Figure 9 for methods 2 and 3. Only the initial de-correlation is of interest, therefore we have used a simple model for R_{22}. A different model could be used to capture the oscillations but this would make the acoustic model much more complicated and have no improvement on the acoustic spectrum predictions.

For method 3 we use the this R_{22} optimized value a_1 (0.85) in the acoustic model and then optimize the other parameters against the acoustic data to find the best prediction. This allows us to find the optimal predictions for the acoustic spectrum while also having a good representation of the turbulence structure. It results in slightly poorer acoustic predictions, with the exception of Multi-Start for $Ma = 0.7$, as noted in Tables 2–4. However, Figure 11 shows that they still give very good predictions for the level of accuracy that we require. Since this method also allows R_{22} to be better represented, overall it is deemed to be better than method 1.

In method 2, we also use this value ($a_1 = 0.85$) in the acoustic model and hand tune the other parameters to find the best prediction. We aimed to find one set of parameters for all acoustic Mach numbers. However, it was found that due to the level shift in the acoustic spectrum, one parameter (l_2) must change for each Mach number. As this method did not require different optimization routines, the results are included in Figures 12–14 and is identical in each. We can see that the predictions are similar to methods 1 and 3. However, hand tuning these parameters is not ideal as it relies on human judgement as to what is a 'good' prediction.

It is easy to see in Figures 12–14 that the three methods of optimization give similar acoustic predictions for all optimization routines, with only Multi-Start giving a noticeable change in prediction for $Ma = 0.5$. Note that from Figure 8 only PSO and MP-AIDEA consistently converged to a minimum objective function value ($fval = 9.2$), Multi-Start displayed a more widely varying value, possibly due to the number of starting points chosen (500). Figure 7 shows that the parameters found for Multi-Start also vary widely on each run of the routine, hence it is less likely that acoustic predictions found using Multi-Start are optimal. On the other hand, Figures 5 and 6 show that the parameters found by PSO and MP-AIDEA respectively, only vary slightly.

We can conclude that our acoustic spectrum model for the trailing-edge noise problem is 'parametrically flat', i.e., the objective function is not noticeably sensitive to the variation of parameters (l_2, l_3, a_1, τ_0) within their specified range. Both evolutionary optimization routines that we have used in this paper have given good results and almost identical predictions. The time taken in optimising method 1 naturally took longer than method 3 since an extra parameter was being found. In general, it was found that method 1 was faster using PSO, however method 3 was faster using MP-AIDEA. However, for the number of starting points that was chosen, Multi-Start was fastest of all, and since the objective function value was still smaller than method 2, if there was a time constraint which restricted the use of evolutionary algorithms, it would be worthwhile to use Multi-Start rather than hand-tuning the problem.

7. Conclusions

Aeroacoustic models for turbulence interaction problems will always require a set of parameters that define the rate of temporal and spatial de-correlation. For the trailing-edge noise problem as illustrated in Figure 1, the acoustic spectrum (Equation (2)) is proportional to the Fourier transform of the streamwise-independent transverse velocity correlation function, $R_{22}(\tilde{u}-u, \tilde{v}-v; \omega)$ via Equations (3), (4) and (6) and depends on $\mathcal{O}(1)$ parameters (l_2, l_3, a_1, τ_0). The latter can be chosen by hand-tuning, however this will not result in a 'mathematically' optimal choice of parameters, and could result in under/over prediction of the acoustic spectrum (as illustrated in Figures 12–14). It also adds to the time taken in assessing its predictive capability. This paper highlights how optimization routines, both evolutionary and non-evolutionary, can be used to determine what the optimal parameters are, resulting in slightly better acoustic predictions (see Figures 12–14). When determining turbulence-associated parameters in acoustic models for which more than the 3 state variables considered in this paper (i.e., Ma, θ, ϕ) are required, such as in predictive models of the full jet-installation noise signature, the use of a numerical optimization routine will become even more beneficial.

Author Contributions: Conceptualization, S.S. and M.Z.A.; methodology, S.S. and M.Z.A.; software, S.S. and E.M.; formal analysis, S.S.; investigation, S.S. and M.Z.A.; writing—original draft preparation, S.S. and M.Z.A.; writing—review and editing, S.S., M.Z.A. and E.M.; visualization, S.S. and M.Z.A. All authors have read and agreed to the published version of the manuscript.

Funding: This research was funded by EPSRC DTP (grant ref: EP/R513349/1).

Institutional Review Board Statement: Not applicable.

Informed Consent Statement: Not applicable.

Data Availability Statement: Not applicable.

Acknowledgments: The authors would like to thank Ioannis Kokkinakis from the University of Strathclyde for his help in the development of the acoustic model. M.Z.A. would like to thank the University of Strathclyde for financial support from the Chancellor's Fellowship.

Conflicts of Interest: The authors declare no conflict of interest.

References

1. Cumpsty, N.; Heyes, A. *Jet Propulsion: A Simple Guide to the Aerodynamics and Thermodynamic Design and Performance of Jet Engines*; Cambridge University Press: Cambridge, UK, 2015.
2. Peake, N.; Parry, A.B. Modern Challenges Facing Turbomachinery Aeroacoustics. *Annu. Rev. Fluid Mech.* **2012**, *44*, 227–248. [CrossRef]
3. Dobrzynski, W. Almost 40 Years of Airframe Noise Research: What Did We Achieve? *J. Aircr.* **2010**, *47*, 353–367. [CrossRef]
4. Olsen, W.; Boldman, D. Trailing edge noise data with comparison to theory. In Proceedings of the 12th Fluid and Plasma Dynamics Conference, AIAA Paper 79-1524, Williamsburg, VA, USA, 23–25 July 1979.
5. Wang, M.E. Wing effect on jet-noise propagation. *J. Aircr.* **1981**, *18*, 295–302. [CrossRef]
6. Head, R.W.; Fisher, M.J. Jet/surface interaction noise: Analysis of farfield low frequency augmentation of jet noise due to the presence of a solid shield. In Proceedings of the 3rd Aeroacoustics Conference, Palo Alto, CA, USA, 20–23 July 1976.
7. Southern, I.S. Exhaust noise in flight: The role of acoustic installation effects. In Proceedings of the 6th Aeroacoustics Conference, Hartford, CT, USA, 4–6 June 1980.
8. Bridges, J.E. Noise from Aft Deck Exhaust Nozzles—Differences in experimental embodiments. In Proceedings of the 52nd Aerospace Sciences Meeting, AIAA Paper 2014-0876, Harbor, MD, USA, 13–17 January 2014.
9. Bridges, J.E.; Brown, C.A.; Bozak, R. Experiments on Exhaust Noise of Tightly Integrated Propulsion Systems. In Proceedings of the 20th AIAA/CEAS Aeroacoustics Conference, AIAA Paper 2014-2904, Atlanta, GA, USA, 16–20 June 2014.
10. Goldstein, M.E.; Afsar, M.Z.; Leib, S.J. Non-homogeneous Rapid-distortion theory on transversely sheared flows. *JFM* **2013**, *736*, 532–569. [CrossRef]
11. Goldstein, M.E.; Leib, S.J.; Afsar, M.Z. Generalized rapid distortion theory on transversely sheared mean flows with physically realizable upstream boundary conditions: Application to the trailing-edge problem. *JFM* **2017**, *824*, 477–512. [CrossRef]

2. Goldstein, M.E.; Leib, S.J.; Afsar, M.Z. Rapid distortion theory on transversely sheared mean flows of arbitrary cross-section. *JFM* **2019**, *881*, 551–584. [CrossRef]
3. Afsar, M.Z.; Leib, S.J.; Bozak, R.F. Effect of de-correlating turbulence on the low frequency decay of jet-surface interaction noise in sub-sonic unheated air jets using a CFD-based approach. *JSV* **2017**, *386*, 177–207. [CrossRef]
4. Afsar, M.Z.; Stirrat, S.A.; Kokkinakis, I.W. Investigation of fast GPU-based algorithms for jet-surface interaction noise calculations. In Proceedings of the 2020 AIAA Aviation and Aeronautics Forum and Exposition, Virtual, Online, 15–19 June 2020; AIAA 2020-2564.
5. Rego, L.; Avallone, F; Ragni, D.; Casalino, D. Jet-installation noise and near-field characteristics of jet–surface interaction. *JFM* **2020**, *895*, A2. [CrossRef]
6. Afsar, M.Z.; Goldstein, M.E.; Fagan, A. Enthalpy-Flux/Momentum-Flux Coupling in the Acoustic Spectrum of Heated Jets. *AIAA J.* **2011**, *49*, 2522–2532 [CrossRef]
7. Karabasov, S.A.; Afsar, M.Z.; Hynes, T.P.; Dowling, A.P.; McMullan, W.A.; Pokora, C.D.; Page, G.J.; McGuirk, J.J. Jet Noise: Acoustic Analogy informed by Large Eddy Simulation. *AIAA J.* **2010**, *48*, 1312–1325 [CrossRef]
8. Harper-Bourne, M. Jet Noise Measurements. In Proceedings of the 9th AIAA/CEAS Aeroacoustics Conference, Hilton Head, SC, USA, 12–14 May 2003; AIAA-2003-3214.
9. Gunzburger, M. Adjoint equation-based methods for control problems in incompressible, viscous flows. *Flow Turbul. Combust.* **2000**, *65*, 249–272. [CrossRef]
10. Vanderplaats, G.N. *Multidiscipline Design Optimization*; Vanderplaats Research and Development, Inc.: Monterey, CA, USA, 2007.
11. Arora, J.S. *Introduction to Optimum Design*, 2nd ed.; Elsevier Academic Press: Amsterdam, The Netherlands, 2004.
12. Haftka R.T.; Gürdal, Z. *Elements of Structural Optimization*, 3rd ed.; Springer: Dordrecht, The Netherlands, 1993.
13. Snyman, J.A. *Practical Mathematical Optimization*; Springer: Berlin/Heidelberg, Germany, 2005.
14. Venter, G. Review of Optimization Techniques. In *Encyclopedia of Aerospace Engineering*; Wiley & Sons: Hoboken, NJ, USA, 2010.
15. Neumaier, A. Complete search in continuous global optimization and constraint satisfaction. *Acta Numer.* **2004**, *13*, 271–370. [CrossRef]
16. Korolev, Y.M.; Karabasov, S.A.; Toropov, V.V. Automatic Optimizer vs. Human Optimizer for Low-Order Jet Noise Modeling. In Proceedings of the 21st AIAA/CEAS Aeroacoustics Conference, Dallas, TX, USA, 22–26 June 2015.
17. Toropov, V.V. Simulation approach to structural optimization. *Struct. Optim.* **1989**, *1*, 37–46. [CrossRef]
18. Toropov, V.V.; Filatov, A.A.; Polynkin, A.A. Multiparameter structural optimization using FEM and multipoint explicit approximations. *Struct. Optim.* **1993**, *6*, 7–14. [CrossRef]
19. Holland, J.H. *Adaption in Natural and Artificial Systems*; University of Michigan Press: Ann Arbor, MI, USA, 1975.
20. Kennedy, J.; Eberhart, R.C. Particle swarm optimization. In Proceedings of the 1995 IEEE International Conference on Neural Networks, Perth, Australia, 27 November–1 December 1995; pp. 1942–1948.
21. Storn, R.; Price, K. Differential evolution—A simple and efficient heuristic for global optimization over continuous spaces. *J. Glob. Optim.* **1997**, *11*, 341–359. [CrossRef]
22. Martí, R. Multi-Start Methods. In *Handbook of Metaheuristics. International Series in Operations Research & Management Science*; Springer: Boston, MA, USA, 2003; Volume 57. [CrossRef]
23. Cox, S.E.; Haftka, R.T.; Baker, C.A.; Grossman, B.; Mason, W.H.; Watson, L.T. A comparison of global optimization methods for the design of a high-speed civil transport . *J. Glob. Optim.* **2001**, *21*, 415–432. [CrossRef]
24. Haim, D.; Giunta, A.A.; Holzwart, M.M.; Mason, W.H.; Watson, L.T.; Haftka, R.T. Comparison of optimization softare packages for an aircraft multidisciplinary design optimization problem. *Des. Optim.* **1999**, *1*, 9–23.
25. Di Carlo, M.; Vasile, M.; Minisci, E. Adaptive multi-population inflationary differential evolution. *Soft Comput.* **2020**, *24*, 3861–3891. [CrossRef]
26. Bridges, J. Effect of Heat on Space-Time Correlations in Jets. In Proceedings of the 12th AIAA/CEAS Aeroacoustics Conference, Cambridge, MA, USA, 8–10 May 2006; AIAA-2006-2534.
27. Brown, C.A. Jet-surface interaction test: Far-field noise results. *ASME J. Engng Gas Turbines Power* **2010** *135*, 071201–071201-7. [CrossRef]
28. Mezura-Montes, E.; Coello, C.A.C. Constraint-handling in nature-inspired numerical optimization: Past, present and future. *Swarm Evol. Comput.* **2011**, *1*, 173–194. [CrossRef]
29. Pedersen, M.E. *Good Parameters for Particle Swarm Optimization*; Hvass Laboratories: Luxembourg, 2010.
30. Das, S.; Suganthan, P.N. Differential Evolution: A Survey of the State-of-the-Art. *IEEE Trans. Evol. Comput.* **2011**, *15*, 4–31. [CrossRef]
31. Vasile, M.; Minisci. E.; Locatelli, M. An Inflationary Differential Evolution Algorithm for Space Trajectory Optimization. *IEEE Trans. Evol. Comput.* **2011**, *15*, 267–281. [CrossRef]
32. Wales, D.J.; Doye, J.P. Global optimization by basin-hopping and the lowest energy structures of Lennard-Jones clusters containing up to 110 atoms. *J. Phys. Chem. A* **1997**, *101*, 5111–5116. [CrossRef]

Article

Detailed Study on the Behavior of Improved Beam T-Junctions Modeling for the Characterization of Tubular Structures, Based on Artificial Neural Networks Trained with Finite Element Models

Francisco Badea [1], Jesus Angel Perez [2] and Jose Luis Olazagoitia [1],*

- [1] Department of Industrial Engineering and Automotive, Nebrija University, Pirineos 55, 28040 Madrid, Spain; Fbadea@nebrija.es
- [2] Department of Construction and Manufacturing Engineering, Mechanical Engineering Area, University of Oviedo, 33203 Gijón, Spain; perezangel@uniovi.es
- * Correspondence: jolazago@nebrija.es

Abstract: The actual behavior of welded T-junctions in tubular structures depends strongly on the topology of the junction at the joint level. In finite element analysis, beam-type elements are usually employed due to their simplicity and low computational cost, even though they cannot reproduce the joints topologies and characteristics. To adjust their behavior to a more realistic situation, elastic elements can be introduced at the joint level, whose characteristics must be determined through costly validations. This paper studies the optimization and implementation of the validation data, through the creation of an optimal surrogate model based on neural networks, leading to a model that predicts the stiffness of elastic elements, introduced at the joint level based on available data. The paper focuses on how the neural network should be chosen, when training data is very limited and, more importantly, which of the available data should be used for training and which for verification. The methodology used is based on a Monte Carlo analysis that allows an exhaustive study of both the network parameters and the distribution and choice of the limited data in the training set to optimize its performance. The results obtained indicate that the use of neural networks without a careful methodology in this type of problems could lead to inaccurate results. It is also shown that a conscientious choice of training data, among the data available in the problem of choice of elastic parameters for T-junctions in finite elements, is fundamental to achieve functional surrogate models.

Keywords: T-junctions; neural networks; finite elements analysis; surrogate; beam improvements; beam T-junctions models; artificial neural networks (ANN) limited training data

Citation: Badea, F.; Perez, J.A.; Olazagoitia, J.L. Detailed Study on the Behavior of Improved Beam T-Junctions Modeling for the Characterization of Tubular Structures, Based on Artificial Neural Networks Trained with Finite Element Models. *Mathematics* **2021**, *9*, 943. https://doi.org/10.3390/math9090943

Academic Editor: David Greiner

Received: 29 March 2021
Accepted: 19 April 2021
Published: 23 April 2021

Publisher's Note: MDPI stays neutral with regard to jurisdictional claims in published maps and institutional affiliations.

Copyright: © 2021 by the authors. Licensee MDPI, Basel, Switzerland. This article is an open access article distributed under the terms and conditions of the Creative Commons Attribution (CC BY) license (https://creativecommons.org/licenses/by/4.0/).

1. Introduction

The utilization of finite element modeling for the simulation of structural components has become a fundamental part of modern engineering, being utilized in the majority of industrial fields.

Despite the advances of the finite element analysis (FEA) software and the increased calculation capabilities, for the modeling of tubular structures in which the length of the profiles is significantly bigger than the width and thickness of the sections, the beam type elements are still widely utilized despite their limitations and the fact that they cannot reproduce the characteristics of the joints neither from the geometrical point of view nor from the behavioral one. Figure 1 presents the equivalent beam type element T-junction that would be utilized for the simulation of the original T1 and T2 junctions.

The joint configuration of these T-junctions (T1 and T2) has a direct influence on the behavior of the structures, determining significant behavioral differences depending on the type and direction of the load. Despite this reality, the beam type element equivalent model will always be composed of three nodes and three beam type elements. This way all the geometrical characteristics of the joint get lost into an infinitely rigid node, which represents a significant shortcoming of these elements in the correct characterization of T-junctions.

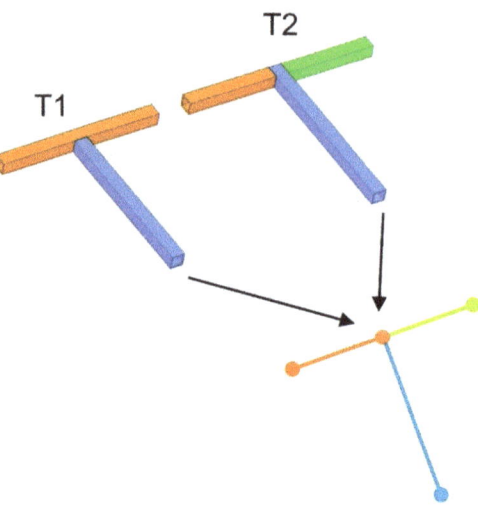

Figure 1. T-junctions equivalent beam type element model.

This issue could be avoided by utilizing shell or volume type elements. However, these alternatives are feasible only for small models, which is not the case when common tubular structures such as buses upper modules, bridges, support structures, or similar big structures need to be modeled. The complexities of the modeling process along with the increase in computational resources due to the increment of nodes make these alternatives unfeasible for general industrial use.

The magnitude of this issue is illustrated in Table 1, in which a comparative analysis is performed based on the number of nodes and degrees of freedom for a simple T-junction of 1 m by 1 m having a standard square hollow profile of 40 mm × 40 mm × 4 mm presented in Figure 2. The table presents the comparative analysis of the T-junction, modeled using beam, shell and volume element types, for comparative purposes, the three models where meshed using 4 mm first order elements. The resulting number of nodes are a consequence of the meshing dimension and the degrees of freedom are calculated taking into consideration the mathematical formulation of the elements. In the case of the beam and shell elements we have 6 degrees of freedom (DOF) per node and in the case of the volume type elements we have 3 DOF per node. A comparative evaluation criterion taking into consideration the complexity of the modeling process and the precision of the results, based on the experience of the authors for modeling structures. The value is calculated as the product between the DOFs and the complexity of the modeling process along with the precision of the results (Table 1).

Table 1. Comparative analysis between characteristics of T-junctions modeled with beam, shell, and volume type elements.

Element Type	No. of Nodes	DOFs	Complexity of Modeling (1–3)	Results Precision (1–3)	Evaluation Coefficient
Beam	910	5460	1	3	16,380
Shell	19,520	117,120	2	2	468,480
Volume	36,144	108,432	3	1	325,296

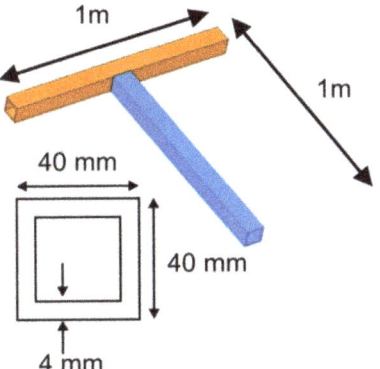

Figure 2. Analyzed T-junction with the cross section detail.

From the previously presented results, it can be seen that, despite their limitations, the beam type elements represent the most feasible alternative when analyzing tubular structures in which the relation of the length and the profile dimensions is quite significant. Due to the above, it can be concluded that the calculation of complex tubular structure (buses, bridges, and railway carriages) is significantly simplified by using beam-type elements. This is why, these elements have been widely recommended [1–3] for modeling large tubular structures.

However, despite being recommended and widely utilized in the industry, the beam formulation is simply not designed to take into account and reproduce the behavior of real joints, since all the elements that reach the same joint are joined in a single infinitely rigid node [4–6]. Furthermore, beam-type elements cannot represent the topological characteristics of the joints and do not differentiate between joints that are apparently similar, but with different topologies. Therefore, there is a clear necessity for the improvements of the beam modeled T-junctions.

One of the improvement solutions proposed in order to avoid ignoring the specific characteristics of the joint topologies was the introduction of elastic elements at the joint level when modeling T-junctions. In [4] it was shown that the flexibility of the joints has a determining influence on the behavior of the analyzed structures and could not be ignored. Later, in [5] it was again confirmed that ignoring the influence of joint stiffness on structural behavior was a mistake. In this case, it was shown that, when the stiffness of the elastic elements in the joint exceeds a certain maximum value, the total deformation energy of the vehicle is insensitive to the joint.

Furthermore, in [7], Lee presented the hypothesis that the behavior of welded joints was elastic and linear, obtaining a general methodology for the model of joints. Applying this, in [8], three rotational springs were introduced into the joints with the aim of optimizing vehicle structures. This way, the deformations that could occur in the joints due to torsional or bending moments were considered. In contrast, infinite stiffness values were assumed in the axial directions, this study showed that it was possible to obtain better approximations with this approach and validated modal calculations with experimental modal analyses. The introduction of elastic elements when modeling beam T-junctions has a fundamental advantage since it provides the possibility to modify the overall behavior of the structure while maintaining the mass distribution of the input model, having the downside of requiring the adjustment of the appropriate stiffness values which would lead to the need for expensive experiments.

Another alternative was to model beam T-junctions with variable stiffness in the adjacent regions to the joints, this was proposed in [9] since the rigidity of the joint is much greater than the surrounding area. Later, in [10] the idea was reformulated by using partially rigid beams. Taking advantage of this new concept, in [11] H-shaped structures with rectangular hollow profiles were studied and partially rigid beams were used at their joints, the model was validated through modal correlation with good results. The same methodology was applied in [12] for the optimization of a three-dimensional bus structure formed by hollow rectangular profiles of 40 mm × 40 mm × 3 mm. To do this, it was necessary to perform a sensitivity analysis to identify the joints that had the greatest influence on the overall behavior of the structure, this way, partially rigid beams were progressively introduced in the most important regions. Alternatively, a new technique was proposed in [13]. In this case, the parameters of the elements near the joint were modified. This approach of the problem is similar to the previous one, but also allowing the modification of the mass and rigidity independently.

Given these limitations, some authors have used shell or volume type elements as an alternative for the study of the joints, modeling the rest of the structure with beam type elements, these elements allow the characteristics of the joint to be modeled more accurately In [14] the shell element model was applied to tubular joints and major improvements were reported. In [15] the models of a single beam with shell elements were compared against the same shell and beam modeled and it was shown that hybrid modeling could reduce the number of elements without compromising the results.

The same approach has been applied to the analysis of structural joints in cars. Thus, hybrid modeling has led to satisfactory results [14,16,17] in this field, however, the application of this hybrid technique to larger models (e.g., hollow beam structures in buses) is unattractive, since it would be necessary to specifically model a large number of joints, which leads to a substantial increase in the model preparation time and more computational resources by increasing the number of elements in the model.

In their study [18] developed an alternative beam T-junction model that allowed them to obtain more precise deformation results when utilizing beam type elements, also allowing them to take into consideration different T-junction configurations. Their focus was to improve the results provided by beam type elements when simulating structures that by their characteristics cannot be realistically simulated with anything but beam type elements due to the complexity of the modeling process, the computational requirements or other practical aspects. The authors proposed an alternative beam model in which they introduced a total of six elastic elements at the joint level along with a complete methodology based on finite element methodology (FEM) comparative analysis that allowed them to improve the behavior of the modeled T-junctions in terms of displacements. This proposal, however, required the user to reproduce the complete methodology in order to obtain six stiffness values for each individual modeled T-junction making it somehow not attractive when having to characterize a wide variety of T-junction combinations. Figure 3 presents the alternative beam T-junction model proposed and utilized by [18]. This T-junction introduces a total of six elastic elements at the joint level, three of them behaving as linear springs (represented by the letter k) and noted with the sub index ux, uy, and uz corresponding to (u) displacement and (x, y, z) the axial directions and another three corresponding to (r) rotations along the (x, y, z) axial directions.

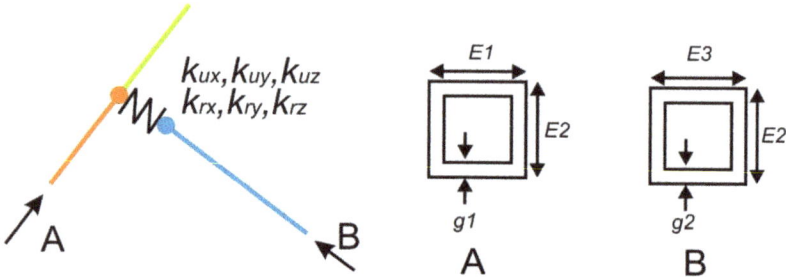

Figure 3. Alternative T-junction model.

This approach was also used by [19], in which multiple simulations utilizing the most common profiles found in buses and coaches upper structures were performed, followed by a statistical analysis in order to obtain regression models for the calculation of the stiffness values for the alternative T-junctions, based on the profile dimensions ($E1$, $E2$, $E3$, $g1$, and $g2$). They utilized a Bayesian Kriging regression model that provided satisfactory results, however, the regression equations obtained were notably complicated having, in some cases, more than 33 terms per equation, making them impractical for everyday use in the wide industry. These regressions were obtained with a MISO (multiple input, single output) approach in which a total of six regression models had to be calculated, one for each k_{ux}, k_{uy}, k_{ux}, k_{rx}, k_{ry}, and k_{rz} stiffness value, leading to the necessity of any potential user to use six complex equations for the estimation of the stiffness values for a single T-junction type.

From the previous studies it can be concluded that it is possible to make an adjustment of the existing data coming from analysis by finite elements in complicated models. In this way, the time and effort needed to carry out new simulations for sets of junction parameters not previously studied can be drastically reduced. Unfortunately, the basic laws that allow for optimal adjustment are not known. In this sense, artificial neural networks (ANN) surrogate models allow, from sufficient known training data, to infer the unknown laws behind the analyzed data. The use of ANNs is not new. There are countless studies that have applied them to the most diverse applications. Specifically, in the last years, the development and implementation of neural networks for the evaluation of multiple processes with a pronounced non-linear behavior has become a reality [20]. ANN have also been utilized along with finite elements such as in [21] where the authors implemented their use with good results in the evaluation of damage detection in bridges. In [22], ANN was used to simplify contact estimation models in ANSYS®. In [23] the use of response surface and ANN models was successfully evaluated against other methods of structural reliability analysis. In addition, ANNs can generate valid surrogated models, based on data obtained from finite elements, to evaluate such complicated issues as noise reduction in braking systems, with very high accuracy [24].

Taking into consideration the benefits of the implementation of neural networks for finite element analysis, the authors considered of great interest the development of a methodology for the estimation of the stiffness values utilizing ANN as an alternative to the utilization of Bayesian Kriging regressions similar to those utilized in [19], that provided satisfactory results although not realistic from the practical point of view due to the complexity of the obtained equations. In fact, ANNs have been used as a valid alternative to such Kriging networks in various articles for the creation of metamodels [25,26] providing very satisfactory results.

This article studies the application of ANN in the creation of surrogate models that allow inferring the information obtained from complicated finite element models, such as those obtained in the analysis of optimal stiffness to predict the behavior of T-welded junctions. The biggest problem when trying to replace finite element modeling with surrogate models is that a lot of training data needs to be obtained for the prediction to be reliable. However, obtaining this data is often very costly in time and effort. There is also uncertainty about how the ANN should be created to learn effectively and be useful for unknown finite element models.

This article presents a detailed study, leading to the presentation of a new methodology for the creation of surrogate models (or metamodels) based on data obtained from finite element calculations. The particularity in this case is that the number of initial data, for the training of the model, is very limited. Few training data can make the model a failure. However, as mentioned, this is usually a common situation in FEM due to the high time of creation of FEM models and their calculation. The possibility of creating models that can learn from these "few" data obtained from FEM calculations, using them in an optimal way, can allow to obtain more precise models and enable the use of ANN to cases with limited data.

The rest of the article is distributed as follows: Section 2 presents the methodology followed for the determination of the ANN topology and the optimal use of the available data; Section 3 describes the results and the analyses obtained, which are discussed in Section 4. Finally, Section 5 presents the conclusions, applicability, and future lines.

2. Methodology

Using the same data from [19] the experimental design for the neural network analysis had a five dimensional input and six dimensional output, where the input values represented the dimensions of the analyzed T-junction profiles ($E1$, $E2$, $E3$, $g1$, and $g2$) and the output values represented the calculated stiffness values for the alternative T-junction model having 3 axial spring (k_{ux}, k_{uy}, k_{uz}) and 3 rotational springs (k_{rx}, k_{ry}, k_{rz}) at the junction level as presented in Figure 3.

This data is based on hollow rectangular profiles commonly utilized in the buses and coaches upper structures, focusing the study on a total of 243 profile combinations for two different T-junction configurations named T1 and T2 presented in Figure 1.

The 243 input values were obtained by getting all of the possible combinations for the five input variables, each variable having 3 levels as presented in Table 2.

Table 2. Input base variable values.

$E1$ (mm)	$E2$ (mm)	$E3$ (mm)	$g1$ (mm)	$g2$ (mm)
40	40	40	2	2
60	60	60	3	3
80	80	80	4	4

As can be seen in the previous table, the input data is staggered and only provides information about the studied process for specific values, leaving significant empty spaces between variables as can be seen in Figure 4, in which the utilized input values for the $E1$–$E3$ characteristics are plotted in a 3D graphic.

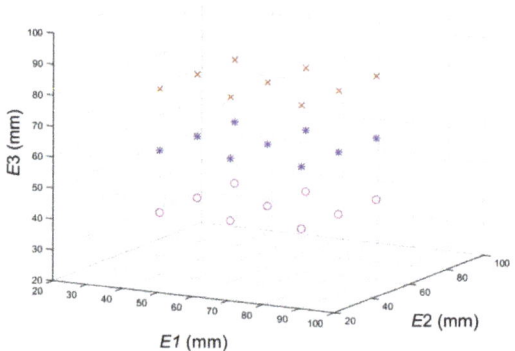

Figure 4. Three dimensional plot for the $E1$–$E3$ utilized values.

Since the base methodology requires comparative FEM analysis between beam and shell modeled T-junctions, for example adding 2 additional thicknesses for the $g1$ and $g2$ variables, would require a total of 675 comparative simulations for each one of the T1 and T2 junctions determining a significant increase in the required computational time. This leads to a situation in which predictive models need to be constructed from a limited amount of staggered input data, in which a proper selection of the neural network training data is of paramount importance. To illustrate that, Figure 5 presents the average error obtained for 50 different neural networks, constructed using different training data sets versus using the same training data set. This error is determined for each neural network as the average difference between the network output values and the target outputs of the validation data obtained from FEM models. As it can be seen, although both cases reach the same average error (0.072), the standard deviation is almost one order of magnitude higher in the case of neural networks constructed using different training data, evidencing the importance of a proper selection of this data.

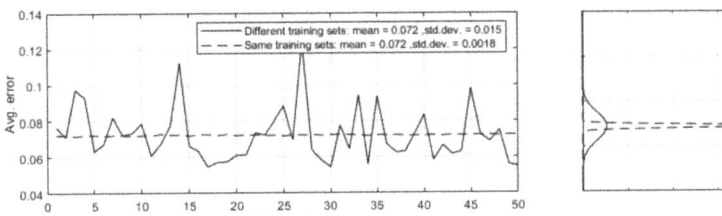

Figure 5. Neural network error dispersion comparison between using same versus different training data.

This behavior motivates a more detailed study of the performance of neural networks with respect of the chosen data set. To carry out this study it was necessary to decide the topology and design of a neural network that could be effectively trained from few input data. In a neural network, the increase of hidden layers and nodes in each layer, requires a lot of training data to avoid the known problem of over fitting. As the number of training data is very limited in our case the network was kept simple, with only one hidden layer and a limited number of nodes (1 to 20) within that layer. In addition, to check the performance of the neural network, the initial data were divided into two subsets. The first one was dedicated to the training of the network (75% of them) and the second one to its validation (25%). Usually, the distribution of training and validation data is around 50% in cases where there are large amounts of data. However, in this case, due to the high cost of obtaining additional training data, it was decided to vary the standard. The rest of the parameters and functionality of the neural network were also kept simple. For the comparison between the outcome between neural networks a limit of 2000 training iterations (epocs) was taken, with a learning rate of 0.1. The activation function of the neurons was the hyperbolic tangent sigmoid, and the training was based on a Levenberg–Marquardt backpropagation algorithm, Figure 6 presents the characteristics of the ANN.

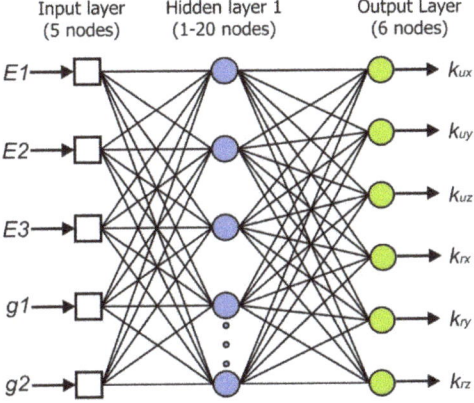

Figure 6. Characteristics of the ANN evaluated.

As mentioned, the limitation of training data has a fundamental influence on the choice of the type of neural network to be used, its topology and its configuration, since it can cause the network not to learn the pattern behind the data, or to learn it too well fitting perfectly to the training data, but offering very poor results in the validation data. This effect is called over fitting and it is critical to avoid it. The goal is to train with as much data as possible and to learn in a robust way and to reduce the error in the validation data.

But, apart from the issue of reasoned choice of network topology, there is the major problem of optimal selection of training data. Since there is so little training data, the information that such data provides about the behavior of the welded T-joints is very valuable. The choice of which data should be used for network training and which data should be used for validation has a critical influence on the result of the analysis. Therefore, the choice of which data will go to 75% of the training set should not be made at random in this type of problem. At this point, the choice of suitable training sets is not evident. For this reason, a thorough study is needed to shed light on this point.

In this sense, next sections show the results of this complete study, where a total of 1000 different training data sets were used to construct the neural network, leading to a total of 20,000 different neural networks for each joint configuration. Input layer consisted of 5 neurons containing input information of the joint dimensions ($E1$, $E2$, $E3$, $g1$, and $g2$, see Figure 3 and Table 2). For the output layer both MISO (multiple input single output) and MIMO (multiple input multiple output) models were initially explored. Both approaches showed very similar results, so it was decided to focus on MIMO models due to the obvious computational cost saving of not having to construct independent neural networks for each output. Thus, output layer consisted of 6 neurons with the 6 DOF spring rates (k_{ux}, k_{uy}, k_{uz}, k_{rx}, k_{ry}, and k_{rz}, see Figure 3). Due to the different behavior of the different joint configurations, independent neural networks were analyzed for T1 and T2 junctions (Figure 1). The following Figure 7 shows a block diagram with the methodology used to construct and check the performance of each neural network. As it is shown, the same input data with the geometric information that was used to construct the beam and shell FEM models and obtain the spring stiffness, is used to feed the neural networks, which are constructed using 1000 randomly chosen trained sets, and 1 to 20 neurons in its hidden layer. For each network, the average error is recorded, which is obtained as the mean absolute difference between the predicted and correct outputs of the validations set (remaining 25% of the data).

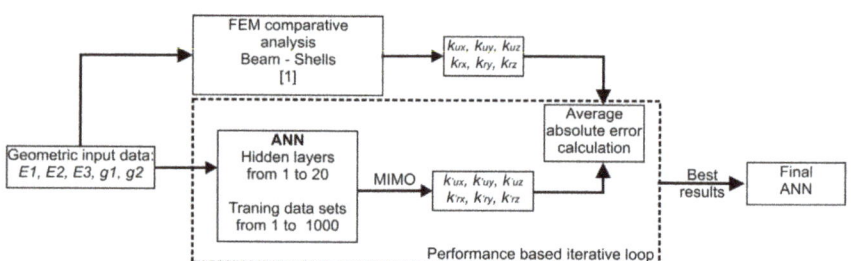

Figure 7. Block diagram of the applied methodology for the selection of the best ANN.

Finally, once the complete methodology is executed, the best performance networks are selected for further analysis in order to validate the selection methodology proposed.

3. Results and Analysis

In the following subsections, the most relevant information, and the obtained results of the application of the methodology will be presented.

3.1. Neural Networks Behavioral Analysis Results

This section shows all relevant findings encountered during the application of the methodology itself. It should be noted that, although of great interest to understand the behavior of the data, the analysis presented in this section in not strictly necessary to run the methodology since it does not affect the network selection, which ultimately depends in the overall average error.

In relation to the performance of the neural network with respect to the number of neurons, Figure 8 shows the average error obtained for each spring and joint configuration, averaged over the 1000 training data sets utilized. As it can be seen, the minimum error values are generally obtained for a number of neurons that ranges approximately between 5 and 10. It can be also noticed that the error level is highly dependent on the spring direction and the junction configuration.

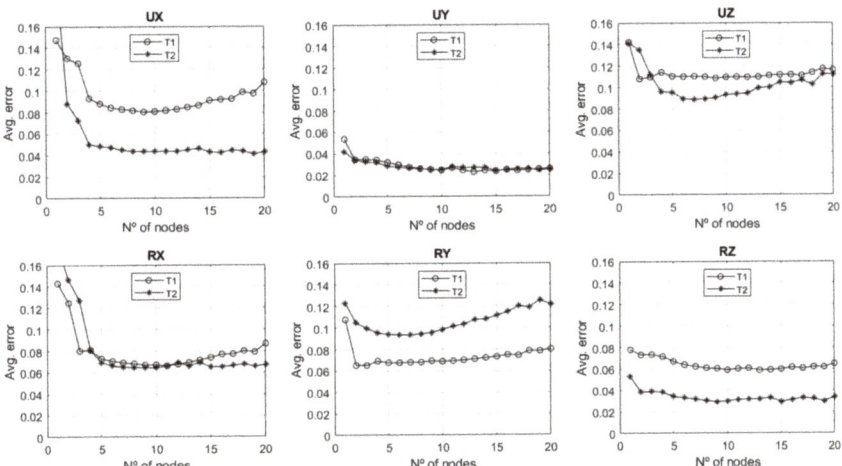

Figure 8. Average errors in comparison to the number of nodes in the ANN (1000 networks).

Although in some cases the average error reach minimum values for more than 10 neurons, it was observed that the variability of these errors increased significantly as the number of neurons is increased, and that some outputs with extremely bad predictions where obtained. In this regard, Figure 9 shows the number of networks observed with an average error above one, identified as outliers. It is clearly noticed that from about 10 neurons, outlier results start to increase. In general, springs with better performance show fewer networks with exceptionally high errors.

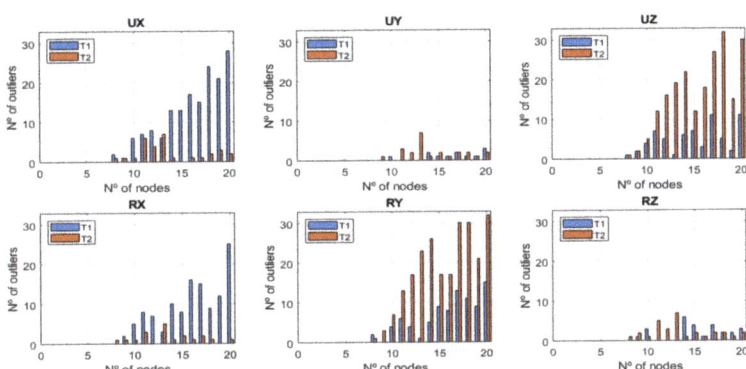

Figure 9. Outlier networks with respect to the number of neurons.

With respect to the performance of the networks regarding the training data, it was noticed a dependency among the number of neurons, i.e., if any network performs optimally for a given number of neurons, it will show the same tendency for the rest number of neurons. To assess this, the 1000 networks were ranked from 1 (least error) to 1000 (highest error) for each direction and number of neurons. If a given training data shows similar rankings for all number of neurons, it would mean that such dependency of the performance with respect to the number of neurons exists. On the other hand, if the rankings from 1 to 20 neurons approach to a uniform distribution, it would indicate not dependency at all. This behavior was quantified by means of the standard deviation of the rankings in Figure 10, for both junction configuration and spring direction. It is noticed that dispersion of rankings is always significantly lower than the one for uniform distribution (σ = 288.6) and that it is especially low for networks that show best or worst rankings, meaning that dependency is especially high in these cases. As expected, behavior differences are observed between junction configuration and spring direction.

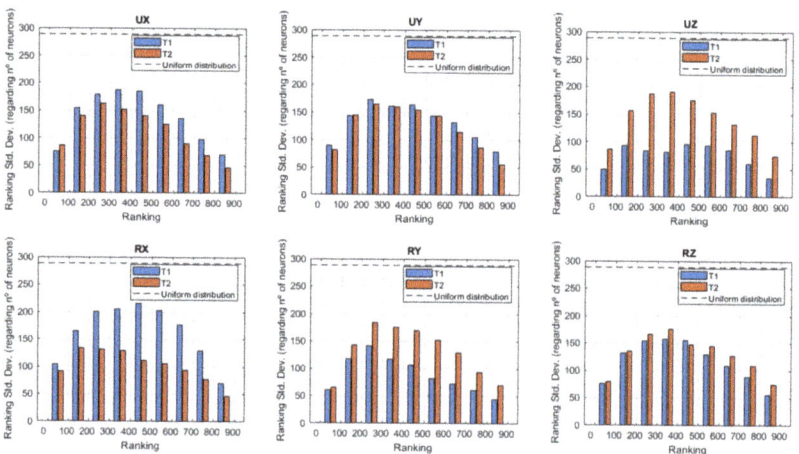

Figure 10. Network ranking dispersion with respect to the number of neurons.

Additionally to the relation of the network performance with respect to the number of neurons, possible dependency with respect to the spring directions were also assessed to check if a given network that performs optimally in one direction, should show the same tendency for the rest of directions. To quantify that in an equivalent manner, standard deviation of the rankings with respect to the spring directions were obtained for each number of neurons, as shown in Figure 11. For the sake of clarity, and since the behavior was found to be very similar for all number of neurons, only results for 1, 5, 10, and 15 neurons are included. In this case, it is observed that for mid rankings, no significant dependency between directions can be stated, since dispersions approaches to the one of uniform distribution. In contrast, extreme rankings show once again lower dispersion, meaning that networks with worst and best results will show dependency between spring directions.

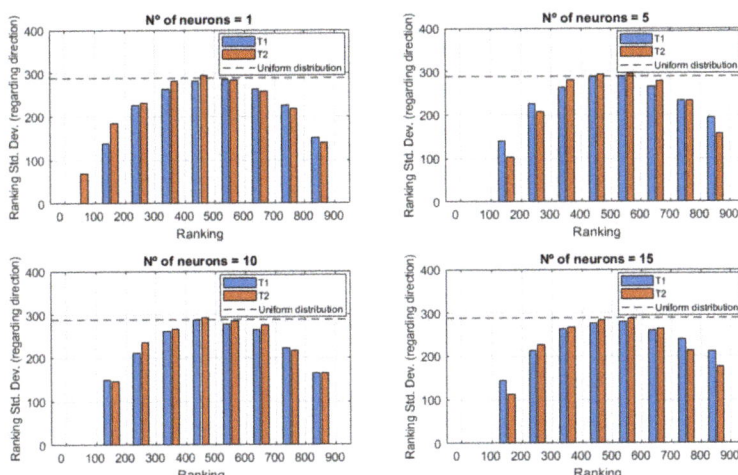

Figure 11. Network ranking dispersion with respect to the spring direction.

Finally, a detailed analysis of the characteristics of the networks that showed the best results was carried out in order to evaluate if there is any affinity among them regarding the input data and variable values. To do so, the 10 best overall ranked networks were selected for both T1 and T2. Later, the input data that was present in at least 8 out of the 10 for all directions were accounted. These data are highlighted for the 243 input data (abscissa), and for each spring direction (ordinate), for both T1 and T2 in Figures 12 and 13, respectively. It can be noticed that there are certain input data that appears systematically in all best performance networks (14 data inputs are shown for T1 in Figure 12, and 13 data inputs are shown for T2 in Figure 13), regardless the spring directions. It is also noticed that such data are different for T1 and T2 junction configurations.

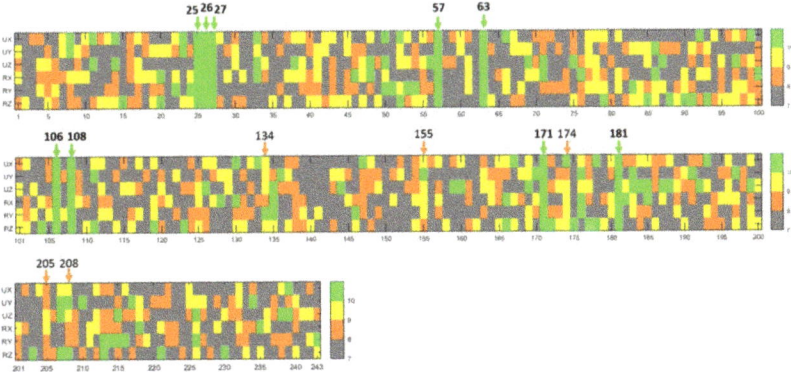

Figure 12. Data frequency graph in the 10 best performance networks for T1. Most used 14 input data are shown.

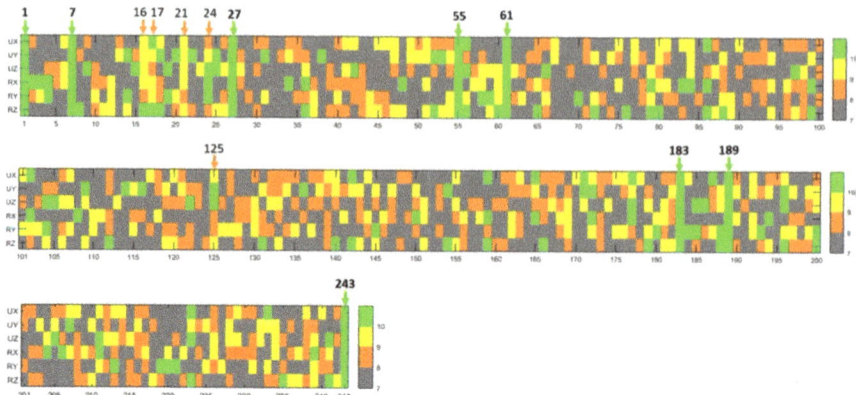

Figure 13. Data frequency graph in the 10 best performance networks for T2. Most used 13 input data are shown.

Furthermore, the values of the geometric variables contained in these input data that appeared systematically in at least 8 out of the 10 best performance networks (14 in Figure 12 and 13 in Figure 13) are analyzed in Figures 14 and 15 for T1 and T2, respectively. The figures show the frequencies of each parameter value. It is noticed a clear tendency to certain values for some of the parameters. Junction configuration T1, for example, takes preferably lower values of $E2$, and higher values for $E3$ and $g1$; meanwhile, junction configuration T2 shows preference for lower values of $E1$ and $E2$, and higher values for $g1$. A priori, no physical meaning to explain this behavior and, its differences between junction configuration, was found.

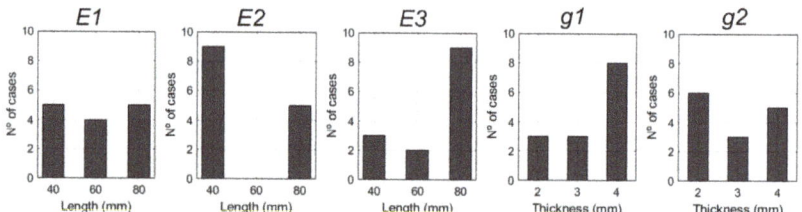

Figure 14. Variables values of frequent input data in best performance networks—T1.

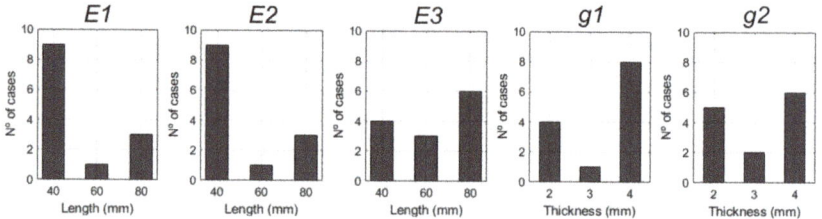

Figure 15. Variables values of frequent input data in best performance networks—T2.

From the evaluation of the 1000 different networks, the best overall for T1 and T2 was selected for validation. Since no performance improvement was observed from around 10 neurons, and in order to avoid isolated high error values, the selection was limited to a maximum of 10 neurons. The selected ones were a 9 and 10 neuron network for T1 and T2, respectively, that performed with overall average errors of 0.039 and 0.035. The corresponding neuron weights are gathered in Tables 3 and 4 (for T1), and in Tables 5 and 6 (for T2).

Table 3. Input layer to hidden layer weights for T1.

		Hidden Layer Neuron								
		1	2	3	4	5	6	7	8	9
Input layer	$E1$	17.68	−0.52	0.38	0.90	−0.38	−0.38	0.05	−0.74	3.50
	$E2$	−1.49	−0.01	−0.02	1.75	0.02	0.83	1.50	0.21	6.05
	$E3$	0.90	−0.15	−0.96	−1.34	0.96	−0.97	1.03	1.61	−4.16
	$g1$	0.79	−0.73	3.25	−0.32	−3.23	−0.48	23.36	0.66	−1.60
	$g2$	−0.30	−0.05	−0.81	−0.28	0.81	0.09	1.15	0.83	0.06

Table 4. Hidden layer to output layer weights for T1.

		Hidden Layer Neuron								
		1	2	3	4	5	6	7	8	9
Output layer	k_{ux}	−0.63	−0.68	−5.86	−286.73	−5.85	−44.07	−0.07	10.83	0.82
	k_{uy}	0.30	−0.44	21.54	−10.59	21.43	42.36	0.21	18.12	−0.01
	k_{uz}	−3.08	−0.16	2.93	6.33	2.94	−245.51	−0.01	22.29	−0.20
	k_{rx}	0.94	−0.10	30.83	−117.97	30.96	103.39	0.01	157.71	0.19
	k_{ry}	−0.43	0.08	3.82	19.53	3.83	−154.39	0.01	21.58	−0.36
	k_{rz}	−0.13	−0.26	22.90	−42.41	22.81	29.46	0.21	37.62	0.07

Table 5. Input layer to hidden layer weights for T2.

		Hidden Layer Neuron									
		1	2	3	4	5	6	7	8	9	10
Input layer	$E1$	−0.05	0.41	−0.21	0.53	0.13	−0.34	0.43	15.70	2.01	−0.13
	$E2$	−18.19	−0.06	0.00	1.01	0.12	0.96	−2.03	2.67	−2.19	−0.12
	$E3$	0.34	−0.08	0.05	−1.18	0.21	−0.65	0.33	−3.50	−48.09	−0.22
	$g1$	−20.36	0.13	0.09	−0.45	0.37	−0.09	−1.42	−0.83	21.88	−0.35
	$g2$	20.00	0.24	−0.72	0.23	0.19	−0.51	2.22	−0.77	−25.75	−0.20

Table 6. Hidden layer to output layer weights for T2.

		Hidden Layer Neuron									
		1	2	3	4	5	6	7	8	9	10
Output layer	k_{ux}	−0.02	91.15	44.78	18.55	−56.35	−92.22	0.26	3.80	−0.01	−77.39
	k_{uy}	−0.06	18.77	26.16	−5.54	143.37	27.15	0.06	−9.37	−0.01	186.32
	k_{uz}	−0.01	−0.59	14.25	−128.62	−34.06	−161.27	−0.20	−31.99	0.04	−49.50
	k_{rx}	0.01	−14.05	−52.31	11.67	−28.38	48.26	0.86	−5.78	−0.01	−43.06
	k_{ry}	0.00	−3.27	3.02	52.74	−24.07	−211.65	−0.35	−93.02	0.03	−34.27
	k_{rz}	−0.06	4.72	15.56	−22.93	149.42	18.68	0.03	−16.70	0.00	194.29

3.2. Analysis of Selected Neural Network

Although global evaluation of the 1000 different training data sets was based on averaged errors and dispersions, a more detailed look on the network estimations becomes necessary to complete the analysis and assess the capability to properly predict the validation outputs. In this regard, Figures 16 and 17 show the correct and estimated outputs of the selected networks for T1 and T2, respectively. A general good fit can be observed. It is noted that there are significant differences in the precision of some spring directions, especially for T2 joint configuration, where UX, UY, RX, and RZ show an average error of approximately one third with respect to the rest of directions. In any case, the worst-case average error keeps below 7% regardless spring directions and junction configurations (6.7% for RY in T2), which already supposes a significant error improvement with respect to the original deviation in beam type structural models, which can reach up to 90% according to [19].

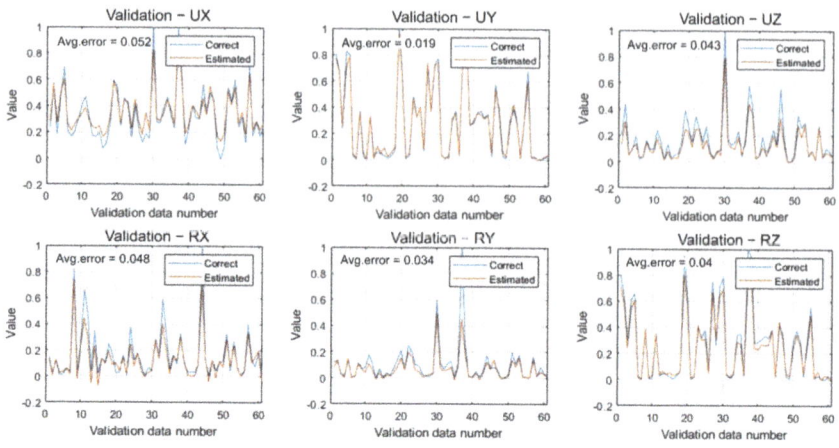

Figure 16. Validation results—T1.

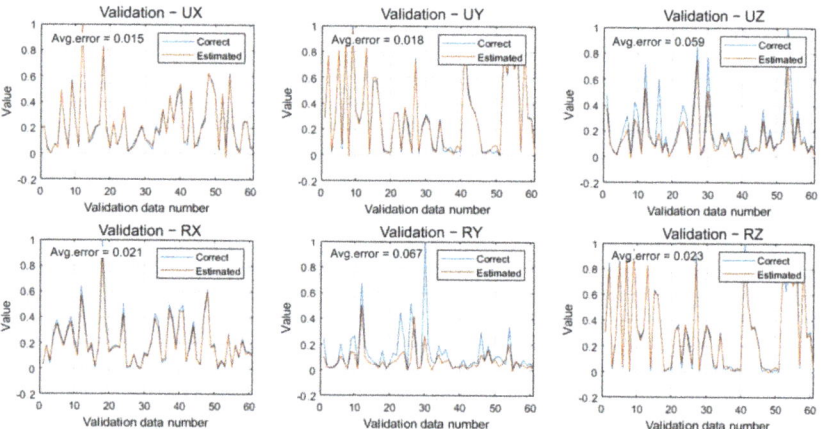

Figure 17. Validation results—T2.

4. Discussion

A neural network-based approach is presented as an alternative to Bayesian Kriging regressions for improving the accuracy of beam type element structural models. Due to the limited amount of input data, an adequate selection of the training set was proved to be fundamental in order to get valid results. In this regard, a methodology that deals with this reality and ensures a valid choice of the training data was also developed.

The notable amount of information generated during the application of the methodology allowed a more profound study of the behavior and performance of the different neural networks constructed. Such study is not strictly necessary to run the methodology since it has no influence on the best network selections. Nevertheless, the information obtained can be of great interest to increase the understanding about the input data and the behavior of the networks, and thus be utilized in futures related studies.

The obtained results show that there is a general increase in the accuracy of the networks when increasing the number of neurons up to 4 to 5. The optimal number of neurons was observed to be dependent on the joint configuration and the elastic element direction, ranging from 6 to 10, the use of more neurons increases the risk of obtaining isolated extremely bad predictions, probably due to over fitting.

It was also found that there is a clear correlation between the accuracy of the network with respect to the number of neurons, meaning that if a certain training set presents high precision for a given number of neurons, it will also present the same tendency for the rest. With respect to the correlation of the accuracy of the networks with respect to the elastic element direction, it was only observed for extreme ranked networks, i.e., only if a given network is one of the best (or the worst) ranked for a specific direction, it will also tend to be one of the best (or the worst) for other directions. Both T1 and T2 joint configurations showed very similar behavior regarding these dependencies.

Additionally, there were located specific input data that appeared systematically in the 10 most accurate networks, regardless the spring direction. Although the same was observed for both joint configurations, the data that appeared systematically in T1 and in T2 were different. It was also found that these data had preference to contain certain variable values in the case of T1, lower values of $E2$ and higher values for $E3$ and $g1$ where observed, while T2 showed preference for lower values of $E1$ and $E2$, and higher values for $g1$.

A priori, no physical meaning was found to explain the behavior of the networks for this data. Consequently, it cannot be stated that similar behavior should be noticed in other structural problems. Future works would be needed to prove if these tendencies apply to data sets of similar problems or could be even extensive to data sets of completely different nature.

After running the 20,000 neural networks of the methodology, the least error networks for T1 and T2 were selected for validation. As it was proven, from about 10 neurons no accuracy improvement was observed at the same time the deviations on isolated predictions increased, so such selection was limited from 1 to 10 neurons. Validation results showed that some spring directions predictions were very close to the reference values, with average errors around 2%. The highest average error was of 6.7% corresponding to the Ry elastic element of the T2 junction.

From the performed study, it can be concluded that neural networks represent a solid alternative to the Bayesian Kriging regressions for the accuracy improvement of structural models constructed with beam type elements, as long as a proper selection of the training data set is assured through the presented methodology.

The present study provides a detailed insight on all the relevant aspects and difficulties that could be encountered when applying neural networks for the estimation of the elastic element stiffness in the alternative beam T-junction model studied. At the same time, it evaluates real aspects such as the generation of ANN with a reduced number of training data, the high sensitivity of the process to the selection of the initial sampling for training, and the deviation ranges that would be considered as acceptable so that any potential user could apply this methodology with a significant amount of confidence.

The interactions that take place at the joint level of T-junctions are complex and depend on the geometry of the junction, the type of loads and their direction. As it was pointed in the literature, these interactions are hard to characterize, and thus the alternative of presenting a complete methodology for the improvement of these T-junctions with a significant degree of confidence suppose a valuable tool for the structural design using finite element models.

5. Conclusions

In this paper the behavior of welded T-junctions in tubular structures was studied with the purpose of improving their behavior by means of artificial neural networks trained with finite element models. The topology of the junction at the joint level determines significant behavioral differences that cannot be taken into consideration with regular beam type elements.

To adjust their behavior to more precise results, elastic elements were introduced at the joint level, characterizing their stiffness utilizing artificial neural networks. In this paper the optimization and implementation of the validation data, through the creation of an optimal surrogate model based on neural networks was presented.

The results led to a model that predicted the stiffness of these elastic elements, in a satisfactory manner. The paper also focuses on how the neural network should be chosen when training data is very limited and, more importantly, which of the available data should be used for training and which for validation.

The results indicated that the use of neural networks without a careful methodology in this type of problems could lead to inaccurate results.

The present work, more generally, is applicable to many other applications where there is insufficient training data (or data that costs a lot of time or money to obtain). This situation occurs in experimental trials, where the manufacture and use of the necessary test equipment to obtain more data is limiting. Furthermore, this situation appears in the computational calculation of complicated FEM or computational fluid dynamics (CFD) models, where a complicated calculation can take several hours and days and obtaining a large pool of data to train a neural network is pure utopia. In this sense, the methodology presented in this paper would be applicable to all these cases where data availability is scarce. The work presented is part of a broader development, where in the future it is expected to apply this methodology to larger problems based on this type of profiles, such as the study of bus structures.

Author Contributions: Conceptualization y methodology, F.B. and J.L.O.; software, F.B. and J.A.P.; validation, J.L.O. and J.A.P.; formal analysis and investigation, J.L.O. and F.B.; resources, J.A.P. and J.L.O.; data curation, J.A.P.; writing—original draft preparation, review, and editing, F.B., J.A.P and J.L.O.; visualization, J.A.P. and F.B.; supervision, J.L.O.; project administration, F.B.; funding acquisition, J.L.O. All authors have read and agreed to the published version of the manuscript.

Funding: This research was partially funded by the AgenciaEstatal de Investigación [grant RETOS 2018-RTI2018-095923-B-C22] and by the Cátedra Global Nebrija-Santander en Recuperación de Energía en el Transporte de Superficie.

Institutional Review Board Statement: Not applicable.

Informed Consent Statement: Not applicable.

Data Availability Statement: The datasets analyzed in this study are available on request from the corresponding author, or publicly available here: [oa.upm.es/33012/].

Conflicts of Interest: The authors declare no conflict of interest.

References

1. Zienkiewicz, O.C.; Taylor, R.L. *The Finite Element Method for Solid and Structural Mechanics*; Butterworth-Heinemann: Oxford, UK, 2014; p. III.
2. Liu, G.R.; Quek, S.S. *The Finite Element Method: A Practical Course*; Butterworth-Heinemann: Oxford, UK, 2003.
3. Hutton, D.V. *Fundamentals of Finite Element Analysis*; McGraw-Hill: New York, NY, USA, 2004; ISBN 71218572 505.
4. Chang, D.C. Effects of flexible connections on body structural response. In Proceedings of the Automotive Engineering Congress and Exposition, Detroit, MI, USA, 25 February 1974.
5. Chon, C.T. Sensitivity of total strain energy of a vehicle structure to local joint stiffness. *AIAA J.* **1987**, *25*, 1391–1395. [CrossRef]
6. Garcia, A.; Vicente, T. Characterization and influence of semi-rigid joints in the buses and coaches structural behavior. In Proceedings of the Bus & Coach Experts Meeting, 33rd International Conference on Vehicle Safety, Keszthely, Hungary, 2–4 September 2002.
7. Lee, K.; Nikolaidis, E. Identification of flexible joints in vehicle structures. *AIAA J.* **1992**, *30*, 482–489. [CrossRef]
8. Moon, Y.M.; Jee, T.H.; Park, Y.P. Development of an automotive joint model using an analytically based formulation. *J. Sound Vib.* **1999**, *220*, 625–640. [CrossRef]
9. Mottershead, J.E.; Friswell, M.I. Model updating in structural dynamics: A survey. *J. Sound Vib.* **1993**, *167*, 347–375. [CrossRef]
10. Ahmadian, H.; Mottershead, J.; Friswell, M. Joint modelling for finite element model updating. In Proceedings of the 14th International Modal Analysis Conference (IMAC), Dearborn, MI, USA, 12–15 February 1996; pp. 591–596.
11. Horton, B.; Gurgenci, H.; Veidt, M.; Friswell, M.I. Finite element model updating of the welded joints in a tubular H-frame. In Proceedings of the 17th International Modal Analysis Conference, Proceedings of the International Modal Analysis Conference—IMAC, Kissimmee, FL, USA, 8–11 February 1999; Society of Experimental Engineers: Bethel, CT, USA, 1999; Volume 2, pp. 1556–1562.

12. Horton, B.; Gurgenci, H.; Veidt, M.; Friswell, M.I. Finite element model updating of a welded space frame. *Proc. Int. Modal Anal. Conf. IMAC* **2000**, *1*, 529–535.
13. Gladwell, G.M.L.; Ahmadian, H. Generic element matrices suitable for finite element model updating. *Mech. Syst. Signal Process* **1995**, *9*, 601–614. [CrossRef]
14. Kim, J.H.; Seok Kim, H.; Woon Kim, D.; Kim, Y.Y. New accurate efficient modeling techniques for the vibration analysis of T-joint thin-walled box structures. *Int. J. Solids Struct.* **2002**, *39*, 2893–2909. [CrossRef]
15. Sreenath, S.; Saravanan, U.; Kalyanaraman, V. Beam and shell element model for advanced analysis of steel structural members. *J. Constr. Steel Res.* **2011**, *67*, 1789–1796. [CrossRef]
16. Donders, S.; Takahashi, Y.; Hadjit, R.; Van Langenhove, T.; Brughmans, M.; Van Genechten, B.; Desmet, W. A reduced beam and joint concept modeling approach to optimize global vehicle body dynamics. *Finite Elem. Anal. Des.* **2009**, *45*, 439–455. [CrossRef]
17. Mundo, D.; Hadjit, R.; Donders, S.; Brughmans, M.; Mas, P.; Desmet, W. Simplified modelling of joints and beam-like structures for BIW optimization in a concept phase of the vehicle design process. *Finite Elem. Anal. Des.* **2009**, *45*, 456–462. [CrossRef]
18. Alcalá, E.; Badea, F.; Martin, Á.; Aparicio, F. Methodology for the accuracy improvement of FEM beam type T-junctions of buses and coaches structures. *Int. J. Automot. Technol.* **2013**, *14*, 817–827. [CrossRef]
19. Romero, F.B.; McWilliams, J.M.; Fazio, E.A.; Izquierdo, F.A. Bayesian kriging regression for the accuracy improvement of beam modeled T-junctions of buses and coaches structures with a methodology based on FEM behavioral analysis. *Int. J. Automot. Technol.* **2014**, *15*, 1027–1041. [CrossRef]
20. Papadopoulos, V.; Soimiris, G.; Giovanis, D.G.; Papadrakakis, M. A neural network-based surrogate model for carbon nanotubes with geometric nonlinearities. *Comput. Methods Appl. Mech. Eng.* **2018**, *328*, 411–430. [CrossRef]
21. Lee, J.J.; Lee, J.W.; Yi, J.H.; Yun, C.B.; Jung, H.Y. Neural networks-based damage detection for bridges considering errors in baseline finite element models. *J. Sound Vib.* **2005**, *280*, 555–578. [CrossRef]
22. Hattori, G.; Serpa, A.L. Contact stiffness estimation in ANSYS using simplified models and artificial neural networks. *Finite Elem. Anal. Des.* **2015**, *97*, 43–53. [CrossRef]
23. Gomes, H.M.; Awruch, A.M. Comparison of response surface and neural network with other methods for structural reliability analysis. *Struct. Saf.* **2004**, *26*, 49–67. [CrossRef]
24. Parra, C.; Olazagoitia, J.L.; Biera, J. Practical tool for the design of brake pads to avoid squeal noise in automotive brake systems. In *SAE Technical Papers, Proceedings of the SAE 2010 Annual Brake Colloquium and Engineering Display, Phoenix, AZ, USA, 10–13 October 2010*; SAE International: Pittsburgh, PA, USA, 2010.
25. Ren, Y.; Bai, G. Comparison of neural network and Kriging method for creating simulation-optimization metamodels. In Proceedings of the 8th IEEE International Symposium on Dependable, Autonomic and Secure Computing, DASC, Chengdu, China, 12–14 December 2009; pp. 815–821.
26. Chowdhury, M.; Alouani, A.; Hossain, F. Comparison of ordinary kriging and artificial neural network for spatial mapping of arsenic contamination of groundwater. *Stoch. Environ. Res. Risk Assess.* **2010**, *24*, 1–7. [CrossRef]

Article

Differential Evolution with Estimation of Distribution for Worst-Case Scenario Optimization

Margarita Antoniou [1,2,*] and Gregor Papa [1,2]

1 Computer Systems Department, Jožef Stefan Institute, Jamova c. 39, SI-1000 Ljubljana, Slovenia; gregor.papa@ijs.si
2 Jožef Stefan International Postgraduate School, Jamova c. 39, SI-1000 Ljubljana, Slovenia
* Correspondence: margarita.antoniou@ijs.si

Abstract: Worst-case scenario optimization deals with the minimization of the maximum output in all scenarios of a problem, and it is usually formulated as a min-max problem. Employing nested evolutionary algorithms to solve the problem requires numerous function evaluations. This work proposes a differential evolution with an estimation of distribution algorithm. The algorithm has a nested form, where a differential evolution is applied for both the design and scenario space optimization. To reduce the computational cost, we estimate the distribution of the best worst solution for the best solutions found so far. The probabilistic model is used to sample part of the initial population of the scenario space differential evolution, using a priori knowledge of the previous generations. The method is compared with a state-of-the-art algorithm on both benchmark problems and an engineering application, and the related results are reported.

Keywords: worst-case scenario; robust; min-max optimization; evolutionary algorithms

1. Introduction

Many real-world optimization problems, including engineering design optimization, typically involve uncertainty that needs to be considered for a robust solution to be found. The worst-case scenario optimization refers to obtaining the solution that will perform best under the worst possible conditions. This approach gives the most conservative solution but also the most robust solution to the problem under uncertainty.

The formulation of the problem that arises is a special case of a bilevel optimization problem (BOP), where one optimization problem has another optimization problem in its constraints [1,2]. In the worst-case scenario case, the maximization of the function in the uncertain space is nested in the minimization in the design space, leading to a min-max optimization problem. Therefore, optimization can be achieved in a hierarchical way.

Min-max optimization has been solved by classical methods such as mathematical programming [3], branch-and-bound algorithms [4] and approximation methods [5]. These methods have limited application as they require simplifying assumptions about the fitness function, such as linearity and convexity.

In recent years, evolutionary algorithms (EAs) have been developed to solve min-max optimization problems. Using EAs mitigates the problem of making specific assumptions about the underlying problem, as they are population-based and they directly use the objectives. In this way, they can handle mathematically intractable problems that do not follow specific mathematical properties.

A very popular approach to solve min-max problems with the EAs is the co-evolutionary approach, where the populations of design and scenario space are co-evolving. In [6], a co-evolutionary genetic algorithm was developed, while in [7], particle swarm optimization was used as the evolution strategy. In such approaches, the optimization search over the design and scenario space is parallelized, reducing significantly the number of function evaluations. In general, they manage to successfully solve symmetrical problems but perform poorly in asymmetrical problems by looping over bad solutions, due to the red queen effect [8]. As the condition of symmetry does not hold in the majority of the problems [definition can be found in Section 2, they become unsuitable for most of the problems.

One approach to mitigate this problem is to apply a nested structure, solving the problem hierarchically as for e.g., in [9], where a nested particle swarm optimization is applied. This leads to a prohibitively increased computation cost, as the design and scenario space is infinite for continuous problems. Min-max optimization problems solved as bilevel problems with bilevel evolutionary algorithms were presented in [10], where three algorithms—the BLDE [11], a completely nested algorithm, the BLEAQ [1], an evolutionary algorithm that employs quadratic approximation in the mappings of the two levels, and the BLCMAES [12], a specialized bilevel CMA-ES—known to perform well in bilevel problems, were tested on a min-max test function and showed good performance in most of the cases but required a high number of function evaluations. A recently proposed differential evolution (DE) with a bottom-boosting scheme that does not use surrogates proved to reach superior accuracy, though the number of functional evaluations (FEs) needed is still relatively high [13].

Using a surrogate model can lower the computational cost. Surrogate models are approximation functions of the actual evaluation and are quicker and easier to evaluate. Surrogate-assisted EAs have been developed for min-max optimization, such as in [14]. In that work, a surrogate model is built with a Gaussian process to approximate the decision variables and the objective value, assuming that evaluating the worst-case scenario performance is expensive. This might be problematic when the real function evaluation is also expensive. In [15], a Kriging-based optimization algorithm is proposed, where Kriging models the objective function as a gaussian process. A newly proposed surrogate-assisted EA applying multitasking can be found in [16], where a radial basis function is trained and used as a surrogate.

As already explained, there are two ways so far to reduce the computational cost when using EAs for min-max problems: the co-evolutionary approach and the use of surrogates, which both come with the disadvantage that either cannot be applied in all the problems or the final solution lacks accuracy.

The DE [17] is one of the most popular EAs because of its efficiency for global optimization. Estimation of distribution algorithm (EDA) is a newer population-based algorithm that relies on estimating the distribution for global convergence, rather than crossover and mutation, and has great convergence [18]. Hybrid DE-EDAs have been proposed to combine the good exploration and exploitation characteristics of each in several optimization problems, such as in [19] for solving a job-shop scheduling problem and in [20] for the multi-point dynamic aggregation problem. EDA with a particle swarm optimization has been developed for bilevel optimization problems, where it served as a hybrid algorithm of the upper-level [21].

In this paper, we propose a DE with EDA for solving min-max optimization problems. The algorithm has a nested form, where a DE is applied for both the design and scenario space optimization. To reduce the computational cost, instead of using surrogates, we estimate the distribution of the best worst solution for the best solutions found so far. Then, this distribution is passed to a scenario space optimization, and a part of the population is sampled from it as a priori knowledge. That way, there is a higher probability that the population will contain the best solution, and there is no need for training a surrogate model. We also limit the search for the scenario space. If one solution found is already worse than the best worst scenario, it is skipped.

The rest of this paper is organized as follows: Section 2 introduces the basic concepts of the worst-case scenario and min-max optimization. A brief description of the general DE and EDA algorithm is provided in Section 3, along with a detailed description of the proposed method. In Section 4, we describe the test functions and the parameter settings used in our experiments. In Section 5, the results are presented and discussed. Finally, Section 6 concludes our paper.

2. Background

In this section, the definitions of the deterministic optimization problem and the worst-case scenario optimization as an instance of robust optimization are presented.

2.1. Definition of Classical Optimization Problem

A typical optimization problem is the problem of minimizing an objective over a set of decision variables subject to a set of constraints. The generic mathematical form of an optimization problem is:

$$\min_x f(x) \tag{1}$$
$$\text{subject to} \quad g(x) \leq 0,$$

where $x \in R^n, x^L \leq x \leq x^U$ is a decision vector of n dimension, $f(x)$ is the objective function and $g(x)$ are the inequality constraints. The global optimization techniques solve this problem, giving a deterministic optimal design. Usually, no uncertainties are considered. This approach, though widely used, is not very useful when a designer desires the optimal solution given the uncertainties of the system. Therefore, robust optimization approaches are applied [22].

2.2. Definition of Worst Case Scenario Optimization Problem

When one seeks the most robust solution under uncertainties, then the worst-case scenario approach is applied. Worst-case scenario optimization deals with minimizing the maximum output in the scenario space of a problem, and it is usually formulated as a min-max problem. The general worst-case scenario optimization problem in its min-max formulation is described as:

$$\min_{x \in X} \max_{y \in Y} f(x, y) \tag{2}$$

where $X \in R^m$ represents the set of possible solutions and $Y \in R^n$ the set of possible scenarios. The problem is a special instance of a bilevel optimization problem (BOP), where one optimization problem (the upper level, UL) has another optimization problem in its constraints (the lower level, LL). The reader can find more about the BOPs in [1]. Here, the UL and LL share the same objective function $f(x, y)$, where UL is optimizing with respect to the variables x of the design space and the lower level is optimizing with respect to the uncertain parameters y of the scenario space. If the upper-level problem is a minimization problem, then the worst-case scenario given by the uncertain variables y of a solution x can be found by maximizing $f(x, y)$. From now on, we will refer to the design space as upper level (UL) and scenario space as lower level (LL) interchangeably. In Figure 1, a general sketch of the min-max optimization problem as bilevel problem is shown, where for every fixed x in the UL, a maximization problem over the scenario space y is activated in the LL.

When the problem holds the following condition:

$$\min_{x \in X} \max_{y \in Y} f(x, y) = \max_{y \in Y} \min_{x \in X} f(x, y)$$

then the problem is symmetrical. Problems that satisfy the symmetrical condition are simpler to solve since the feasible regions of the upper and lower level are independent.

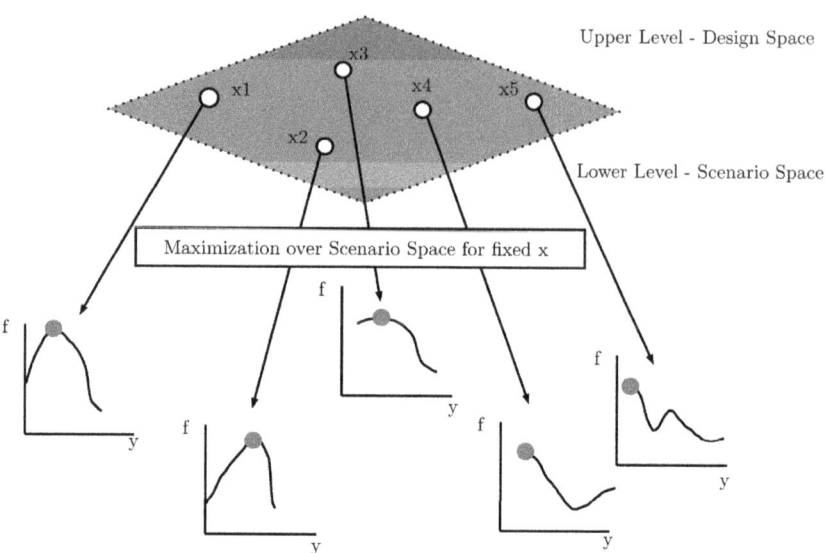

Figure 1. A general sketch of the min-max optimization problem as a bilevel problem, inspired by [16].

3. Algorithm Method

In this section, we briefly describe the differential evolution and estimation of distribution algorithms. Then, we explain the proposed algorithm for obtaining worst-case scenario optimization.

3.1. Differential Evolution (DE)

DE [17] is a population-based metaheuristic search algorithm and falls under the category of evolutionary algorithm methods. Following the standard schema of such methods, it is based on an evolutionary process, where a population of candidate solutions goes through mutation, crossover, and selection operations. The main steps of the algorithm can be seen below:

1. Initialization: A population of $NPop$ individuals is randomly initialized. Each individual is represented by a D dimensional parameter vector, $X_{i,g} = (x^1_{i,g}, x^2_{i,g}, ..., x^D_{i,g})$ where $i = 1, 2, ..., nPop$, $g = 1, 2, ... MaxGen$, where $MaxGen$ is the maximum number of generations. Each vector component is subject to upper and lower bounds X_{min} and X_{max}. The initial values of the ith individual are generated as:

$$X_i = X_{min} + rand(0,1) * (X_{max} - X_{min}) \qquad (3)$$

where rand(0,1) is a random integer between 0 and 1.

2. Mutation: The new individual is generated by adding the weighted difference vector between two randomly selected population members to a third member. This process is expressed as:

$$V_{i,G} = X_{r1,G} + F * (X_{r2,G} - X_{r3,G}) \qquad (4)$$

V is the mutant vector, X is an individual, $r1, r2, r3$ are randomly chosen integers within the range of $[1, NPop]$ and $r1, r2, r3 \neq i$, G corresponds to the current generation, F is the scale factor, usually a positive real number between 0.2 and 0.8. F controls the rate at which the population evolves.

3. Crossover: After mutation, the binomial crossover operation is applied. The mutant individual $V_{i,G}$ is recombined with the parent vector $X_{i,G}$, in order to generate the

offspring $U_{i,G}$. The vectors of the offspring are inherited from $X_{i,G}$ or $V_{i,G}$ depending on a parameter called crossover probability, $C_r \in [0,1]$ as follows:

$$U_{i,G} = \begin{cases} V_{i,G}, & \text{if } rand \leq C_r \text{ or } t = random(i). \\ X_{i,G}, & \text{otherwise}. \end{cases} \quad (5)$$

where $rand \in (0,1)$ is a uniformly generated number, $random(i) \in 1, ..., D$ is a randomly chosen index, which assures that $V_{i,G}$ gives at least one element to $U_{i,G}$. $t = 1, ..., D$ denotes the t-th element of the individual's vector.

4. Selection: The selection operation is a competition between each individual $X_{i,G}$ and its offspring $U_{i,G}$ and defines which individual will prevail in the next generation. The winner is the one with the best fitness value. The operation is expressed by the following equation:

$$X_{i,G+1} = \begin{cases} U_{i,G}, & \text{if } f(U_{i,G}) \leq f(X_{i,G}). \\ X_{i,G}, & \text{otherwise}. \end{cases} \quad (6)$$

The above steps of mutation, crossover, and selection are repeated for each generation until a certain set of termination criteria has been met. Figure 2 shows the basic flowchart of the DE.

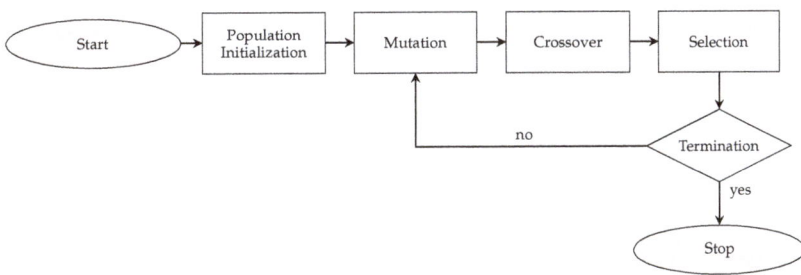

Figure 2. Basic flowchart of the differential evolution algorithm (DE).

3.2. Estimation of Distribution Algorithms (EDAs)

The basic flowchart of the EDA is shown in Figure 3. The general steps of the EDA algorithm are the following:

1. Initialization: A population is initialized randomly.
2. Selection: The most promising individuals $S(t)$ from the population $P(t)$, where t is the current generation, are selected.
3. Estimation of the probabilistic distribution: A probabilistic model $M(t)$ is built from $S(t)$.
4. Generate new individuals: New candidate solutions are generated by sampling from the $M(t)$.
5. Create new population: The new solutions are incorporated into $P(t)$, and go to the next generation. The procedure ends when the termination criteria are met.

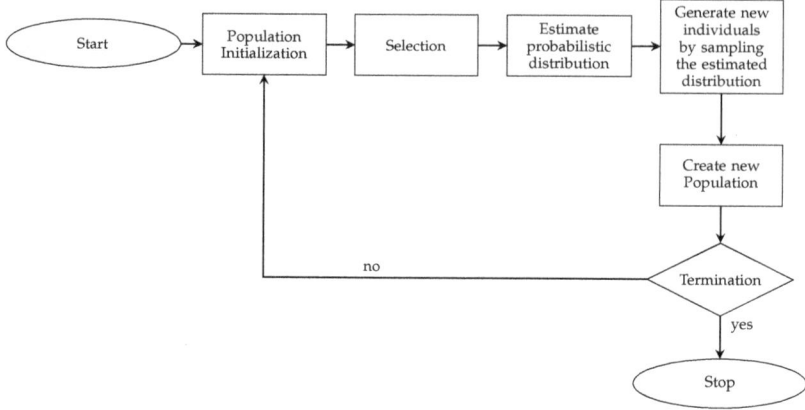

Figure 3. Basic flowchart of the estimation of distribution algorithm (EDA).

3.3. Proposed Algorithm

In the proposed algorithm, we keep the hierarchically nested formulation of a min-max problem, which solves asymmetrical problems. The design space (UL) decision variables are evolving with a DE. For the evaluation of each UL individual, first the scenario space (LL) problem is solved by the DE. This solution is then transferred to the upper level. To reduce the cost, we apply an estimation of distribution mechanism between the decision space search (UL) and the scenario space search (LL). In that way, we use a priori knowledge obtained during the optimization. To further reduce the FEs, we search only for solutions with good worst-case scenarios. If the objective function of a solution X_1 under any scenario is already worse in terms of worst-case performance of the best solution X_2 found so far, there is no need for further exploring X_1 over scenario space. Therefore, the mutant individual's performance is checked under the parent's worst-case scenario, and further explored only when it is better in terms of the fitness function. Figure 4 shows the general framework of the proposed approach. The main steps of the proposed algorithm for the UL:

1. Initialization: A population of size $NPop$ is initialized according to the general DE procedure mentioned in the previous section, where the individuals are representing candidate solutions in the design space X.
2. Evaluation: To evaluate the fitness function, we need to solve the problem in the scenario space. For a fixed candidate UL solution X_i, the LL DE is executed. More detailed steps are given in the next paragraphs. The LL DE returns the solution corresponding to the worst-case scenario for the specific X_i. For each individual, the corresponding best $Y_{best} = argmax_{y \in Y} f(X_i, y)$ solutions are stored, meaning the solution y that for a fixed x maximizes the objective function.
3. Building: The individuals in the population $P(i)$ are sorted as the ascending of the UL fitness values. The best $nPop/2$ are selected. From the best $nPop/2$ individuals, we build the distribution to establish a probabilistic model M_G for the LL solution. The d-dimensional multivariate normal densities to factorize the joint probability density function (pdf) are:

$$F(x, \mu, \Sigma) = \frac{1}{\sqrt{|\Sigma|(2\pi)^d}} e^{-1/2(x-\mu)\Sigma^{-1}(x-\mu)'} \quad (7)$$

where x is the d-dimensional random vector, μ is the d-dimensional mean vector and Σ is the dxd covariance matrix. The two parameters are estimated from the best $nPop/2$ of the population, from the stored lower level best solutions. In that way, in each generation, we extract statistical information about the LL solutions of the previous UL population. The parameters are updated accordingly in each generation, following the general schema of an estimation of distribution algorithm.

4. Evolution: Evolve UL with the steps of the standard DE of mutation, crossover, producing an offspring $U_{i,G}$.
5. Selection: As mentioned above, the selection operation is a competition between each individual $X_{i,G}$ and its offspring $U_{i,G}$. The offspring will be evaluated in the scenario space and sent in LL only if $f(U_{i,G}, Y_{i,G}) \leq f(X_{i,G}, Y_{i,G})$, where $Y_{i,G}$ corresponds to the worst case vector of the parent individual $X_{i,G}$. In that way, a lot of unneeded LL optimization calls will be avoided, reducing FEs. If the offspring is evaluated in the scenario space, the selection procedure in Equation (6) is applied.
6. Termination criteria:
    - Stop if the maximum number of function evaluations $MaxFEs$ is reached.
    - Stop if the improvement of the best objective value of the last $MaxImpGen$ generations is below a specific number.
    - Stop if the absolute difference of the best and the known true optimal objective value is below a specific number.
7. Output: the best worst case function value $f(x^*, y^*)$, the solution corresponding to the best worst-case scenario x^*, y^*

For the LL:

1. Setting: Set the parameters of the probability of crossover CR, the population size $nPop$, the mutation rate F, the sampling probability β.
2. Initialization: Sample $nPop$ individuals to initialize the population. If $\beta \leq random(0, 1)$, then the individual is sampled from the probabilistic model M_{GUL} built in the UL with the Equation (7). The model here is sampled with the $mvnrnd(mu, Sigma)$ built-in function of Matlab, which accepts a mean vector mu and covariance matrix sigma as input and returns a random vector chosen from the multivariate normal distribution with that mean and covariance [23]. Otherwise, it is uniformly sampled in the scenario space according to the Equation (3). Please note that for the first UL generation, β is always 0, as no probabilistic model is built yet. For the following generations, β can range from (0,1) number, where $\beta = 1$ means that the population will be sampled only from the probabilistic model. This might lead the algorithm to be stuck in local optima and to converge prematurely. An example of an initial population generated with the aforementioned method with $\beta = 0.5$ is shown in Figure 5. Magenta asterisk points represent the population generated by the probabilistic model M_{GUL} of the previous UL generation. Blue points are samples uniformly distributed in the search space. In Figure 6, the effect of the probabilistic model on the initial population of LL for f_8 during the optimization is shown. As the iterations increase, the LL members of the populations sampled from the probabilistic distribution reach the promising area that maximizes the function. In the zoomed subplot in each subfigure, one can see that all such members of the population are close to the global maximum, compared to the randomly distributed members.
3. Mutation, crossover, and selection as the standard DE.
4. Termination criteria:
    - Stop if the maximum number of generations $MaxGen$ is reached.
    - Stop if the absolute difference between the best and the known true optimal objective value is below a specific number.
5. Output: the maximum function value $f(x^*, y^*)$, the solution corresponding to the worst-case scenario $y^* = argmax(f(x^*, y))$.

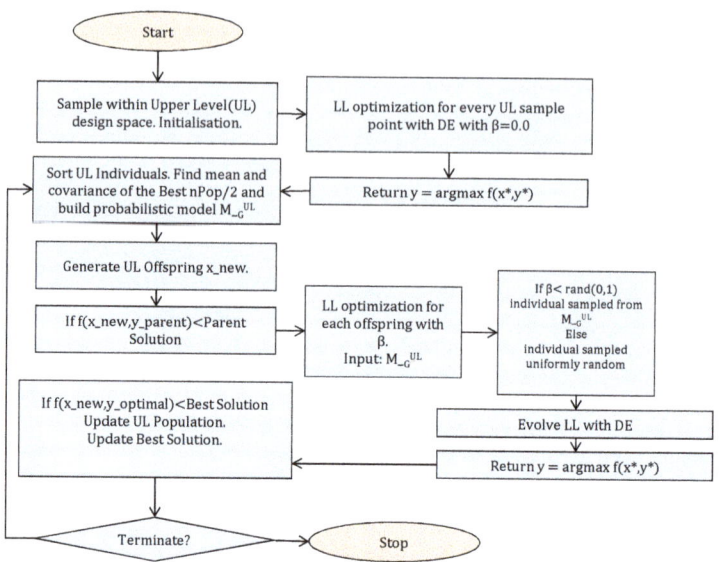

Figure 4. General framework of the proposed algorithm.

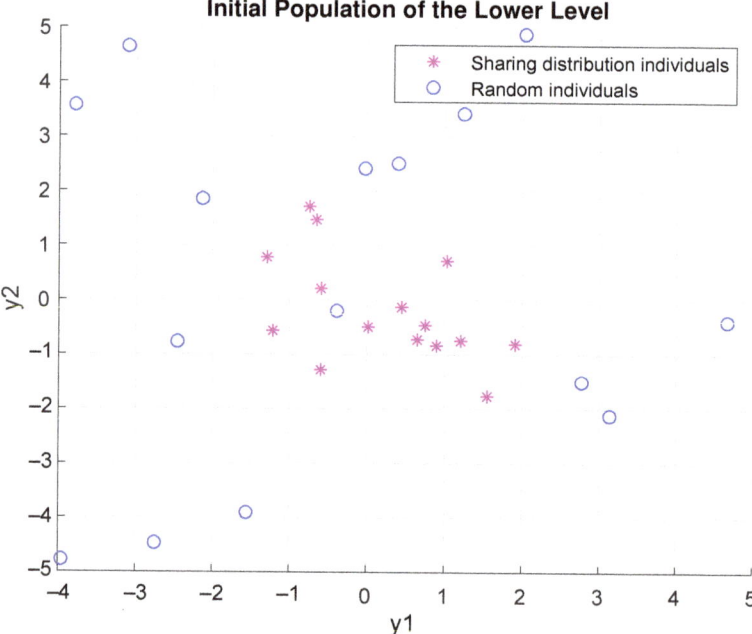

Figure 5. Balancing exploration and exploitation with the sharing distribution mechanism of the LL population. Magenta asterisk points represent the population generated by the distribution of the previous UL generations. Blue points are samples uniformly distributed in the search space. The idea behind this is to keep the "knowledge" already gained in previous generations while also giving the opportunity to the algorithm to search the whole search space. Here with $\beta = 0.5$.

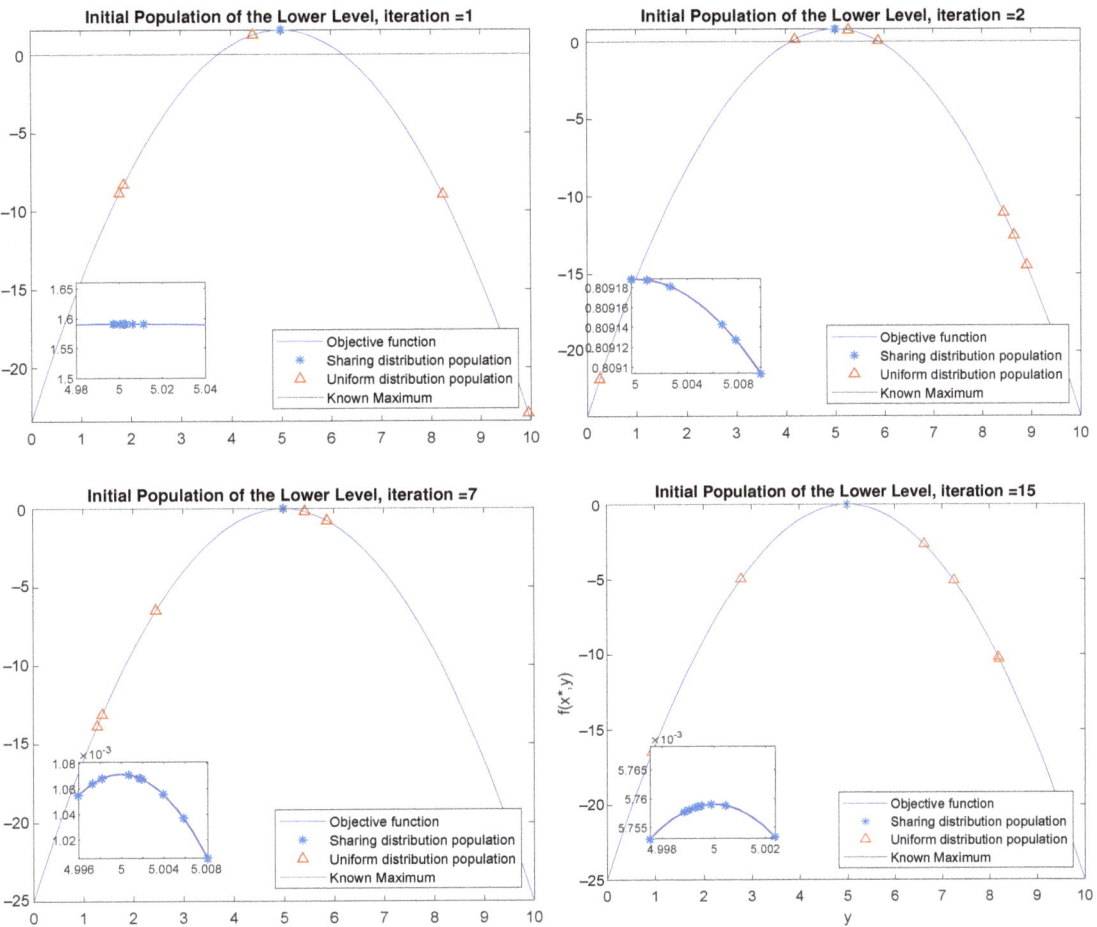

Figure 6. Effect of the probabilistic model on the initial population of LL for f_8. As the iterations increase, the LL members of the population that were produced from the probabilistic model reach the promising area that maximizes the function. The area they are concentrated is shown in the zoomed plots of each plot.

4. Experimental Settings

In this section, we describe the 13 benchmark test functions used for this study and provide the parameter settings for our experiments.

4.1. Test Functions

The performance of the proposed algorithm was tested on 13 benchmark problems of min-max optimization. The problems used are found collected in [15] along with their referenced optimal values. The first 7 problems f_1–f_7 are taken from [24] and they are convex in UL and concave in the LL. The problems described as min-max are:

Test function f_1:

$$\min_{x \in X} \max_{y \in Y} f_1(x,y) = 5(x_1^2 + x_2^2) - (y_1^2 + y_2^2) + x_1(-y_1 + y_2 + 5) + x_2(y_1 - y_2 + 3) \quad (8)$$

with $x \in [-5,5]^2, y \in [-5,5]^2$. The points $x^* = -0.4833, -0.3167$ and $y^* = 0.0833, -0.0833$ are the known solutions of the f_1, and the optimal value is approximated at $f_1(x^*, y^*) = -1.6833$.

Test function f_2:

$$\min_{x \in X} \max_{y \in Y} f_2(x,y) = 4(x_1 - 2)^2 - 2y_1^2 + x_1^2 y_1 - y_2^2 + 2x_2^2 y_2 \qquad (9)$$

with $x \in [-5,5]^2, y \in [-5,5]^2$. The points $x^* = 1.6954, -0.0032$ and $y^* = 0.7186, -0.0001$ are the known solutions of the f_2, and the optimal value is approximated at $f_2(x^*, y^*) = 1.4039$.

Test function f_3:

$$\min_{x \in X} \max_{y \in Y} f_3(x,y) = x_1^4 y_2 + 2x_1^3 y_1 - x_2^2 y_2(y_2 - 3) - 2x_2(y_1 - 3)^3 \qquad (10)$$

with $x \in [-5,5]^2, y \in [-3,3]^2$. The points $x^* = -1.1807, 0.9128$ and $y^* = 2.0985, 2.666$ are the known solutions of the f_4, and the optimal value is approximated at $f_3(x^*, y^*) = -2.4688$.

Test function f_4:

$$\min_{x \in X} \max_{y \in Y} f_4(x,y) = -\sum_{i=1}^{3}(y_i - 1)^2) + \sum_{i=1}^{2}(x_i - 1)^2 + y_3(x_2 - 1) + y_1(x_1 - 1) + y_2 x_1 x_2 \qquad (11)$$

with $x \in [-5,5]^2, y \in [-3,3]^3$. The points $x^* = 0.4181, 0.4181$ and $y^* = 0.709, 1.0874, 0.709$ are the known solutions of the f_4, and the optimal value is approximated at $f_4(x^*, y^*) = -0.1348$.

Test function f_5:

$$\min_{x \in X} \max_{y \in Y} f_5(x,y) = -(x_1 - 1)y_1 - (x_2 - 2)y_2 - (x_3 - 1)y_3 + 2x_1^2 + 3x_2^2 + x_3^2 \qquad (12)$$

with $x \in [-5,5]^3, y \in [-1,1]^3$. The points $x^* = 0.1111, 0.1538, 0.2$ and $y^* = 0.4444, 0.9231, 0.4$ are the known solutions of the f_5, and the optimal value is approximated at $f_5(x^*, y^*) = 1.3453$.

Test function f_6:

$$\min_{x \in X} \max_{y \in Y} f_6(x,y) = -y_1(x_1^2 - x_2 + x_3 - x_4 + 2) + y_2(-x_1 + 2x_2^2 - x_3^2 + 2x_4 + 1) +$$
$$y_3(2x_1 - x_2 + 2x_3 - x_4^2 + 5) + 5x_1^2 + 4x_2^2 + 3x_3^2 + 2x_4^2 - \sum_{i=1}^{3} y_i^2 \qquad (13)$$

with $x \in [-5,5]^4, y \in [-2,2]^3$. The points $x^* = -0.2316, 0.2228, -0.6755, -0.0838$ and $y^* = 0.6195, 0.3535, 1.478$ are the known solutions of the f_6, and the optimal value is approximated at $f_6(x^*, y^*) = 4.543$.

Test function f_7:

$$\min_{x \in X} \max_{y \in Y} f_7(x,y) = 2x_1 x_5 + 3x_4 x_2 + x_5 x_3 + 5y_4^2 + 5y_5^2 - x_4(y_4 - y_5 - 5) +$$
$$x_5(y_4 - y_5 + 3) + \sum_{i=1}^{3}(x_i(y_i^2 - 1)) - \sum_{i=1}^{5} y_i^2 \qquad (14)$$

with $x \in [-5,5]^5, y \in [-3,3]^5$. The points $x^* = 1.4252, 1.6612, 1.2585, -0.9744, -0.7348$ and $y^* = 0.5156, 0.8798, 0.2919, 0.1198, -0.1198$ are the known solutions of the f_7, and the optimal value is approximated at $f_7(x^*, y^*) = -6.3509$.

Test function f_8 [25]:

$$\min_{x \in X} \max_{y \in Y} f_8(x,y) = (x_1 - 5)^2 - (y_1 - 5)^2 \tag{15}$$

with $x \in [0, 10], y \in [0, 10]$. The points $x^* = 5$ and $y^* = 5$ are the known solutions of the f_8, and the optimal value is approximated at $f_8(x^*, y^*) = 0$. This test function is a saddle point function. The function along with the known optimum is plotted in Figure 7, and it serves as an example of a symmetric function.

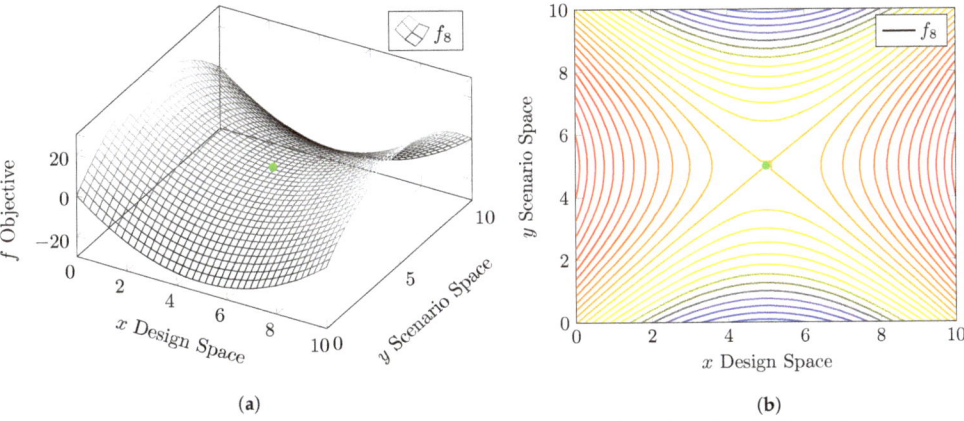

Figure 7. Three-dimensional mesh and contour plots of the symmetrical test function f_8. Green dot corresponds to the known optimum. (**a**) A 3D mesh of the symmetrical test function f_8. (**b**) Contour plot of the symmetrical test function f_8.

Test function f_9 [25]:

$$\min_{x \in X} \max_{y \in Y} f_9(x,y) = \min\{3 - 0.2x_1 + 0.3y_1, 3 + 0.2x_1 - 0.1y_1\} \tag{16}$$

with $x \in [0, 10], y \in [0, 10]$. The points $x^* = 0$ and $y^* = 0$ are the known solutions of the f_9, and the optimal value is approximated at $f_9(x^*, y^*) = 3$. It is a two-plane asymmetrical function. The contour plot and 3-D plot of this function, along with the known optima, are shown in Figure 8 and serves as an example of an asymmetrical function.

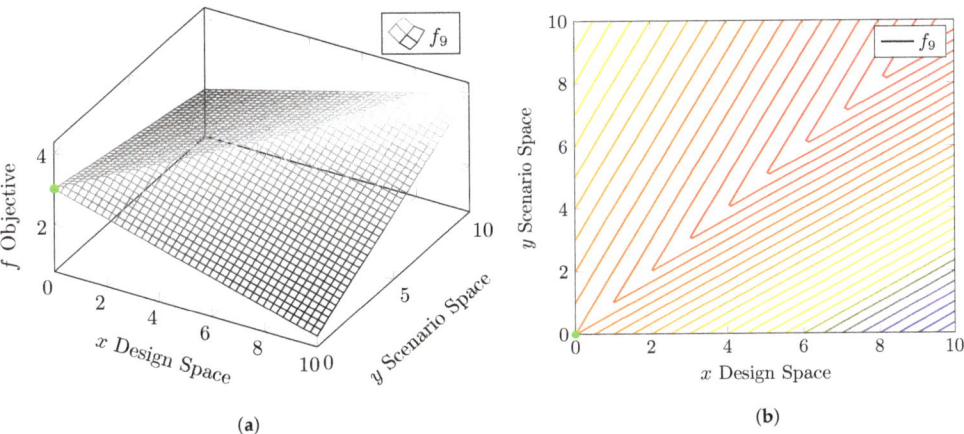

Figure 8. 3D mesh and contour plots of the asymmetrical test function f_9. Green dot corresponds to the known optimum. (**a**) 3D mesh of the asymmetrical test function f_9. (**b**) Contour plot of the asymmetrical test function f_9.

Test function f_{10} [25]:

$$\min_{x \in X} \max_{y \in Y} f_{10}(x,y) = \frac{\sin(x_1 - y_1)}{\sqrt{x_1^2 + y_1^2}} \quad (17)$$

with $x \in [0,10], y \in [0,10]$. The points $x^* = 10$ and $y^* = 2.1257$ are the known solutions of the f_{10}, and the optimal value is approximated at $f_{10}(x^*, y^*) = 0.097794$. It is a damped sinus asymmetrical function.

Test function f_{11} [25]:

$$\min_{x \in X} \max_{y \in Y} f_{11}(x,y) = \frac{\cos(\sqrt{x_1^2 + y_1^2})}{\sqrt{x_1^2 + y_1^2 + 10}} \quad (18)$$

with $x \in [0,10], y \in [0,10]$. The points $x^* = 7.0441$ and $y^* = 10$ or $y^* = 0$ are the known solutions of the f_{11}, and the optimal value is approximated at $f_{11}(x^*, y^*) = 0.042488$. It is a damped cosine wave asymmetrical function.

Test function f_{12} [6]:

$$\min_{x \in X} \max_{y \in Y} f_{12}(x,y) = 100(x_2 - x_1^2)^2 + (1 - x_1)^2 - y_1(x_1 + x_2^2) - y_2(x_1^2 + x_2) \quad (19)$$

with $x \in [-0.5, 0.5] \times [0,1], y \in [0,10]^2$. The points $x^* = 0.5, 0.25$ and $y^* = 0,0$ are the known solutions of the f_{12}, and the optimal value is approximated at $f_{12}(x^*, y^*) = 0.25$.

Test function f_{13} [6]:

$$\min_{x \in X} \max_{y \in Y} f_{13}(x,y) = (x_1 - 2)^2 + (x_2 - 1)^2 + y_1(x_1^2 - x_2) + y_2(x_1 - x_2 - 2) \quad (20)$$

with $x \in [-1,3]^2, y \in [0,10]^2$. The points $x^* = 1,1$ and $y^* = any, any$ are the known solutions of the f_{13}, and the optimal value is approximated at $f_{13}(x^*, y^*) = 1$.

4.2. Parameter Settings

The parameter setting used for all the experiments of this study are shown in Table 1. The population size depends on the dimensionality of the problem, where for the UL $max(n_x + n_y, 5) * 2$ is used and for the LL $max(n_y, 5) * 2$, where n_x, n_y is the dimensionality of the UL and LL, respectively.

Table 1. Control parameters used in the reported results.

	Upper-Level	Lower-Level
Population size	$max(n_x + n_y, 5) * 2$	$max(n_y, 5) * 2$
Crossover	0.9	0.9
Mutation	uniformly (0.2, 0.8)	uniformly (0.2, 0.8)
Desired Accuracy	1×10^{-5}	1×10^{-5}
Maximum Number of Generations	-	10
Maximum Number of Function Evaluations	5000	-
Maximum Number of Improvement Generations	30	-
Least Improvement	1×10^{-5}	-

All the simulations were undertaken on an Intel (R) Core (TM) i7-7500 CPU @ 2.70 GHz, 16 GB of RAM, and the Windows 10 operating system. The code and the experiments were implemented and run in Matlab R2018b.

5. Experimental Results and Discussion

5.1. Effectiveness of the Probabilistic Sharing Mechanism

To evaluate the effectiveness of the probabilistic sharing mechanism of the proposed algorithm, we compare three different instances that correspond to three different β values. The first algorithmic instance has $\beta = 0$, meaning that the estimation of distribution in the optimization procedure is not activated, and the algorithm becomes a traditional nested DE. This instance serves therefore as the baseline. The second algorithmic instance corresponds to $\beta = 0.5$, where half of the initial population of the LL is sampled from the probabilistic model. Last, for the third algorithmic instance, we set a value of $\beta = 0.8$, testing the ability of the algorithm when 80% of the initial population of the LL is sampled from the probabilistic model.

Due to the inherent randomness of the EAs, repeated experiments are held to assess a statistical analysis of the performance of the algorithm. We report results of 30 independent runs, which is the minimum number of samples used for statistical assessment and tests. In Table 2, the statistical results of the 30 runs of the different instances of the algorithm are reported. More specifically, we report the mean, median, and standard deviation of the accuracy of the objective function. We calculate the accuracy as the absolute differences between the best objective function values provided by the algorithms and the known global optimal objective values of each test function. This is expressed as

$$Acc = |f' - f^*| \tag{21}$$

where f' and f^* are the best and the true optimal values, respectively.

Table 2. Accuracy comparison of the different instances of the algorithm over the 30 runs.

Problems		$\beta = 0$	$\beta = 0.5$	$\beta = 0.8$
f_1	Mean	3.45×10^{-1}	2.77×10^{-5}	2.29×10^{-5}
	Median	9.49×10^{-2}	$\mathbf{3.33 \times 10^{-5}}$	$\mathbf{3.33 \times 10^{-5}}$
	Std	6.55×10^{-1}	3.43×10^{-5}	1.53×10^{-5}
	p-value	≤ 0.05	NA	>0.05
	Median FEs	20,115	28,300	46,535
f_2	Mean	1.11×10^{-1}	1.25×10^{-3}	1.28×10^{-4}
	Median	4.96×10^{-2}	$\mathbf{5.53 \times 10^{-6}}$	$\mathbf{5.79 \times 10^{-6}}$
	Std	5.38×10^{-1}	4.27×10^{-3}	5.43×10^{-4}
	p-value	≤ 0.05	NA	>0.05
	Median FEs	20,665	16,180	17,140
f_3	Mean	1.64×10^{0}	2.27×10^{-3}	2.35×10^{-2}
	Median	9.51×10^{-1}	$\mathbf{1.86 \times 10^{-5}}$	$\mathbf{2.47 \times 10^{-5}}$
	Std	2.29×10^{0}	1.30×10^{-2}	8.35×10^{-2}
	p-value	≤ 0.05	NA	>0.05
	Median FEs	27,535	39,830	46,785
f_4	Mean	3.49×10^{-1}	2.93×10^{-5}	1.64×10^{-3}
	Median	2.27×10^{-1}	$\mathbf{2.03 \times 10^{-5}}$	$\mathbf{3.39 \times 10^{-5}}$
	Std	4.49×10^{-1}	4.41×10^{-5}	8.88×10^{-3}
	p-value	≤ 0.05	NA	>0.05
	Median FEs	19,940	26,478	40,516
f_5	Mean	6.23×10^{-2}	9.95×10^{-4}	8.43×10^{-6}
	Median	2.63×10^{-2}	2.99×10^{-4}	$\mathbf{8.55 \times 10^{-7}}$
	Std	1.05×10^{-1}	4.18×10^{-3}	5.08×10^{-5}
	p-value	≤ 0.05	>0.05	NA
	Median FEs	38,694	78,444	97,506

Table 2. Cont.

Problems		$\beta = 0$	$\beta = 0.5$	$\beta = 0.8$
f_6	Mean	2.16×10^{-1}	1.96×10^{-3}	1.19×10^{-2}
	Median	1.62×10^{-1}	$\mathbf{7.86 \times 10^{-6}}$	6.33×10^{-6}
	Std	2.67×10^{-1}	6.74×10^{-3}	6.50×10^{-2}
	p-value	≤ 0.05	>0.05	NA
	Median FEs	55,740	**69,798**	77,356
f_7	Mean	5.57×10^{-1}	7.90×10^{-2}	7.90×10^{-2}
	Median	4.76×10^{-1}	$\mathbf{7.90 \times 10^{-2}}$	7.90×10^{-2}
	Std	3.75×10^{-1}	1.34×10^{-4}	9.54×10^{-6}
	p-value	≤ 0.05	NA	≤ 0.05
	Median FEs	143,580	**360,460**	541,940
f_8	Mean	9.27×10^{-6}	3.03×10^{-6}	3.34×10^{-6}
	Median	6.16×10^{-6}	$\mathbf{1.17 \times 10^{-6}}$	2.12×10^{-6}
	Std	1.99×10^{-5}	3.24×10^{-6}	3.23×10^{-6}
	p-value	≤ 0.05	NA	>0.05
	Median FEs	9120	8150	**8070**
f_9	Mean	0.00×10^0	0.00×10^0	2.96×10^{-3}
	Median	$\mathbf{0.00 \times 10^0}$	$\mathbf{0.00 \times 10^0}$	$\mathbf{0.00 \times 10^0}$
	Std	0.00×10^0	0.00×10^0	1.62×10^{-2}
	p-value	NaN	NA	>0.05
	Median FEs	**3435**	3935	3715
f_{10}	Mean	7.54×10^{-6}	8.03×10^{-8}	8.74×10^{-4}
	Median	$\mathbf{2.86 \times 10^{-7}}$	2.98×10^{-7}	2.95×10^{-7}
	Std	3.60×10^{-5}	4.96×10^{-7}	4.79×10^{-3}
	p-value	NA	>0.05	>0.05
	FEs	**4435**	3995	3880
f_{11}	Mean	5.11×10^{-3}	3.62×10^{-3}	1.43×10^{-2}
	Median	$\mathbf{1.79 \times 10^{-3}}$	2.95×10^{-4}	1.14×10^{-2}
	Std	7.44×10^{-3}	8.06×10^{-3}	1.25×10^{-2}
	p-value	>0.05	NA	≤ 0.05
	FEs	**21,965**	30,480	33,485
f_{12}	Mean	3.72×10^{-1}	5.21×10^{-1}	7.05×10^{-1}
	Median	$\mathbf{2.25 \times 10^{-1}}$	4.77×10^{-1}	7.43×10^{-1}
	Std	5.98×10^{-1}	5.23×10^{-1}	1.42×10^{-1}
	p-value	NA	≤ 0.05	≤ 0.05
	Median FEs	**14,945**	15,795	28,210
f_{13}	Mean	3.99×10^{-2}	2.42×10^{-2}	1.98×10^{-1}
	Median	6.51×10^{-2}	$\mathbf{1.82 \times 10^{-4}}$	1.07×10^{-5}
	Std	7.29×10^{-1}	1.21×10^{-1}	1.04×10^0
	p-value	≤ 0.05	>0.05	NA
	Median FEs	22,430	**56,880**	61525

In order to compare the instances, the non-parametric statistical Wilcoxon signed-rank test [26] was carried out at the 5% significance, where for each test function, the best instance in terms of median accuracy used a control algorithm against the other two. The reported ≤ 0.05 means that it rejects the null hypothesis and the two samples are different, while >0.05 means the opposite. The best algorithm in terms of median accuracy is shown in bold. We also report the median of the total number of function evaluations. In bold are the lowest median FEs corresponding to the best algorithmic instance in terms of median accuracy. As we can see, the proposed method outperforms the baseline in most of the test functions. More specifically, the second and the third instances are significantly better than the first in the test functions $f_1 - f_8$ and f_{13}. For these test functions, the results of these two instances do not differ significantly, therefore there is a tie. What we can note though, is that instance 2 repeatedly requires fewer FEs to reach the same results. Therefore, it performs better in terms of computation expense. For test function f_9, all the instances are performing equally in terms of median accuracy, while the baseline instance reports less FEs. The third instance is best in test functions f_5-f_6. For test function f_7, there is a tie between the second and third instance. The first instance performs better in test functions f_{10} and f_{12}, while for f_{11}, the first and second instance outperforms the third. In many cases, the baseline algorithmic case reports a low number of FEs. These cases, where it does not reach the desired accuracy, indicate premature convergence, when the "least improvement" termination criterion is activated and the algorithm is terminated before reaching the maximum number of evaluations. In 11 out of 13 test functions, instance 2 outperforms at least one instance or performs equally, which makes selecting a $\beta = 0.5$ a safe choice.

In Figure 9, the success rate of each algorithmic instance and each test function is reported. As a success rate, we define here the percentage of the number of runs where the algorithm reached the desired accuracy of the total runs for each test function. It is interesting to note that the baseline first instance did not at all reach the desired accuracy in 9 out of 13 test problems. The performance of the algorithm improves dramatically by the use of the estimation of distribution. On the other hand, the instance with $\beta = 0.5$ reaches the desired accuracy for at least one run in 11 out of 13 problems and, instance with $\beta = 0.8$ in 10 out of 13 problems. The second instance reaches the accuracy of 100% for asymmetrical functions f_9 and f_{10}. For test functions f_7 and f_{12}, none of the algorithms reach the desired accuracy in the predefined number of FEs. f_7 is one of the test problems with higher dimensionality, and a higher number of function evaluations might be needed in order to reach higher accuracy.

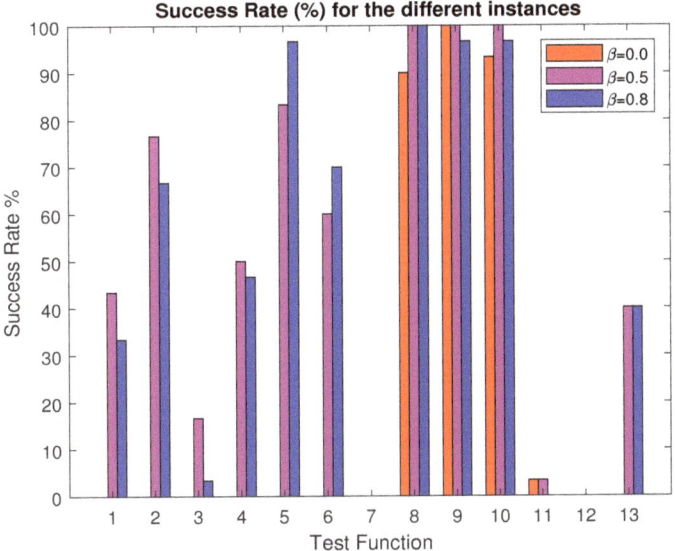

Figure 9. Barchart of the success rate (%) of each algorithmic instance and each test function. The red color corresponds to the instance where $\beta = 0.0$, magenta $\beta = 0.5$ and blue $\beta = 0.8$.

In Figure 10, the convergence plots of the accuracy of the upper level for each algorithmic instance and test function are shown. The red color corresponds to the instance where $\beta = 0.0$, magenta $\beta = 0.5$ and blue $\beta = 0.8$. The bumps that can be spotted in the convergence are probably because of inaccurate solutions of the worst-case scenario. This can be mostly seen in Figure 10g, for test function f_7, where the convergence seems to go further than the desired accuracy. In Figure 10j for f_{10}, algorithmic instance 2 and 3 seem to converge in even earlier generations, in contrast to the baseline first algorithmic instance.

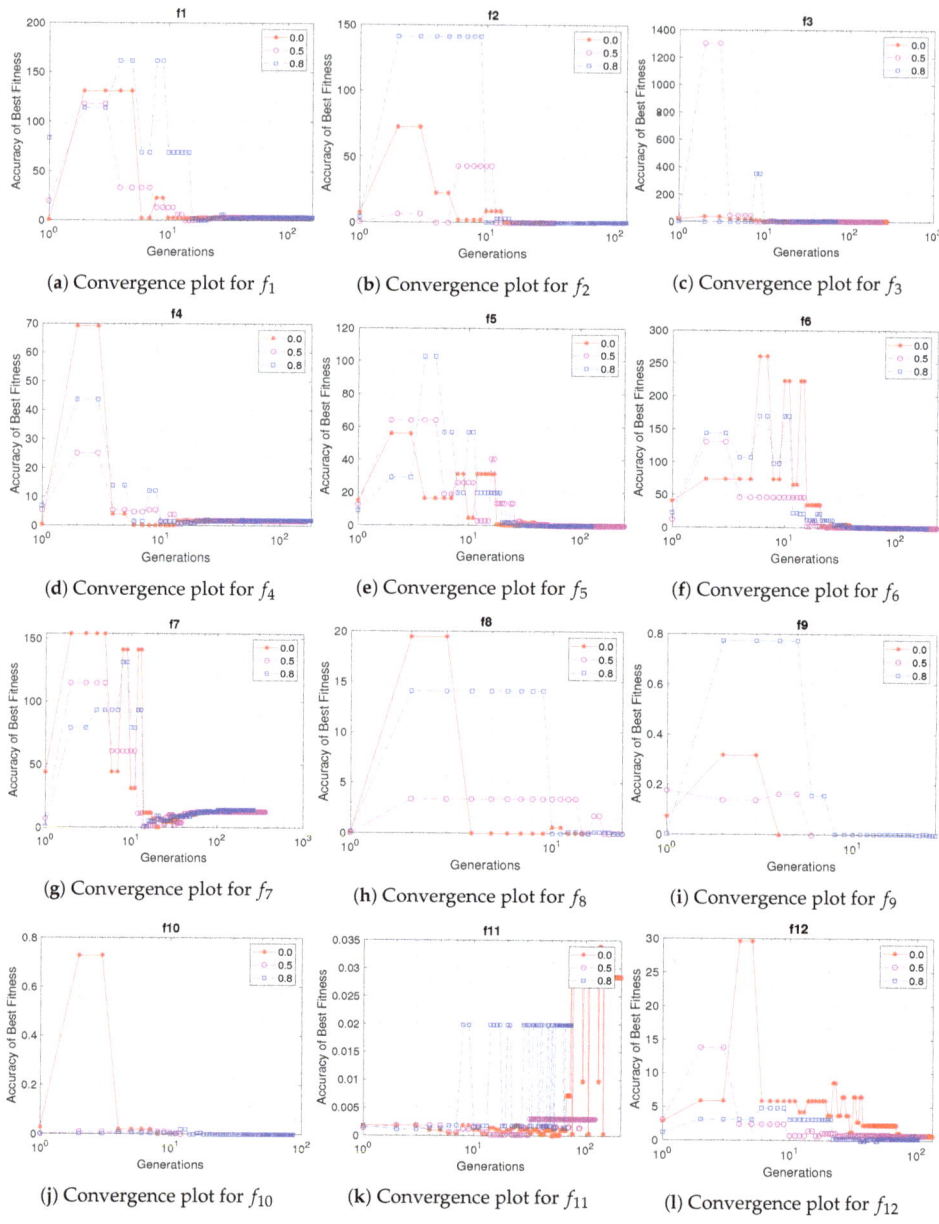

(a) Convergence plot for f_1
(b) Convergence plot for f_2
(c) Convergence plot for f_3
(d) Convergence plot for f_4
(e) Convergence plot for f_5
(f) Convergence plot for f_6
(g) Convergence plot for f_7
(h) Convergence plot for f_8
(i) Convergence plot for f_9
(j) Convergence plot for f_{10}
(k) Convergence plot for f_{11}
(l) Convergence plot for f_{12}

Figure 10. Cont.

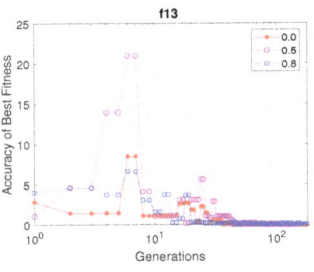

(m) Convergence plot for f_{13}

Figure 10. Fitness accuracy convergence of the upper level of the median run for all the test functions and algorithm instances. The red color corresponds to the instance where $\beta = 0.0$, magenta $\beta = 0.5$ and blue $\beta = 0.8$. Generations axes is in logarithmic scale.

5.2. Comparison with State-of-the-Art Method MMDE

In this subsection, we compare the proposed method with one state-of-the-art min-max EA. The MMDE [13] employs a differential evolution algorithm along with a bottom-boosting scheme and a regeneration strategy to detect best worst-case solutions. The MMDE showed statistically significant superior performance against a number of other min-max EAs, so we only compared with the MMDE. For the comparative experiments, the following settings are applied. For the proposed method, the DE parameters of UL are the same as in Table 1, while for the LL, the population size was set to $max(n_y, 5) * 3$ and $\beta = 0.5$. For the MMDE, the proposed settings from the reference paper are used and are crossover $CR = 0.5$ and mutation $F = 0.7$. The MMDE also has two parameters K_s and T that control the number of FEs in the bottom-boosting scheme and partial-regeneration strategy. Here, they are set to 190 and 10, respectively, as in the original settings. To have a fair comparison, the termination criterion for both algorithms is only the total number of FEs and set to 10^4. Since the number of FEs is limited, an additional check was employed for the proposed method, where if a new solution of the UL is already found in the previous population, then it is not passed to the lower level, since the worst-case scenario is already known. The algorithms are run 30 times on test functions $f8 - f13$. For comparing the two methods, we use the mean square error (MSE) of the obtained solutions in the design space (UL) to the true optimum, a metric commonly used for comparing min-max algorithms. More specifically, the MSE is calculated:

$$MSE(X_{best}, X_{opt}) = \frac{1}{D_X} \sum_{n=1}^{D_X} (x_{best}^n - x_{opt}^n)^2 \qquad (22)$$

where X_{best} is the best solution found by the algorithm and X_{opt} the known optimal solution, while D_X is the dimensionality of the solution. In Table 3, we report the mean, median and standard deviation of the mean square error (MSE). In Figure 11, these values are illustrated as boxplots. The Wilcoxon signed-rank test [26] was conducted at the 5% significance, and we report if the p-value rejects or not the null hypothesis. The proposed method outperforms the MMDE for the test functions f_8 and f_{10}, while it performs equally good on asymmetrical test function f_9. On the test functions f_{11}–f_{13}, the MMDE performs better than the proposed method.

Table 3. MSE Comparison with MMDE over 30 runs and 10^4 FEs.

Problems		$\beta = 0.5$	MMDE
f_8	Mean	2.0234×10^{-5}	2.6618×10^{-5}
	Median	$\mathbf{1.8269 \times 10^{-7}}$	9.5487×10^{-6}
	Std	7.5060×10^{-5}	6.1841×10^{-5}
	p-value	NA	≤ 0.05
f_9	Mean	2.5849×10^{-1}	3.3719×10^{-3}
	Median	$\mathbf{0.0000 \times 10^0}$	$\mathbf{0.0000 \times 10^0}$
	Std	9.6251×10^{-1}	1.1081×10^{-2}
	p-value	NA	>0.05
f_{10}	Mean	1.0029×10^0	5.1712×10^{-1}
	Median	$\mathbf{0.0000 \times 10^0}$	$\mathbf{0.0000 \times 10^0}$
	Std	2.8499×10^0	2.4408×10^0
	p-value	NA	≤ 0.05
f_{11}	Mean	3.3428×10^{-1}	7.9495×10^{-4}
	Median	5.5485×10^{-2}	$\mathbf{8.8027 \times 10^{-5}}$
	Std	8.7588×10^{-1}	1.4876×10^{-3}
	p-value	≤ 0.05	NA
f_{12}	Mean	8.1786×10^{-3}	1.1339×10^{-5}
	Median	9.6258×10^{-5}	$\mathbf{2.1344 \times 10^{-6}}$
	Std	1.9804×10^{-2}	3.2300×10^{-5}
	p-value	≤ 0.05	NA
f_{13}	Mean	5.0537×10^{-2}	5.5425×10^{-3}
	Median	1.9716×10^{-2}	$\mathbf{2.7037 \times 10^{-3}}$
	Std	7.7093×10^{-2}	7.7943×10^{-3}
	p-value	≤ 0.05	NA

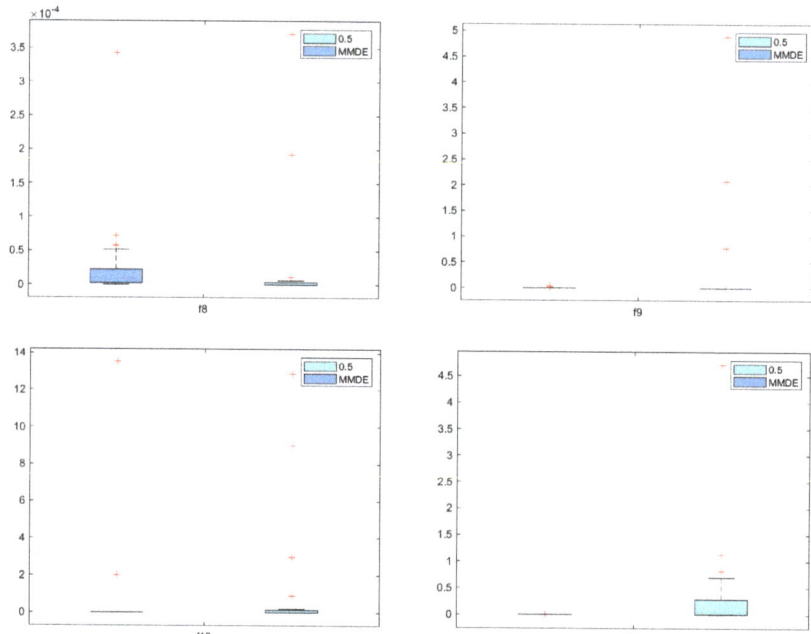

Figure 11. Cont.

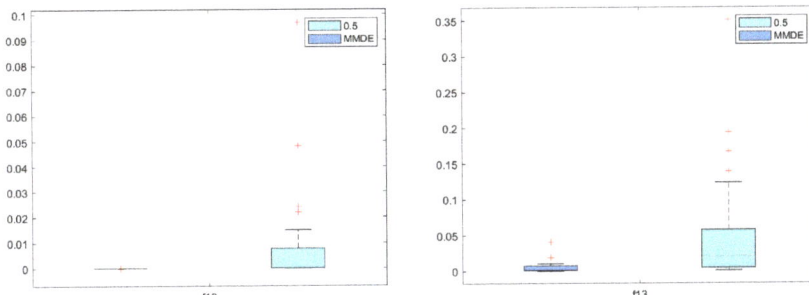

Figure 11. Boxplots of MSE values of the proposed method and MMDE over 30 runs for the test functions f_8–f_{13}.

5.3. Engineering Application

To further investigate the performance of the proposed method, we solved a simple engineering application that also serves a benchmark, taken from [27]. It refers to the optimal design of a vibration absorber (Figure 12) for a structure where uncertainties occur in the forcing frequency. A structure with mass m_1 is subjected to a force and an unknown frequency. Through a viscous damping effect, a smaller structure of mass m_2 is employed to compensate for the oscillations caused by this disturbance. The design challenge is to figure out how to make this damper robust to the worst force frequency. The objective function is the normalized maximum displacement of the main structure and is expressed as [27]:

$$J = \frac{1}{Z}\sqrt{(1-\beta_{freq}^2/T^2 + 4*(\zeta_2\beta_{freq}/T)^2} \tag{23}$$

where

$$Z^2 = [\beta_{freq}^2(\beta_{freq}^2-1)/T^2 - \beta_{freq}^2(1+\mu) - 4\frac{\zeta_1\zeta_2\beta_{freq}^2}{T} + 1]^2 \\ + 4[\zeta_1\beta_{freq}^3/T^2 + \frac{\zeta_2\beta_{freq}^3(1-\mu) - \zeta_2\beta_{freq}}{T} - \zeta_1\beta_{freq}]^2 \tag{24}$$

The fixed parameters for the specific problem are $\mu = 0.1$ and $\zeta_1 = 0.1$. The decision variables in the design space are ζ_2 and T, while variable β_{freq} is the decision variable in the scenario space against which the design should be robust against. The problem can be written as a min-max problem:

$$\min_{x \in X}\max_{y \in Y} J(x,y) = \min_{\zeta_2,T}\max_{\beta_{freq}} J(\zeta_2, T, \beta_{freq}) \tag{25}$$

with $\zeta_2 \in [0,1]$, $T \in [0,1]$ and $\beta_{freq} \in [0,2.5]$. The points $x^* = (\zeta_2^*, T^*) = 0.1986, 0.8619$ and $y^* = \beta_{freq}^* = 1.043$ are the known solutions of the J, and the optimal value is approximated at $J(x^*, y^*) = 2.6227$ as reported in [28]. We run the problem 30 independent times for both the proposed method and the MMDE algorithm with the same parameter settings as in the previous subsection. In Table 4, we report the mean, median and standard deviation of the obtained accuracy and MSE for the proposed method and the MMDE. Both algorithms perform well at approximate the known global optima with an accuracy of $\times 10^{-2}$ and MSE $\times 10^{-4}$. The statistical test showed that the proposed method performs equally well with the MMDE for the engineering application.

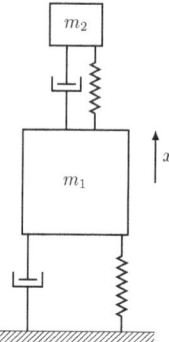

Figure 12. Vibration absorber.

Table 4. Statistical comparison with MMDE over 30 runs and 10^4 FEs for the engineering application.

Problems		$\beta = 0.5$	MMDE
$Acc(J_{minmax})$	Mean	1.7100×10^{-1}	1.0472×10^{-1}
	Median	$\mathbf{6.4784 \times 10^{-2}}$	9.9247×10^{-2}
	Std	2.6084×10^{-1}	6.0458×10^{-2}
	p-value	NA	>0.05
$MSE(x)$	Mean	1.4668×10^{-2}	7.6533×10^{-4}
	Median	6.4275×10^{-4}	$\mathbf{3.6815 \times 10^{-4}}$
	Std	5.1909×10^{-2}	7.6206×10^{-4}
	p-value	>0.05	NA

6. Conclusions

In this work, we propose an algorithm for solving worst-case scenario optimization as a min-max problem. The algorithm employs a nested differential evolution with an estimation of the distribution between the two levels to enhance the efficiency of solving the problems in terms of both accuracy and computational cost. A probabilistic model is built from the best worst-case solutions found so far and is used to generate samples as an initial population of the lower level DE to speed up the convergence. First, the efficiency is investigated by comparing the nested algorithm with different probabilities of using the probabilistic model on 13 test functions of various dimensions and characteristics. To further investigate the performance of the algorithm, it is compared with the MMDE, one state-of-the-art algorithm known to perform well on these problems on both benchmark functions and on an engineering application. The results show that, most times, the proposed method performs better or equal to the MMDE.

In future work, the method could be tested with different population-based EAs in UL or LL, as it is independent of the evolutionary strategy. The parameter, β, that defines the probability that the probabilistic model will be used, could be adapted during the optimization. Last, the method can be tested on higher dimensional test functions and/or engineering applications.

Author Contributions: Conceptualization, M.A.; methodology, M.A.; software, M.A.; validation, M.A.; formal analysis, M.A.; investigation, M.A.; writing—original draft preparation, M.A.; writing—review and editing, M.A. and G.P.; visualization, M.A.; supervision, G.P.; project administration, G.P.; funding acquisition, G.P. All authors have read and agreed to the published version of the manuscript.

Funding: This work was supported by the European Commission's H2020 program under the Marie Skłodowska-Curie grant agreement No. 722734 (UTOPIAE) and by the Slovenian Research Agency (research core funding No. P2-0098).

Data Availability Statement: The code and the related results are available on request from the corresponding author.

Acknowledgments: The authors would like to thank Xin Qiu for providing the source code of the MMDE.

Conflicts of Interest: The authors declare no conflict of interest.

Abbreviations

The following abbreviations are used in this manuscript:

DE	differential evolution
EDA	estimation of distribution algorithm
UL	upper level
LL	lower level
BOP	bilevel optimization problem
FEs	function evaluations

References

1. Sinha, A.; Lu, Z.; Deb, K.; Malo, P. Bilevel optimization based on iterative approximation of multiple mappings. *J. Heuristics* **2017**, *26*, 1–35. [CrossRef]
2. Antoniou, M.; Korošec, P. Multilevel Optimisation. *Optimization under Uncertainty with Applications to Aerospace Engineering*; Springer: Cham, Switzerland, 2021; pp. 307–331.
3. Feng, Y.; Hongwei, L.; Shuisheng, Z.; Sanyang, L. A smoothing trust-region Newton-CG method for minimax problem. *Appl. Math. Comput.* **2008**, *199*, 581–589. [CrossRef]
4. Montemanni, R.; Gambardella, L.M.; Donati, A.V. A branch and bound algorithm for the robust shortest path problem with interval data. *Oper. Res. Lett.* **2004**, *32*, 225–232. [CrossRef]
5. Aissi, H.; Bazgan, C.; Vanderpooten, D. Min–max and min–max regret versions of combinatorial optimization problems: A survey. *Eur. J. Oper. Res.* **2009**, *197*, 427–438. [CrossRef]
6. Barbosa, H.J. A coevolutionary genetic algorithm for constrained optimization. In Proceedings of the 1999 Congress on Evolutionary Computation-CEC99 (Cat. No. 99TH8406), Washington, DC, USA, 6–9 July 1999; IEEE: Piscataway, NJ, USA; 3, pp. 1605–1611.
7. Shi, Y.; Krohling, R.A. Co-evolutionary particle swarm optimization to solve min-max problems. In Proceedings of the 2002 Congress on Evolutionary Computation. CEC'02 (Cat. No. 02TH8600), Honolulu, HI, USA, 12–17 May 2002; IEEE: Piscataway, NJ, USA; Volume 2, pp. 1682–1687.
8. Vasile, M. On the solution of min-max problems in robust optimization. In Proceedings of the EVOLVE 2014 International Conference, A Bridge between Probability, Set Oriented Numerics, and Evolutionary Computing, Beijing, China, 1–4 July 2014.
9. Chen, R.B.; Chang, S.P.; Wang, W.; Tung, H.C.; Wong, W.K. Minimax optimal designs via particle swarm optimization methods. *Stat. Comput.* **2015**, *25*, 975–988. [CrossRef]
10. Antoniou, M.; Papa, G. Solving min-max optimisation problems by means of bilevel evolutionary algorithms: A preliminary study. In Proceedings of the 2020 Genetic and Evolutionary Computation Conference Companion, Cancun, Mexico, 8–12 July 2020; pp. 187–188.
11. Angelo, J.S.; Krempser, E.; Barbosa, H.J. Differential evolution for bilevel programming. In Proceedings of the 2013 IEEE Congress on Evolutionary Computation, Cancun, Mexico, 20–23 June 2013; IEEE: Piscataway, NJ, USA; pp. 470–477.
12. He, X.; Zhou, Y.; Chen, Z. Evolutionary bilevel optimization based on covariance matrix adaptation. *IEEE Trans. Evol. Comput.* **2018**, *23*, 258–272. [CrossRef]
13. Qiu, X.; Xu, J.X.; Xu, Y.; Tan, K.C. A new differential evolution algorithm for minimax optimization in robust design. *IEEE Trans. Cybern.* **2017**, *48*, 1355–1368. [CrossRef] [PubMed]
14. Zhou, A.; Zhang, Q. A surrogate-assisted evolutionary algorithm for minimax optimization. In Proceedings of the IEEE Congress on Evolutionary Computation, Barcelona, Spain, 18–23 July 2010; IEEE: Piscataway, NJ, USA; pp. 1–7.
15. Marzat, J.; Walter, E.; Piet-Lahanier, H. Worst-case global optimization of black-box functions through Kriging and relaxation. *J. Glob. Optim.* **2013**, *55*, 707–727. [CrossRef]
16. Wang, H.; Feng, L.; Jin, Y.; Doherty, J. Surrogate-Assisted Evolutionary Multitasking for Expensive Minimax Optimization in Multiple Scenarios. *IEEE Comput. Intell. Mag.* **2021**, *16*, 34–48. [CrossRef]
17. Storn, R.; Price, K. Differential evolution–a simple and efficient heuristic for global optimization over continuous spaces. *J. Glob. Optim.* **1997**, *11*, 341–359. [CrossRef]
18. Larranaga, P. A review on estimation of distribution algorithms. In *Estimation of Distribution Algorithms*; Springer: Boston, MA, USA, 2002; pp. 57–100.
19. Zhao, F.; Shao, Z.; Wang, J.; Zhang, C. A hybrid differential evolution and estimation of distribution algorithm based on neighbourhood search for job shop scheduling problems. *Int. J. Prod. Res.* **2016**, *54*, 1039–1060. [CrossRef]

20. Hao, R.; Zhang, J.; Xin, B.; Chen, C.; Dou, L. A hybrid differential evolution and estimation of distribution algorithm for the multi-point dynamic aggregation problem. In Proceedings of the Genetic and Evolutionary Computation Conference Companion, Kyoto, Japan, 15–19 July 2018; pp. 251–252.
21. Wang, G.; Ma, L. The estimation of particle swarm distribution algorithm with sensitivity analysis for solving nonlinear bilevel programming problems. *IEEE Access* **2020**, *8*, 137133–137149. [CrossRef]
22. Gorissen, B.L.; Yanıkoğlu, İ.; den Hertog, D. A practical guide to robust optimization. *Omega* **2015**, *53*, 124–137. [CrossRef]
23. Available online: https://www.mathworks.com/help/stats/mvnrnd.html (accessed on 3 August 2021).
24. Rustem, B.; Howe, M. *Algorithms for Worst-Case Design and Applications to Risk Management*; Princeton University Press: Princeton, NJ, USA, 2009.
25. Jensen, M.T. A new look at solving minimax problems with coevolutionary genetic algorithms. In *Metaheuristics: Computer Decision-Making*; Springer: Boston, MA, USA, 2003; pp. 369–384.
26. Wilcoxon, F. Individual comparisons by ranking methods. *Biom. Bull.* **1945**, *1*, 80–83. [CrossRef]
27. Marzat, J.; Walter, E.; Piet-Lahanier, H. A new expected-improvement algorithm for continuous minimax optimization. *J. Glob. Optim.* **2016**, *64*, 785–802. [CrossRef]
28. Brown, B.; Singh, T. Minimax design of vibration absorbers for linear damped systems. *J. Sound Vib.* **2011**, *330*, 2437–2448. [CrossRef]

Article

Evolutionary Design of a System for Online Surface Roughness Measurements

Valentin Koblar [1,2] and Bogdan Filipič [2,3,*]

1 Kolektor Group d.o.o., Vojkova Ulica 10, SI-5280 Idrija, Slovenia; valentin.koblar@kolektor.com
2 Jožef Stefan International Postgraduate School, Jamova Cesta 39, 1000 Ljubljana, Slovenia
3 Department of Intelligent Systems, Jožef Stefan Institute, Jamova Cesta 39, 1000 Ljubljana, Slovenia
* Correspondence: bogdan.filipic@ijs.si

Abstract: Surface roughness is one of the key characteristics of machined components as it affects the surface quality and, consequently, the lifetime of the components themselves. The most common method of measuring the surface roughness is contact profilometry. Although this method is still widely applied, it has several drawbacks, such as limited measurement speed, sensitivity to vibrations, and requirement for precise positioning of the measured samples. In this paper, machine vision, machine learning and evolutionary optimization algorithms are used to induce a model for predicting the surface roughness of automotive components. Based on the attributes extracted by a machine vision algorithm, a machine learning algorithm generates the roughness predictive model. In addition, an evolutionary algorithm is used to tune the machine vision and machine learning algorithm parameters in order to find the most accurate predictive model. The developed methodology is comparable to the existing contact measurement method with respect to accuracy, but advantageous in that it is capable of predicting the surface roughness online and in real time.

Keywords: quality control; roughness measurement; machine vision; machine learning; evolutionary algorithm; parameter optimization

1. Introduction

Demands for increased productivity and product quality in highly competitive industries, such as the automotive industry, have necessitated the use of online systems for inspecting the quality of massively produced parts. One of the quality measures that is especially challenging for online examination is surface roughness of machined parts. Surface roughness is defined as an amplitude value measuring the vertical heights of the surface deviations from a reference line [1]. Inadequate surface roughness of machined parts can significantly affect the functionality of a product and can lead to a premature failure. Moreover, measurement of surface roughness in production can reduce machining costs, since the machining parameters, such as machining speed and the period between the changes of machining tools, can be appropriately chosen.

The most widely used method of surface roughness measuring is contact profilometry. This method uses a stylus type device that correlates displacements induced by surface irregularities to the surface roughness of the inspected specimen. The method is standardized and has been widely used in industrial laboratories and manufacturing industry [2]. The technology of contact profilometry is well developed and can provide measurements of surface roughness within the accuracy of a micrometer. However, this method has several drawbacks. Since the stylus tip must be brought into contact with the measured specimen, the measured surface can be altered by scratches. Moreover, this method is time-consuming and sensitive to vibrations, and therefore not suitable for online measurements in high-volume production processes. More details about stylus-based roughness measurements and their advantages and shortcomings can be found in [3].

To overcome the drawbacks of contact methods, several non-contact methods, such as optical profilometry, scanning electron microscopy, atomic force microscopy, and laser scanning microscopy, have been developed. These methods can provide very accurate measurements of surface roughness and are becoming increasingly popular, also in the automotive industry [4]. However, the methods still require the preparation of adequate samples, are sensitive to vibrations and the measuring apparatuses are expensive. Consequently, none of these methods can be used for online and real-time surface roughness measurements.

This paper presents the development of a machine vision system for roughness evaluation of graphite commutator mounting holes. The graphite commutators are components of electric motors used in automotive fuel pumps. The final phase in the graphite commutator production is the precise turning of the commutator mounting hole to achieve an adequate hole inner diameter and surface roughness. Both characteristics, the diameter and roughness, are important for reliable operation of a fuel pump. Several online methods for measuring the inner diameter of holes are applicable; however, online roughness measurement of the hole surface roughness represents a major challenge.

Specifically, the work proposes combining machine vision (MV), machine learning (ML) and optimization methods to build a predictive model capable of determining the mounting hole roughness. The MV algorithm extracts the attributes from the commutator mounting hole surface that are used by ML to build a roughness predictive model. However, MV and ML methods depend on numerous parameters that notably affect the outcome and are hard to set to their optimum values. To overcome this limitation, an optimization algorithm is used to set the MV and ML algorithm parameters.

The paper is further organized as follows. Section 2 presents the related work in MV-based systems for measurement of surface roughness. The design and development of the online surface roughness measurement system are presented in Section 3. Section 4 describes the optimization methodology for automated tuning of MV and ML algorithm parameters in the development process. Section 5 describes the experimental setup and validation procedure used in the development. The experimental results are discussed in Section 6. Finally, Section 7 concludes the paper with a summary of findings and ideas for future work.

2. Related Work

The initial experiment with a setup similar to the one presented in this paper, combining MV, ML and optimization methods was carried out in [5]. The differential evolution (DE) [6] algorithm was used to search for optimal MV parameter settings, such as binary threshold and filter parameter values. Based on the attributes extracted from 300 images of the commutator mounting holes, the ML algorithm was employed to build classification and regression predictive models. The study found that in comparison to the domain expert this methodology always finds better MV parameter settings. In the classification task, the methodology was able to find a classification model of 100% accuracy in very few examined generations, while the regression task proved to be more demanding.

Much research and development has been carried out in the field of prediction and control of surface roughness using MV. Regarding the way of calculating the roughness parameters, these methods can be divided into analytical methods [7–11], where parameters extracted from images are correlated to the measured roughness by a mathematical function, and methods engaging artificial intelligence (AI) [12–19] to build the roughness predictive models.

Shahabi and Ratnam [7] studied vision-based roughness measurements in a turning process. They used back-light illumination to extract the line profiles of turned workpieces. By varying the parameters on the lathe, such as the turning speed and feed rate, they produced workpieces with various roughness values. They showed that after applying the smoothing filter and performing linear regression data fitting, the extracted edge profile of the workpiece can be directly correlated to the average surface roughness parameter

R_a. The maximum difference of R_a between the MV-based estimate and the roughness measured by the conventional stylus method was 10%.

Jeyapoovan and Murugan [8] developed an MV-based roughness measurement method using Euclidean and Hamming distances of the surface features to determine the value of the roughness parameter R_a. The Euclidean distance is a distance between two points in a plane or space, while the Hamming distance represents a distance between two items by the number of mismatches among their pairs of variables. These two parameters were then compared to the values of the parameters in the database of specimen images that were measured using a stylus instrument. The authors observed that the values of the Euclidean and Hamming distances were very low for surfaces with similar surface roughness values. Therefore, the roughness values can be successfully classified using these two parameters.

Nithyanantham and Suresh [9] demonstrated that using the optical surface roughness parameter G_a, the algebraic average of an image's gray levels results in a strong correlation between G_a and R_a. After applying a geometric search technique that enhanced the edges detected in the images, the correlation coefficient between the parameters G_a and R_a was significantly improved and was higher than 0.92.

Jibin and Arunachalam [10] studied the illumination compensation techniques for surface roughness evaluation using MV. The acquired images of ground samples machined at different parameter values were used for illumination compensation utilizing image filtration techniques. Based on these images, the authors calculated the correlation between the extracted surface texture parameters and the reference measurements carried out by an optical profiler. The results of the study showed that by using additional lightning, filtration techniques and statistical methods, the extracted texture parameters are highly correlated to the measured roughness values. Therefore, such a system can be an integral part of any grinding system to inspect the machined components.

Patel and Kiran [11] used the correlation approach to calculate the roughness parameters for end-milled parts. The authors used the contrast, energy, entropy and homogeneity features of the captured images to calculate the correlation with the reference measurements of the roughness parameter R_a obtained by a surface profilometer. The authors gained the best results using the correlation of image energy feature and roughness parameter R_a, where the maximum relative error was 8%.

More advanced methodologies for vision-based roughness measurements incorporate AI methods. These methods are able to find more complex and consequently more accurate models for the evaluation of surface roughness. Fadare and Oni [12] presented a methodology that uses an artificial neural network (ANN) to predict the roughness values. In contrast to the previously described analytical methods, several features are extracted from images using the fast Fourier transform (FFT) analysis. Based on these features and the tool wear index (TWI), a predictive ANN model was trained. The output of the ANN model was the optical surface roughness parameter G_a, which was then correlated with the R_a parameter value measured on the reference pieces. The authors reported that the proposed MV system using the ANN model has acceptable accuracy for online monitoring of surface roughness.

Instead of an ANN, Ravikumar et al. [13] used the algorithm for induction of decision trees called C4.5. The classification model was built based on the histogram features extracted from sample images. Since the decision tree can only classify the given instances into different quality classes, the authors determined three quality classes the instances belonged to. These classes were defined as acceptable workpieces, workpieces with scratches and workpieces with major defects. The result of the classification model was validated and compared to the manually determined classes. The misclassification of the decision tree model was estimated to 8.6%.

Samtaş [14] used an ANN to train a predictive model for surface roughness estimation after the face milling operation. The reference workpieces were firstly measured by the surface roughness profilometer. Afterwards, images of the reference workpieces were captured and processed by an MV algorithm. Next, each image was converted to a binary image and represented by a matrix of "0" and "1", and further transformed to a single dimensional array which had a length of the number of pixels in the image. The ANN was then trained to match the arrays of the measured workpieces with the arrays of the reference workpieces in order to predict the roughness values. The author reported to achieve the confidence of the roughness prediction above 99%.

Elangovan et al. [15] studied the prediction of surface roughness using vibration signals in a turning process. The data for roughness prediction consisted of the cutting parameters, the flank wear and the captured vibration signal parameters. Based on these data and using a ML regression algorithm, a model for predicting the roughness parameter R_a was built. Several combinations of input attributes were studied; however, the best results were gained after applying the principal component analysis (PCA) [20]. The reported root mean square error (RMSE) was about 0.35.

In a paper by Simunovic et al. [16], an adaptive neuro-fuzzy inference system (ANFIS) for roughness assessment was proposed. In the experiment, the input variables were represented by the face milling machining parameters: spindle speed, feed per tooth, and depth of cut. In addition, for every set of the input variables the roughness parameter R_a was measured. Based on the attributes extracted from the captured grayscale images, fuzzy rules mapping the grayscale image attributes to roughness parameter values were generated. The authors reported high accuracy in determining the roughness value, which is reflected in a low normalized root mean square error (NRMSE) value of 6.98%.

An alternative method for roughness measurement was presented by Yi et al. [17]. The authors proposed a visual method where light from the red and green color block is projected at a predetermined angle to the grinding workpiece surface. From the color difference (CD), i.e., the difference in the values of the red and green components of each point, the authors calculated the correlation between the CD value and the roughness parameter R_a. For this purpose, they used a support vector machine (SVM) [21]. The reported accuracy calculated as a relative difference between the measured and the predicted roughness values was over 90%.

Morales Tamayo et al. [18] used an ANN model to predict the steel surface roughness in the dry turning process of stainless steel. The researchers produced the specimens by varying the cutting parameters during the turning process. These parameters were then used as an input for the ANN model to predict the surface roughness parameter R_a. The results were analyzed by calculating the mean absolute error (MAE) and R^2 value between the reference and predicted values of the R_a parameter. The minimum reported MAE was 2.87% and the maximum achieved R^2 value 99%. Based on these results, the authors claim that this methodology can be used to predict the surface roughness in dry turning of steel.

Recently, Lin et al. [19] presented surface roughness modeling for machined parts considering the cutting parameters and machining vibration in the end-milling process. Predictive models were developed using multiple regression analysis and ANN modeling. In addition to the cutting parameters, the authors also measured the machining vibration and used it as an input parameter for the ANN model. Utilizing the built ANN model, they predicted the surface roughness parameter R_a and compared it to the reference measurements. The comparison between the prediction performance of the multiple regression and ANN models revealed that the latter achieved higher prediction accuracy. Based on the RMSE and mean absolute percentage error (MAPE) values, the authors state that the ANN predictive model can serve as base for an on-line surface roughness measurement system.

According to the reviewed literature, we can state that there is no unique method suitable for online MV-based roughness measurements. In contrast to our application, where the roughness of the inner hole surface has to be measured, in most previous studies, roughness was measured on a flat surface or at the outer diameter of workpieces. The inner diameter of a commutator mounting hole amounts to only a few millimeters, what makes our application especially challenging.

As already mentioned, we initially treated the roughness determination problem as a classification task which was to distinguish acceptable and unacceptable commutators, and as a regression task where the roughness parameter R_z was predicted [5]. Since the regression task has proved to be much more demanding than the classification task, this work further extends the scope of the research for predicting the R_z parameter value.

3. Online Surface Roughness Measurement System

The proposed system for online prediction of the commutator mounting hole roughness operates in the following steps (see Figure 1):

- Online capturing of images;
- Preprocessing of images;
- Extraction of attributes from the preprocessed images;
- Prediction of commutator mounting hole roughness based on the ML model.

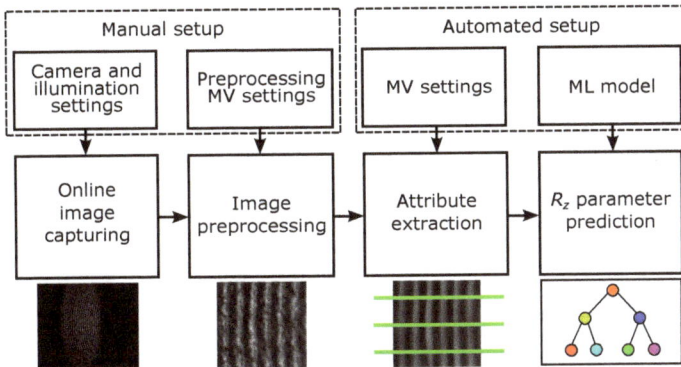

Figure 1. Steps of predicting the mounting hole roughness.

A preparatory step in designing the online roughness prediction system was the selection of representative specimens. In addition, the roughness value of each specimen was measured to obtain the reference roughness values. These were later used to assess the prediction accuracy. A detailed description of each step of the proposed approach is presented in Sections 3.1–3.4.

The parameter settings of image capturing and image preprocessing steps were determined by a domain expert, based on a trial-and-error method. The criterion in the online image capturing step was to find the camera-illumination setup, where the features of the commutator mounting hole surface were emphasized the most. In the image preprocessing step, the hole surface region was extracted from the original captured image. The MV operators were selected in a way that the MV algorithm always extracts the most informative region, regardless of its absolute position in the original image.

Extraction of attributes from an image and roughness prediction based on the extracted attributes are the crucial steps in designing the online roughness measurement system. The inputs are the MV algorithm settings and the ML model obtained during the optimization process in the development phase of the proposed system. In order to obtain the most appropriate MV and ML algorithm settings, the optimization procedure presented in Section 4 was carried out.

3.1. Data Preparation

The design of the online system for roughness measurements started with selecting representative samples and performing the reference roughness measurements. Surface roughness can be determined by several parameters that are categorized into amplitude parameters, spacing parameters, and hybrid parameters [22]. The commutator concerned in this study has a hole roughness defined by the parameter R_z; hence, this parameter was used as a roughness measure in the experiments. R_z represents the height difference between the maximum peak height and the maximum valley depth of a line profile on a predetermined sampling length.

The samples were selected from a recalled batch of commutators with inadequate hole roughness values. The dataset contained 700 instances, which is significantly higher than the dataset in our initial study [5]. The dataset was split into the training set (630 instances) and hold-out set (70 instances) used for the result evaluation. The hold-out set was selected manually and represented 10% of all available instances. To achieve a representative distribution of instances with regard to the roughness parameter R_z, all 700 instances were sorted by the R_z parameter value in ascending order and every tenth instance was moved to the hold-out set (systematic sampling).

The reference roughness measurements were performed by the contact profilometer Mitutoyo Surftest SJ-210. To reduce the measurement error originating from the previously described stylus sensitivity, the reference R_z values were calculated as an average of three measurements. Commutators with the roughness parameter value $R_z \leq 16$ μm are considered acceptable, while the ones with $R_z > 16$ μm unacceptable. The distribution of instances with regard to the R_z value is shown in Figure 2.

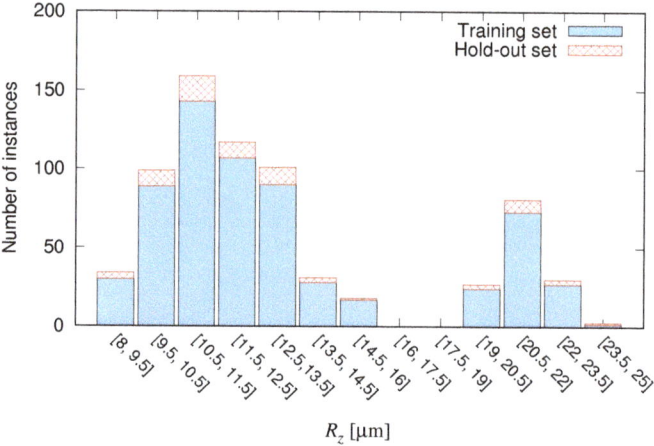

Figure 2. Distribution of instances with respect to the value of R_z.

3.2. Image Capturing and Preprocessing

The inner diameter of the commutator mounting hole concerned in this study is only 6 mm. This fact represented a major challenge in capturing the hole surface images. However, after the validation of several camera-illumination setups, the setup shown in Figure 3 was established.

Grayscale images of all 700 commutator mounting holes were manually captured and labeled with the corresponding reference roughness values. The 8-bit grayscale image of the mounting hole surface has a resolution of 2592 × 1944 pixels. However, to be able to extract the attributes that are correlated with the R_z roughness value, additional preprocessing of the images has to be performed. The purpose of preprocessing is to extract only the portion of the image where the hole surface treatment is clearly visible. The sequence of the MV operators used in the image preprocessing step is shown in Figure 4. The operators, their parameter settings and their sequence in the preprocessing step were determined based on the expert knowledge, gained in the development of similar MV applications.

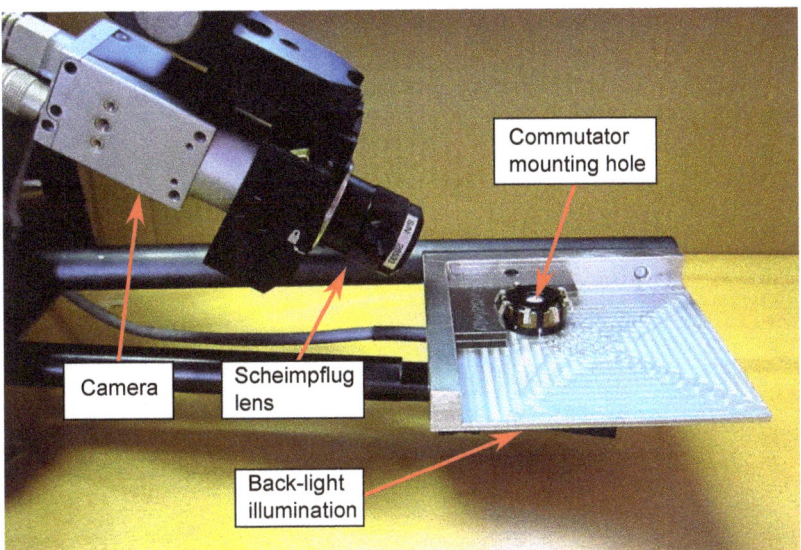

Figure 3. Camera-illumination setup.

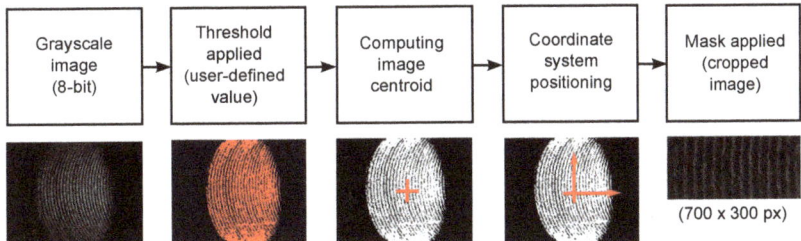

Figure 4. Image preprocessing algorithm.

The initial step in preprocessing the images is binarization, where a binary threshold is applied to the 256-grayscale image. In the next step, the resulting binary image is used to calculate the image intensity centroid. Since the camera-illumination setup is designed in such a way that the hole surface image contains the highest proportion of high-intensity pixels, the calculated coordinates of intensity centroid are always positioned at about the same location of the mounting hole, regardless of its absolute position in the image. The coordinate system for precise positioning of an image mask is then applied to the calculated centroid position. Finally, the extraction mask positioned at the coordinate system origin is applied to the image, and the region of the mask size (700 × 300 pixels), i.e., the region of interest (ROI), is extracted from the original grayscale image. An example of the image with marked and extracted ROI can be seen in Figure 5.

3.3. Attribute Extraction

The attribute extraction algorithm consists of four image operators that require four parameters to be set. Similarly as in the image preprocessing, the operators and their sequence were determined manually, based on the expert knowledge and experience. The operators and their sequence in the attribute extraction algorithm are outlined in Figure 6.

Figure 5. Captured image with marked ROI (**top**) and extracted part of the image (**bottom**).

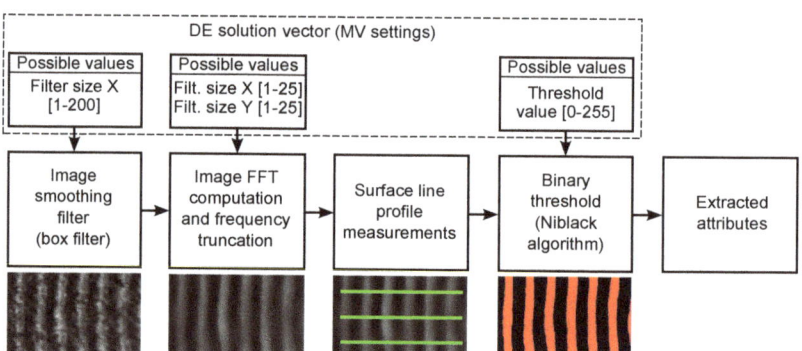

Figure 6. The attribute extraction algorithm.

First, a box filter with a varying kernel size $\{1, 2, \ldots, 200\}$ is applied to the image. Due to the inhomogeneities of the thermoset material used in commutator body manufacturing, some "salt-and-pepper" noise is present in the image. The box filter reduces this noise and emphasizes the features describing the mounting hole roughness. Next, the FFT filtering is applied, truncating a certain portion of high frequencies and thus additionally emphasizing the features that result from the machine treatment of the hole. The FFT filter enables to set the filter kernel size in both X and Y directions in the image, in the range of $\{1, 2, \ldots, 25\}$. Afterwarsd, line profile measurements analog to the contact profilometer measurements are performed on the grayscale image and the image attributes are extracted. An additional set of attributes is extracted after applying the Niblack binarization algorithm [23]. The threshold of the Niblack algorithm is set in the range of $\{1, 2, \ldots, 255\}$ and outputs a binary image consisting of the stripes representing the "peaks" and "valleys" on the commutator hole surface (see Figure 6).

In total, 24 numerical attributes describing the properties of the commutator mounting hole surface were extracted from each image. Specifically, 20 attributes were extracted from the grayscale images and four from the binary images. The grayscale attributes were describing the highest and the lowest grayscale value of a pixel in a selected line profile, the number of grayscale peaks and valleys along the line profile, and the mean grayscale value of all the peaks and valleys. In addition, four attributes representing the roughness parameter estimates were calculated from the grayscale line profile. These were R_t, R_a, and R_z (ISO and DIN variants). Here, R_t (maximum profile height) is determined as

$$R_t = r^{\max} - r^{\min}, \tag{1}$$

where $r^{\max}$ is the highest grayscale value representing the highest peak and $r^{\min}$ the lowest grayscale value representing the lowest valley. R_a (arithmetic mean of profile values) is calculated as

$$R_a = \frac{1}{n} \sum_{1}^{n} r_i, \tag{2}$$

where n is the number of values in the line profile and r_i the i-th value in the profile. R_z^{ISO} is obtained as

$$R_z^{\text{ISO}} = \frac{1}{5} \sum_{j=1}^{5} (r_j^{\max} - r_j^{\min}), \tag{3}$$

where $r_j^{\max}$ and $r_j^{\min}$ represent the highest peak and the lowest valley in the j-th profile, respectively. Similarly, R_z^{DIN} is determined as

$$R_z^{\text{DIN}} = \frac{1}{N} \sum_{j=1}^{N} (r_j^{\max} - r_j^{\min}), \tag{4}$$

where N is the number of considered profiles. Details on these roughness parameters can be found in [3].

Moreover, the maximum, the minimum and the average peak and valley grayscale values were calculated. Lastly, two additional grayscale attributes, the maximum value of a grayscale image signal and its corresponding index, were extracted by the FFT algorithm. Finally, the four attributes extracted from the binary images were the percentage of pixels representing the peaks, the percentage of pixels representing the valleys, and the average peak and valley width in the image.

3.4. Roughness Prediction

The task in this research was to predict the value of the roughness parameter R_z considering the attributes extracted from the images. An MV algorithm with given settings (Figure 6) was applied to the commutator mounting hole images to create a dataset of attributes. In the roughness prediction step, the ML predictive model was applied to the extracted attributes dataset, and the value of the roughness parameter R_z is predicted. For R_z prediction, algorithms for building regression trees and ensembles of regression trees were used. The reason for using the regression trees was in that they can be interpreted and their implementation in the online roughness measurement system is not overly complex.

Besides the MV parameter settings, the ML parameter settings influence the ML prediction accuracy too. In order to find suitable MV and ML algorithm settings, the optimization procedure presented in Section 4 was applied. The goal of this procedure was to find the ML model with the highest prediction accuracy.

The chosen evaluation metric was the root relative squared error (RRSE), which has already been used in [5]. It measures the error of the induced ML model in comparison to the error of a simple predictor which ignores the predictions and always outputs the average of the actual values. It is defined as

$$\text{RRSE} = \sqrt{\frac{\sum_{i=1}^{n}(p_i - a_i)^2}{\sum_{i=1}^{n}(a_i - \bar{a})^2}} \quad (5)$$

where n is the number of instances, p_i the predicted value of i-th instance, a_i the actual value of i-th instance, and $\bar{a}$ the average actual value. The final results of the experiments were additionally assessed with respect to the mean absolute error

$$\text{MAE} = \frac{1}{n}\sum_{i=1}^{n}|p_i - a_i| \quad (6)$$

that is informative for practical considerations as it is expressed in µm.

4. Optimization of Algorithm Parameters

In contrast to manual search for suitable MV and ML algorithm parameter settings, the task can be formulated as an optimization problem where the goal is to find the MV and ML algorithm settings that minimize the roughness prediction error. The optimization procedure that produces the most accurate predictive model is shown in Figure 7.

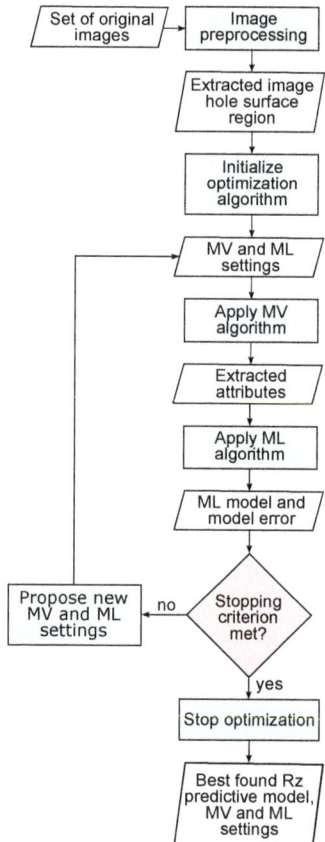

Figure 7. Optimization procedure that searches for the best MV and ML algorithm settings and outputs the best R_z predictive model.

The input to the optimization procedure are images of the commutator mounting hole surfaces. First, the images are preprocessed to extract the part of the image where the mounting hole is located. During the first run of the optimization procedure, the initial MV and ML algorithm settings are set by the optimization algorithm. Based on these settings, the MV algorithm extracts the attributes from the images and creates an attribute file. The number of instances in the file corresponds to the number of input images. Next, the attribute file together with the previously initialized ML settings is passed to the ML algorithm. The output of the ML algorithm is a roughness predictive model and its objective function value, that is the RRSE prediction error. While the stopping criterion is not met, the optimization algorithm keeps generating new populations of solutions. The procedure of generating a new population depends on the employed optimization algorithm. In our case, solutions represent the MV and ML parameter settings, which are passed to the MV and ML algorithms and are in the next iteration evaluated by the optimization algorithm. The optimization procedure stops when a predefined number of solution evaluations is completed. The output of the optimization procedure is the best found roughness predictive model with the corresponding MV and ML algorithm parameter settings.

5. Experimental Setup and Validation Procedure

5.1. Setup

The software environment used in the experiments consisted of MV algorithms for image preprocessing and attribute extraction, the open-source data mining tool Weka [24], and the optimization algorithm jDE [25]. The components were integrated through an interface written in the C++ programming language.

The MV algorithms were implemented using the Open Computer Vision (OpenCV) library [26] and utilizing the CUDA parallel computing platform and programming module [27]. CUDA supports the use of graphical processing units (GPUs) for accelerated algorithm execution.

Weka was selected as a data mining tool since it is easy to call from the C++ interface. Two regression algorithms available in Weka were used to generate the roughness predictive models:

- M5P [28] for building regression trees, and
- RandomForest (RF) [29] for constructing forests of random trees.

Both algorithms involve various parameters that influence the training of predictive models and, consequently, the predictive model accuracy. In this study, the ML algorithm parameters were subject to an automated optimization procedure and, therefore, the best ML parameter values were found by the optimization algorithm.

The following parameters of the M5P algorithm and their values were considered in the optimization procedure:

- M, the minimum number of instances per leaf in the tree, $\{1, 2, \ldots, 20\}$;
- N, use of tree pruning, $\{true, false\}$;
- U, use of smoothing in predictions, $\{true, false\}$.

The parameters and their values for the RF algorithm were as follows:

- Depth, the maximum depth of the tree, $\{1, 2, \ldots, 150\}$;
- K, the number of attributes to randomly investigate, $\{1, 2, \ldots, 25\}$;
- I, the number of iterations, $\{20, 21, \ldots, 200\}$;
- B, randomly breaking the ties when several attributes are equally good, $\{true, false\}$.

The optimization algorithm jDE was used to search the MV and ML parameter decision spaces. jDE is a variant of differential evolution (DE) [6], where only the population size and the stopping condition need to be set manually, while the differential weight and the crossover rate are set through self-adaptation [25]. In all the experiments, the population size was set to 50 and 100 generations were examined, resulting in 5000 evaluated solutions per optimization run. The population size and the number of generations were determined empirically by monitoring the solution improvement over generations.

The decision space size in the case of optimizing the MV and M5P algorithm parameters was $2.6 \cdot 10^9$ (number of possible MV and ML algorithm parameter settings, which is equal to the number of possible values multiplied over all parameters). In the case of optimizing the MV and RF algorithm parameters, it was even larger, i.e., $4.3 \cdot 10^{13}$.

5.2. Validation Procedure

To build the most accurate roughness predictive model, the optimization was performed over the MV parameters, and the M5P and RF algorithm parameters. The optimization procedure was run ten times for each ML algorithm. Based on the extracted dataset of attributes, for each solution, namely ML and MV parameter settings, a regression predictive model was built. The prediction error, i.e., the RRSE value, of each built predictive model was calculated using 10-fold cross-validation (CV). This error was used as the predictive model accuracy estimate and was the optimization objective to be minimized. In Figure 8, the RRSE values averaged over ten optimization runs are denoted with suffix "OPT". The error during the optimization runs could also be assessed using the hold-out set, but as shown in [30], in general, minimizing the error estimated with single CV also minimizes other error estimates.

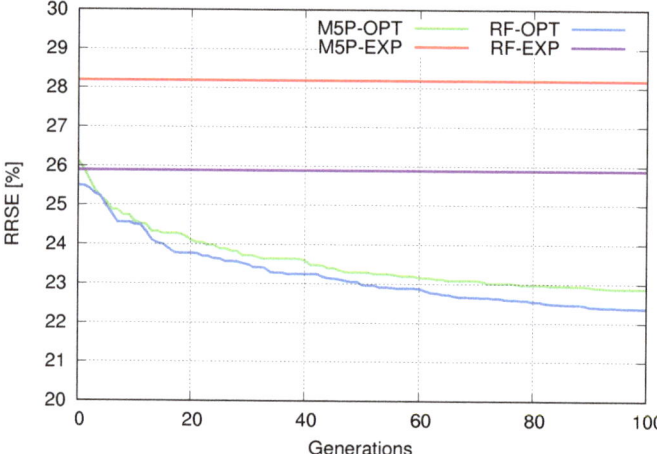

Figure 8. Roughness prediction error (RRSE) estimates for regression trees, random forests and the expert-defined setting. Error estimates for the optimized regression trees and random forests are averaged over ten runs and marked with suffix "OPT". Error estimates for the expert defined setting are marked with "EXP". For details see Section 5.2.

After each run of the optimization procedure, the accuracy of the best found predictive model was assessed using the hold-out set. In addition, the same RRSE assessment was performed for the M5P and RF predictive models built using the expert-defined MV settings. In this case, the expert settings were defined just for the MV parameters, while the parameters of the ML algorithms were set to their default values. Since a single predictive model is constructed in this way, the related result in Figure 8 is represented by a straight line and marked with the suffix "EXP".

6. Results and Discussion

Observing the progress of optimization in terms of RRSE over generations for different algorithms and setups shown in Figure 8, one can draw several conclusions on the resulting predictive models and their accuracy.

First, the manual setup of the MV algorithm parameters does not result in optimal parameter settings. Comparison of the final RRSE values for the manual and optimized settings shows a major difference in the accuracy of the roughness predictive models. The results clearly show that the optimization of the MV parameter settings increases the prediction accuracy.

Next, using different ML algorithms results in predictive models of quite different prediction accuracy, i.e., average RRSE 28.2% for M5P and 25.9% for RF. Recall that the expert parameter settings of the MV algorithm in these runs were kept constant and the ML algorithm parameters were set to their default values. Given the fact that the default values of ML algorithm parameters are not optimized for a specific ML task, it can be expected that the optimization of the ML algorithm parameters improves the accuracy of the predictive model.

Finally, the comparison of the RRSE values averaged over the optimization runs for the M5P and RF algorithms shows that RF achieves better prediction results. In these runs, both the MV and ML algorithm parameters are subject to optimization. As a result, the average RRSE for M5P is 22.9% and for RF 22.4%.

Table 1 shows the comparison of the predictive models validation results between the expert and optimized MV parameters settings. The models were validated by 10-fold CV and by the hold-out set, and averaged over ten runs of each algorithm. The best average prediction accuracy was achieved by the "RF-OPT" algorithm. However, based on the comparison of the RRSE estimates for each ML algorithm, we can observe that the hold-out set validation always yields lower RRSE value than the 10-fold CV. This may be due to the performed systematic sampling of the instances in the hold-out set and relatively small size of the hold-out set (70 instances). The related MAE values were proportional to the RRSE results.

Table 1. Prediction error estimates.

Algorithm	10-Fold CV		Hold-Out Set	
	RRSE [%]	MAE [μm]	RRSE [%]	MAE [μm]
M5P-EXP	28.2	0.82	26.1	0.84
RF-EXP	25.9	0.80	23.9	0.83
M5P-OPT	22.9	0.74	22.5	0.71
RF-OPT	22.4	0.71	21.9	0.70

To better understand the difference between the expert-defined MV settings and MV settings found by the optimization algorithm, we compared the MV algorithm output images. The differences between the original image and the images processed using the expert-defined and the jDE-optimized MV parameter settings are shown in Figure 9. The images processed using the expert-defined and the jDE-optimized MV settings are in comparison to the original image filtered and smoothed. They are very similar from the human eye perspective. However, based on the differences in the output image, a predictive model with a substantially better prediction accuracy is built in the latter case. This indicates that even small differences in image preprocessing arising from different MV parameter settings can result in improved prediction accuracy.

In addition, we analyzed the most informative attributes appearing in the predictive models. The analysis was performed in Weka [24] for the M5P and RF algorithms. The most informative attributes were always selected from the grayscale image attributes mostly describing the geometrical properties of the commutator mounting hole. These properties result from the final treatment of the commutator mounting hole. Regardless of the used ML algorithm, the five most frequent attributes were the number of detected valleys in the image, the number of detected peaks in the image, the lowest grayscale value of a pixel in the valley, and the minimum and the average valley width. Recall that the roughness parameter R_z, which we are trying to predict, represents the maximum difference between the peak height and the valley depth along the measured line profile. Based on this knowledge, the connection between the geometrical properties extracted from the image and the R_z parameter can be interpreted. However, other attributes were also used to build the predictive models, but their selection varied depending on the used algorithm and specific run.

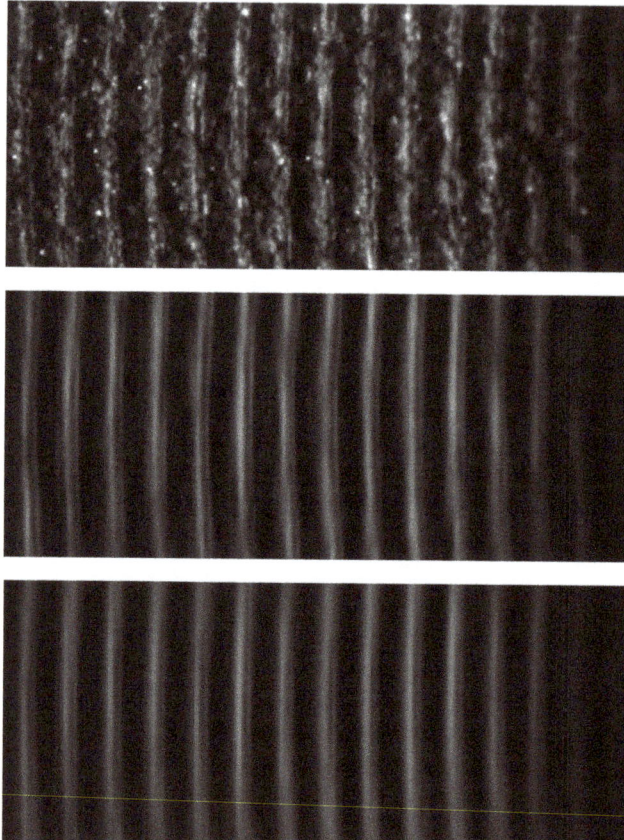

Figure 9. Comparison of the processed images using the expert and the optimized settings: the original extracted image (**top**), the image processed with expert-defined MV parameter settings (**center**), and the image processed with MV and ML parameters set by the optimization algorithm (**bottom**).

To verify the prediction results of the M5P and RF algorithms using the optimized settings, the best model found by each algorithm was identified and applied to predict the roughness of all 700 instances. This was done separately for the training set and the hold-out set. The results of prediction for the two learning algorithms are shown in Figures 10 and 11, respectively.

The results of the M5P regression tree show that the accuracies on the training set and the hold-out set are similar. The estimated RRSE of the best regression tree found on the training set was 21.2%, while the RSSE achieved on the hold-out set was 22.5%. The MAE of this predictive model assessed on the hold-out set was 0.71 μm, which is an acceptable result for practical application.

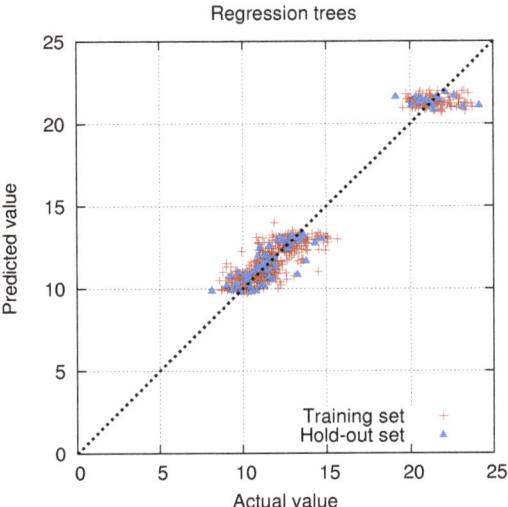

Figure 10. Comparison of the measured and predicted values of R_z for the training and hold-out set, using the best found regression tree model.

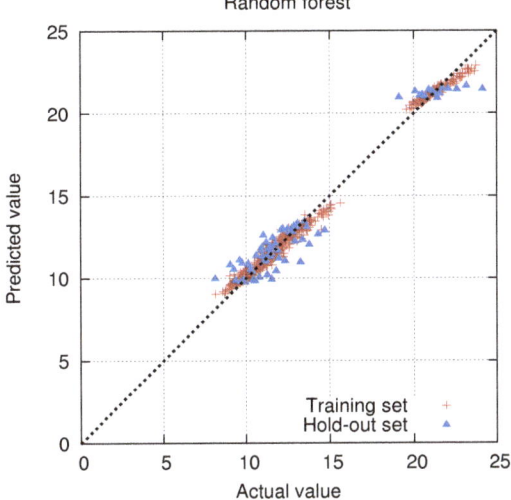

Figure 11. Comparison of the measured and predicted values of R_z for the training and hold-out sets, using the best found random forest model.

Comparison of the prediction accuracy of the M5P and RF models shows that the spread of prediction error is much lower in the case of RF. The best found RF model has the RRSE measured on the training set equal to 8.6%, while the estimated RRSE of the same model on the hold-out set is 22.1%. The MAE of this model assessed on the hold-out set is 0.70 µm.

In addition, we compared the accuracy of the best found predictive model and the accuracy of the existing contact method. Recall that the reference value of the roughness parameter R_z was calculated as an average of three measurements performed with a contact profilometer. The MAE of these measurements was calculated as a difference between a randomly selected one of the three measurements and the average value of these three measurements. The resulting value was 0.60 μm. The MAE of the best RF model was 0.70 μm, which is comparable to the accuracy of the contact measurements. In addition, it was confirmed by a customer that the method is appropriate to perform the statistical process control (SPC) in the commutator production.

The proposed MV-based roughness measurement method has, in comparison to the existing contact method, several advantages. Performing a single measurement with a contact profilometer takes at least 10 seconds, while a complete MV-based measurement, which consists of capturing the image, extracting the attributes and predicting the roughness value, is performed in approximately one second. Since the MV-based method is very efficient and enables contactless measurements, it is suitable for online implementation. In addition, if the method is used in SPC, it enables to perform a higher number of roughness measurements per batch, resulting in a higher reliability of SPC.

7. Conclusions

This paper presents a novel method of measuring the surface roughness of specific machined parts for the automotive industry. The method is based on the MV quality control that enables online and real-time roughness measurements. In addition to MV, the methodology combines ML and evolutionary optimization to build an accurate model for predicting the R_z roughness parameter. The evolutionary optimization algorithm searches for appropriate MV and ML parameter settings to produce a predictive model of acceptable accuracy.

During the development of the MV-based roughness measurement system, two ML algorithms were tested: an algorithm for building the regression trees and a random forest algorithm. The random forest algorithm proved to be more repeatable and accurate on average than the regression tree algorithm; however, the best solutions found by both algorithms were comparable. During the MV and ML parameter optimization, the prediction error was assessed by 10-fold cross-validation. After the optimization, the accuracy of the final predictive models was tested on a hold-out set of previously unseen instances. The validation showed that the found predictive models achieved comparable accuracy on training and hold-out datasets. In addition, it was confirmed, that the optimization methodology is beneficial in setting of the MV parameters for reliable quality control.

The best found RF predictive model has the RRSE value of 22.1%, resulting in the absolute mean prediction error of 0.70 μm. This result is satisfactory and comparable to the accuracy of the SPC contact roughness measurement systems currently installed in the commutator production. However, the developed methodology enables to perform the roughness measurement on the production line and control the quality of the turning process online and in real-time.

The optimization methodology presented in this work can be applied to any MV algorithm to tune its settings and build a predictive model. Nonetheless, the MV operators and their sequence used in the optimization procedure were determined manually, relying on the expert knowledge and experience. Therefore, they may not be optimally selected, and consequently, the prediction error of the best found regression model, in the case of using alternative MV operators and their sequence, could be even lower. Accordingly, our future work will focus on upgrading the presented methodology with automated MV algorithm construction where expert assistance will no longer be needed.

Author Contributions: Conceptualization, V.K. and B.F.; methodology, V.K. and B.F.; software, V.K.; validation, V.K. and B.F.; formal analysis, V.K. and B.F.; investigation, V.K. and B.F.; resources, V.K. and B.F.; data curation, V.K.; writing—original draft preparation, V.K.; writing—review and editing, B.F.; visualization, V.K. and B.F.; supervision, B.F.; project administration, B.F.; funding acquisition, B.F. All authors have read and agreed to the published version of the manuscript.

Funding: The authors acknowledge the financial support from the Slovenian Research Agency (research core funding No. P2-0209). In addition, this work is part of a project that has been funded by the ARTEMIS Joint Undertaking under Grant Agreement No. 332913, and the Slovenian Ministry of Economic Development and Technology under Grant Agreements No. C2130-13-090110 and

C2130-13-090111. It is also part of a project that has received funding from the European Union's Horizon 2020 research and innovation program under Grant Agreement No. 692286.

Institutional Review Board Statement: Not applicable.

Informed Consent Statement: Not applicable.

Data Availability Statement: Not applicable.

Conflicts of Interest: The authors declare no conflict of interest.

References

1. Hamed, A.M.; Saudy, M. Computation of surface roughness using optical correlation. *Pramana* **2007**, *68*, 831–842. [CrossRef]
2. Thomas, T. *Rough Surfaces*; Imperial College Press: London, UK, 1999. [CrossRef]
3. Whitehouse, D.J. *Handbook of Surface and Nanometrology*; CRC Press: Boca Raton, FL, USA, 2010. [CrossRef]
4. Blunt, L.; Jiang, X. (Eds.) *Advanced Techniques for Assessment Surface Topography: Development of a Basis for 3D Surface Texture Standards "Surfstand"*; Butterworth-Heinemann: Oxford, UK, 2003. [CrossRef]
5. Koblar, V.; Pečar, M.; Gantar, K.; Tušar, T.; Filipič, B. Determining surface roughness of semifinished products using computer vision and machine learning. In Proceedings of the 18th International Multiconference Information Society (IS 2015), Ljubljana, Slovenia, 28 September–14 October 2015; Volume A, pp. 51–54.
6. Storn, R.; Price, K. Differential evolution: A simple and efficient heuristic for global optimization over continuous spaces. *J. Glob. Optim.* **1997**, *11*, 341–359. [CrossRef]
7. Shahabi, H.H.; Ratnam, M.M. Noncontact roughness measurement of turned parts using machine vision. *Int. J. Adv. Manuf. Technol.* **2010**, *46*, 275–284. [CrossRef]
8. Jeyapoovan, M.T.; Murugan, M. Surface roughness classfication using image processing. *Measurement* **2013**, *46*, 2065–2072. [CrossRef]
9. Nithyanantham, N.; Suresh, P. Evaluation of cast iron surface roughness using image processing and machine vision system. *ARPN J. Eng. Appl. Sci.* **2016**, *11*, 1111–1116.
10. Jibin, G.J.; Arunachalam, N. Illumination compensated images for surface roughness evaluation using machine vision in grinding process. *Procedia Manuf.* **2019**, *34*, 969–977. [CrossRef]
11. Patel, D.R.P.; Kiran, M. Vision based prediction of surface roughness for end milling. *Mater. Today Proc.* **2020**, *44*, 792–796. [CrossRef]
12. Fadare, D.A.; Oni, A.O. Development and application of a machine vision system for measurement of surface roughness. *ARPN J. Eng. Appl. Sci.* **2009**, *4*, 30–37.
13. Ravikumar, S.; Ramachandran, K.I.; Sugumaran, V. Machine learning approach for automated visual inspection of machine components. *Expert Syst. Appl.* **2011**, *38*, 3260–3266. [CrossRef]
14. Samtaş, G. Measurement and evaluation of surface roughness based on optic system using image processing and artificial neural network. *Int. J. Adv. Manuf. Technol.* **2014**, *73*, 353–364. [CrossRef]
15. Elangovan, M.; Sakthivel, N.; Saravanamurugan, S.; Nair, B.; Sugumaran, V. Machine learning approach to the prediction of surface roughness using statistical features of vibration signal acquired in turning. *Procedia Comput. Sci.* **2015**, *50*, 282–288. [CrossRef]
16. Simunovic, G.; Svalina, I.; Simunovic, K.; Saric, T.; Havrlisan, S.; Vukelic, D. Surface roughness assessing based on digital image features. *Adv. Prod. Eng. Manag.* **2016**, *11*, 93–104. [CrossRef]
17. Yi, H.; Liu, J.; Ao, P.; Lu, E.; Zhang, H. Visual method for measuring the roughness of a grinding piece based on color indices. *Opt. Express* **2016**, *24*, 17215–17233. [CrossRef] [PubMed]
18. Morales Tamayo, Y.; Beltrán Reyna, R.F.; López Bustamante, R.J.; Zamora Hernández, Y.; López Cedeño, K.; Terán Herrera, H.C. Comparison of two methods for predicting surface roughness in turning stainless steel AISI 316L. *Ingeniare. Revista Chilena de Ingeniería* **2018**, *26*, 97–105. [CrossRef]
19. Lin, Y.C.; Wu, K.D.; Shih, W.C.; Hsu, P.K.; Hung, J.P. Prediction of surface roughness based on cutting parameters and machining vibration in end milling using regression method and artificial neural network. *Appl. Sci.* **2020**, *10*, 3941. [CrossRef]
20. Reris, R.; Brooks, J.P. Principal component analysis and optimization: A tutorial. In Proceedings of the 14th INFORMS Computing Society Conference, Richmond, VA, USA, 11–13 January 2015; pp. 212–225.
21. Grinblat, G.L.; Uzal, L.C.; Verdes, P.F.; Granitto, P.M. Nonstationary regression with support vector machines. *Neural Comput. Appl.* **2015**, *26*, 641–649. [CrossRef]
22. Gadelmawla, E.; Koura, M.; Maksoud, T.; Elewa, I.; Soliman, H. Roughness parameters. *J. Mater. Process. Technol.* **2002**, *123*, 133–145. [CrossRef]
23. Trier, O.D.; Jain, A.K. Goal-directed evaluation of binarization methods. *IEEE Trans. Pattern Anal. Mach. Intell.* **1995**, *17*, 1191–1201. [CrossRef]
24. Hall, M.; Frank, E.; Holmes, G.; Pfahringer, B.; Reutemann, P.; Witten, I.H. The WEKA data mining software: An update. *ACM SIGKDD Explor. Newsl.* **2009**, *11*, 10–18. [CrossRef]

25. Brest, J.; Greiner, S.; Bošković, B.; Mernik, M.; Žumer, V. Self-adapting control parameters in differential evolution: A comparative study on numerical benchmark problems. *IEEE Trans. Evolut. Comput.* **2006**, *10*, 646–657. [CrossRef]
26. OpenCV: Open Source Computer Vision. Available online: https://opencv.org/ (accessed on 5 June 2018).
27. NVIDIA. CUDA: Parallel Computing Platform and Programming Model. Available online: https://developer.nvidia.com/cuda-zone/ (accessed on 14 May 2018).
28. Wang, Y.; Witten, I.H. Inducing model trees for continuous classes. In Proceedings of the 9th European Conference on Machine Learning, Prague, Czech Republic, 23–25 April 1997; pp. 128–137.
29. Breiman, L. Random Forests. *Mach. Learn.* **2001**, *45*, 5–32. [CrossRef]
30. Tušar, T.; Gantar, K.; Koblar, V.; Ženko, B.; Filipič, B. A study of overfitting in optimization of a manufacturing quality control procedure. *Appl. Soft Comput.* **2017**, *59*, 77–87. [CrossRef]

Article

Fractional Order PID Controller Design for an AVR System Using Chaotic Yellow Saddle Goatfish Algorithm

Mihailo Micev [1], Martin Ćalasan [1,*] and Diego Oliva [2,3,4]

1. Faculty of Electrical Engineering, University of Montenegro, 81000 Podgorica, Montenegro; mihailom@ucg.ac.me
2. Depto. De Ciencias Computacionales, Universidad de Guadalajara, CUCEI, Av. Revolucion 1500, Guadalajara, 44430 Jal, Mexico; diego.oliva@cucei.udg.mx
3. IN3-Computer Science Dept., Universitat Oberta de Catalunya, 08018 Castelldefels, Spain
4. School of Computer Science & Robotics, Tomsk Polytechnic University, 634050 Tomsk, Russia
* Correspondence: martinc@ucg.ac.me; Tel.: +382-67615237

Received: 29 June 2020; Accepted: 16 July 2020; Published: 18 July 2020

Abstract: This paper presents a novel method for optimal tunning of a Fractional Order Proportional-Integral-Derivative (FOPID) controller for an Automatic Voltage Regulator (AVR) system. The presented method is based on the Yellow Saddle Goatfish Algorithm (YSGA), which is improved with Chaotic Logistic Maps. Additionally, a novel objective function for the optimization of the FOPID parameters is proposed. The performance of the obtained FOPID controller is verified by comparison with various FOPID controllers tuned by other metaheuristic algorithms. A comparative analysis is performed in terms of step response, frequency response, root locus, robustness test, and disturbance rejection ability. Results of the simulations undoubtedly show that the FOPID controller tuned with the proposed Chaotic Yellow Saddle Goatfish Algorithm (C-YSGA) outperforms FOPID controllers tuned by other algorithms, in all of the previously mentioned performance tests.

Keywords: Automatic Voltage Regulation system; Chaotic optimization; Fractional Order Proportional-Integral-Derivative controller; Yellow Saddle Goatfish Algorithm

1. Introduction

The quality of electrical energy is the main demand from consumers in the power system. Since the indicators of the quality are voltage and frequency, these parameters must be maintained at the desired level at every moment. Generally, in every power system, the frequency depends on the active power flow, while the reactive power flow has a greater impact on the voltage level. Additionally, any deviation of the voltage from the nominal value requires the flow of the reactive power, which automatically increases line losses. The fluctuations of the voltage can be repressed using various devices: serial and parallel capacitor banks, synchronous compensators, tap-changing transformers, reactors, Static VAr Compensators (SVC), and Automatic Voltage Regulators (AVR) [1,2]. This paper deals with AVR systems.

The AVR represents the main control loop for the voltage regulation of the synchronous generator (SG), which is the main unit for producing electrical energy in the whole power system. Concretely, the control of the terminal voltage of the synchronous generator is achieved by adjusting its' exciter voltage. Although the main task of the AVR is to provide stable voltage level at the generator's terminals, it is also very important in improving the dynamic response of the terminal voltage. Regardless of the fact that the control theory developed many modern control techniques, the traditional PID controller is still the most used in the AVR systems. In general, in this paper, the optimal tuning of the controller is considered.

Enhancing the performance of the PID controller for AVR systems is possible by using fractional calculus. Fractional order PID controller (FOPID) is the general form of the PID controller that uses fractional order of derivatives and integrals, instead of integer order. Moreover, FOPID can provide a better transient response and is more robust and stable compared to the conventional (known as integer order controller) PID controller [3]. Due to the previously mentioned advantages of the FOPID, this paper deals with this type of controller.

The optimal design of the FOPID controller implies determining the parameters to satisfy defined optimization criteria (or fitness/objective function). In the available literature, the most used method for optimal tuning of the FOPID controller is based on metaheuristic algorithms [4–12]. Particle Swarm Optimization (PSO) and Genetic Algorithm (GA) are applied in [4,5,8,11] to determine the optimal values of the FOPID parameters. For the same purpose, D. L. Zhang et al. proposed an Improved Artificial Bee Colony Algorithm (CNC-ABC) [6]. Also, it can be found that the authors used Chaotic Ant Swarm (CAS) [7], Multi-Objective Extremal Optimization (MOEO) [9], Cuckoo Search (CS) [10], and Salp Swarm Optimization (SSO) algorithm [12] to determine unknown parameters of the FOPID. Besides the FOPID, many existing studies deal with the optimization of the parameters of the classical PID controller for the AVR systems [13–22].

Another very important aspect of the optimization process that needs to be particularly reviewed is the choice of the fitness function. Previously mentioned algorithms introduce a huge variety of fitness functions that take into account time-domain (rise time, settling time, overshoot, and steady-state error), as well as frequency domain parameters (gain margin, phase margin, gain crossover frequency, and so on). One of the most common error-based functions is Integrated Absolute Error (IAE) [6]. Another commonly used time-domain criterion is Zwee Lee Gaing's function originally proposed in [13] for PID controller tuning and applied in [7,10] for optimal tunning of the FOPID controller. Ortiz-Quisbert et al. used the complex function that tends to minimize only time-domain parameters: overshoot, settling time, and maximum voltage signal derivative [11]. One of the most interesting approaches in a fitness function definition is combining error-based functions with time-domain parameters, as presented in [8,9,11]. Concretely, an interesting approach minimizes the Integrated Time Squared Error (ITSE) of the output voltage, energy of the control signal, and ITSE of the load disturbance [8]. The objective function in [9] is composed of IAE, steady-state error, and settling time, while in [11], the objective is to minimize not only IAE, steady-state error, and settling time as previously mentioned, but also the overshoot and the control signal energy. Fitness function that consists only of the frequency domain parameters is proposed in [5] and tends to maximize phase margin and the gain crossover frequency. The trade-off between different frequency domain parameters (phase margin and gain margin) and time-domain parameters (overshoot, rise time, settling time, steady-state error, IAE, and control signal energy) is formulated as an objective function in [4].

Although a large number of FOPID tuning techniques have been proposed in the available literature, the optimal design of the FOPID controller can still be improved by further research. To that end, this paper proposes a novel design approach of the FOPID controller. The contributions of this work are highlighted as follows:

- Firstly, the recently proposed Yellow Saddle Goatfish Algorithm (YSGA) [23] is merged with Chaos Optimization Algorithm [24] in order to obtain novel Chaotic Yellow Saddle Goatfish Algorithm (C-YSGA). Original YSGA can improve the optimization process in terms of accuracy and convergence in comparison to several state-of-the-art optimization methods. The improvement is proven by applying this method on five engineering problems, while the comparison with other methods is carried out by using 27 well-known functions [23]. Additionally, in this paper, the superiority of the original YSGA over several other metaheuristic techniques will be demonstrated on the particular optimization problem. Moreover, an improvement of the YSGA by adding Chaotic Logistic Mapping is introduced. The purpose of merging two algorithms is to additionally improve the convergence speed of the YSGA algorithm. Therefore, the original optimization algorithm for optimal tuning of the FOPID controller will be presented in this paper.

- Afterward, the new objective function that tends to optimize time-domain parameters has been proposed. It is demonstrated that the usage of the proposed objective function provides significantly better results than the other functions proposed in the literature.
- Such an obtained FOPID controller has been compared with those tuned by different optimization algorithms in terms of transient response quality. The conducted analysis clearly demonstrates the superiority of the FOPID controller tuned by C-YSGA.
- Finally, different uncertainties have been introduced to the system in order to examine its behavior. Precisely, the robustness test that implies changing the AVR system parameters is carried out. Also, the ability of the system to cope with the different disturbances (control signal disturbance, load disturbance, and measurement noise) is investigated. During all of the mentioned tests, the FOPID controller tuned by C-YSGA shows significantly better performances compared to the FOPID controller, whose parameters are optimized by the other algorithms considered in the literature.

The organization of this paper is as follows. A brief overview of the AVR system, along with the performance analysis, is provided in Section 2. Section 3 demonstrates the basics of the fractional-order calculus, which is needed for simulating a FOPID controller. Afterward, a compact and wide overview of the available literature related to FOPID parameters optimization is given in Section 4. Section 5 shows the mathematical formulation of the novel C-YSGA algorithm that is presented in this paper. The results of the simulation are given in Section 6. Conclusions are provided in Section 7.

2. Description of the AVR System

The primary function of an AVR system is to maintain the terminal voltage of the generator at a constant level through the excitation system. However, due to the different disturbances in the power system, a synchronous generator does not always work at the equilibrium point. Such oscillations around the equilibrium state can cause deviations of the frequency and the voltage, which can be very harmful to the overall stability of the power system. In order to enhance the dynamic stability of the power system, as well as to provide quality energy to the consumers, excitation systems equipped with AVR are employed. Because of such an important role, the design of an AVR system is a crucial and challenging task.

A typical AVR system consists of the following components:

- controller,
- amplifier,
- exciter,
- generator, and
- sensor.

The object that needs to be controlled in this control scheme is a synchronous generator, whose terminal voltage is measured and rectified by the sensor. An error signal, which presents the difference between the desired and the measured voltage value, is formed in the comparator. One of the main components in the AVR scheme that needs to be chosen carefully is the controller. Based on the error signal and the appropriate control algorithm selected, the controller defines the control signal. Very often the controller is realized as a microcontroller unit, whose output power is deficient. Due to this, the existence of the amplifier is necessary in order to increase the power of the control signal. Finally, an amplified signal is used to control the excitation system of the synchronous generator, and therefore, to define the terminal voltage level. The scheme of such a described system is depicted in Figure 1.

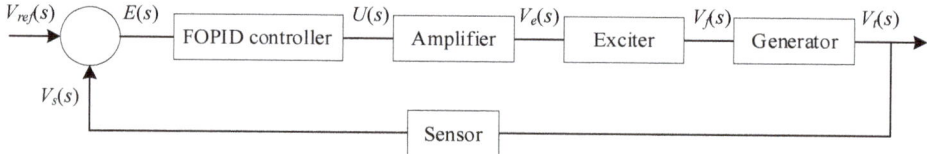

Figure 1. Scheme of the AVR.

In the available literature [4–22], the components of the AVR (except the controller) are presented as the first-order transfer function, which is composed of gain and time constant. Table 1 gives a compact review of the transfer functions and the range of each parameter.

Table 1. Transfer functions of each AVR system component.

Component	Transfer Function	Range of the Parameters
Amplifier	$K_A/(1 + sT_A)$	$10 \leq K_A \leq 400$, $0.02\,s \leq T_A \leq 0.1\,s$
Exciter	$K_E/(1 + sT_E)$	$1 \leq K_E \leq 10$, $0.4\,s \leq T_E \leq 1\,s$
Generator	$K_G/(1 + sT_G)$	$0.7 \leq K_G \leq 1$, $1\,s \leq T_G \leq 2\,s$
Sensor	$K_S/(1 + sT_S)$	$1 \leq K_S \leq 2$, $0.001\,s \leq T_S \leq 0.06\,s$

In the previous table, K_A, K_E, K_G, and K_S stand for gains of an amplifier, exciter, generator, and sensor, respectively, while T_A, T_E, T_G, and T_S are time constants of the amplifier, exciter, generator, and sensor. Values that are considered in this paper are $K_A = 10$, $K_E = 1$, $K_G = 1$, $K_S = 1$, $T_A = 0.1$, $T_E = 0.4$, $T_G = 1$, and $T_S = 0.01$ [5–22]. It is important to mention that the gain of the generator K_G depends on the load of the generator. Namely, K_G can take a value from 0.7 (non-loaded generator) to 1 (nominal loaded generator).

Before including the controller in the system analysis, it is necessary to carry out the analysis of the system in the absence of the controller. To that end, the step response of the AVR system without the controller is given in Figure 2. In order to demonstrate the behavior of the system in the different cases of the load, simulations are conducted for different values of parameter K_G (0.7, 0.8, 0.9, and 1).

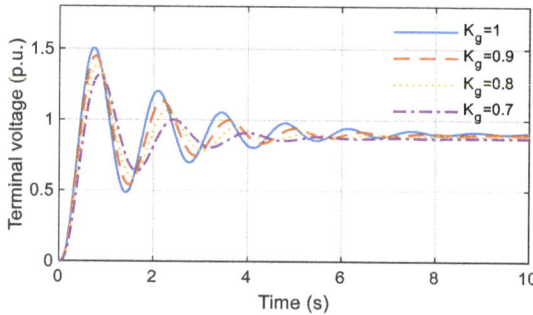

Figure 2. Step response of the AVR system without the controller.

Despite the transient response (time-domain) analysis, it is also very important to take a look at the frequency response of the system. Frequency characteristics or Bode diagrams of the open-loop system can provide information about margins of the stability of the closed-loop system. Precisely, it is important to determine the values of the gain margin and the phase margin, which both need to be positive in order to have a stable system. For the different values of the parameter K_G, frequency responses are shown in Figure 3.

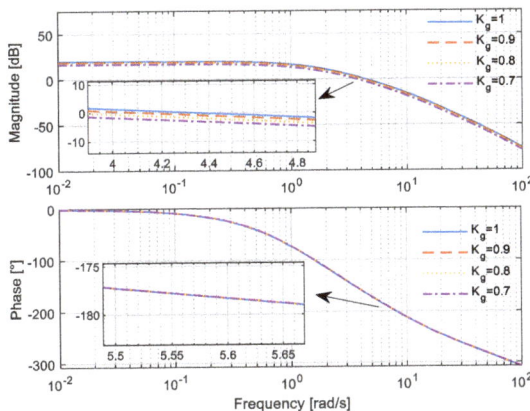

Figure 3. Bode diagrams of the open-loop system.

Another essential characteristic of the system, and the main indicator of stability, is the root locus of the closed-loop system. Root locus gives information about the location of the poles of the closed-loop system. As it is well known from control theory, for the stable system, all the poles must be located in the left half-plane. The graphical representation of the location of the poles is given in Figure 4.

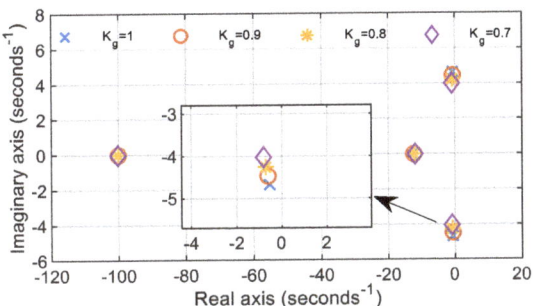

Figure 4. Root locus of the AVR system.

Based on the previously presented figures, all-important time domain and frequency domain parameters can be computed. Namely, Table 2 shows transient response indices-rise time (t_r), settling time (t_s), overshoot in percentage (OS), steady-state error (E_{ss}), frequency response parameters-gain margin (G_m) and phase margin (P_m), as well as the poles of the closed-loop system.

Table 2. Values of the transient response and frequency response parameters.

Parameter	$K_G = 1$	$K_G = 0.9$	$K_G = 0.8$	$K_G = 0.7$
Overshoot (%)	65.214	61.3825	55.9051	50.4818
Rise time (s)	0.2613	0.2755	0.2945	0.3171
Settling time (s)	7.0192	6.5237	5.4086	4.9012
Steady-state error (p.u.)	0.0881	0.102	0.1108	0.1249
Closed-loop system poles	$-0.51 \pm 4.66i$	$-0.6 \pm 4.46i$	$-0.69 \pm 4.25i$	$-0.79 \pm 4.01i$
	-12.48	-12.31	-12.12	-11.92
	-99.97	-99.97	-99.97	-99.97
Gain margin (dB)	4.61	5.53	6.55	7.71
Phase margin (°)	16.1	19.56	23.56	28.26

Root locus and frequency characteristics prove that the AVR system is stable, but the margins of the stability are low due to the poles that are very close to the imaginary axis of the complex plane. Also, large values of the overshoot, the settling time, and the steady-state error indicate that the transient response of the AVR system in the absence of the controller is feeble. In fact, the steady-state error varies from 8% to 12% (depending on the load of the generator), which means the AVR cannot complete the main task—maintaining the voltage level at the reference value. All of the aforementioned deficiencies can be eliminated by adding the controller into the system. According to the available literature, the most used control strategies for AVR systems are based on classical or integer-order PID controller, as well as the generic version of the PID controller that is called Fractional-Order PID controller (FOPID).

Integer-order PID controller is presented by the following transfer function:

$$PID(s) = K_p + \frac{K_i}{s} + K_d s, \tag{1}$$

where K_p is the proportional gain, K_i is the integral gain, and K_d is the derivative gain. Integer-order is a specific case of the PID controller, where the integral and the derivative are first order.

The general type of the PID controller is called Fractional-Order PID controller and is presented using the following transfer function:

$$FOPID(s) = K_p + \frac{K_i}{s^\lambda} + K_d s^\mu, \tag{2}$$

where λ and μ represent the order of the integral and of the derivative, respectively. As the name of the FOPID indicates, these two numbers can be any real numbers (not strictly integers). The aforementioned facts make the FOPID the most general form of the PID controller. Specific forms of FOPID controllers are PID ($\lambda = 1$, $\mu = 1$), PI ($\lambda = 1$, $\mu = 0$), PD ($\lambda = 0$, $\mu = 1$) and P controller ($\lambda = 0$, $\mu = 0$), as illustrated in Figure 5.

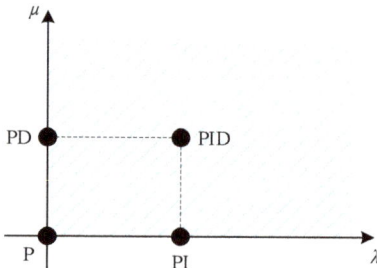

Figure 5. Graphical representation of the different PID controllers.

3. About the Fractional Order Calculus

Regarding the problem of the fractional-order calculus, many different approaches have been proposed. According to [3], the most commonly used definitions of the fractional-order calculus are Grunwald–Letnikov, Riemann–Liouville, and Caputo definition.

Grundwald–Letnikov's approach defines the ath order derivative of the function f in the limits from a to t as follows:

$$D^\alpha \Big|_a^t = \lim_{h \to 0} \frac{1}{h^\alpha} \sum_{r=0}^{[\frac{t-a}{h}]} (-1)^r \binom{n}{r} f(t - rh), \tag{3}$$

where h stands for the time step, and the operator [·] takes only the integer part of the argument. The variable n must satisfy the condition $n-1 < \alpha < n$, while the binomial coefficients are defined by:

$$\binom{n}{r} = \frac{\Gamma(n+1)}{\Gamma(r+1)\Gamma(n-r+1)}, \qquad (4)$$

where the definition of the Gamma function is well known:

$$\Gamma(x) = \int_0^\infty t^{x-1}e^{-t}dt. \qquad (5)$$

Riemann and Liouville proposed the definition of the fractional-order derivative that avoids using limit and sum, but uses integer-order derivative and integral, as follows:

$$_aD^\alpha_t = \frac{1}{\Gamma(n-\alpha)}\left(\frac{d}{dt}\right)^n \int_a^t \frac{f(\tau)}{(t-\tau)^{\alpha-n+1}}d\tau. \qquad (6)$$

Another definition of fractional order derivative is proposed by M. Caputo and is defined by the following Equation:

$$_aD^\alpha_t = \frac{1}{\Gamma(n-\alpha)} \int_a^t \frac{f^{(n)}(\tau)}{(t-\tau)^{\alpha-n+1}}d\tau. \qquad (7)$$

However, the aforementioned formal definitions show the lack of applicability in real-time implementation (digital implementation on a computer) [24]. In order to exceed such a problem, A. Oustaloup proposed the recursive approximation of the fractional-order derivative [25]. Such an approach is trendy among the large number of authors that deal with optimal tuning of the FOPID controller [4,5,7,8,11]. Moreover, in practical implementations of the fractional-order calculus, it can be seen that the Oustaloup's idea dominates over the formal definitions. Because of that, in this paper, Oustaloup's recursive approximation will be used to model fractional-order derivatives and integrals. Mathematical approximation of the αth order derivative (s^α) is given by Equation (8):

$$s^\alpha \approx \omega_h^\alpha \prod_{k=-N}^{N} \frac{s + \omega_k'}{s + \omega_k''}, \qquad (8)$$

where the zeros and the poles are defined as follows:

$$\omega_k = \omega_b \left(\frac{\omega_h}{\omega_b}\right)^{\frac{(k+N+(1+\alpha)/2)}{(2N+1)}}, \quad \omega_k' = \omega_b \left(\frac{\omega_h}{\omega_b}\right)^{\frac{(k+N+(1-\alpha)/2)}{(2N+1)}}. \qquad (9)$$

Before applying the given recursive filter, it is necessary to define the number N that determines the order of the filter (order is $2N + 1$) and the frequency range of the approximation $\{\omega_b, \omega_h\}$. In this study, the order of the filter is chosen to be 9 ($N = 4$), and the selected frequency range is $\{10^{-4}, 10^4\}$ rad/s.

It is imperative to mention that Equation (8) is valid only for $\alpha \in (0, 1)$. Thus, in the case the fractional-order α is higher than 1, it is necessary to conduct a simple mathematical manipulation. Precisely, fractional-order α can be separated as follows:

$$s^\alpha = s^n s^\delta, \ \alpha = n + \delta, \ n \in \mathbb{Z}, \ \delta \in (0,1). \qquad (10)$$

Afterward, Oustaloup's recursive approximation is applied only on s^δ, since s^n is already an integer-order derivative.

4. Overview of the Literature

The problem of optimal design of the FOPID controller means the determination of the parameters K_p, K_i, K_d, λ, and μ so that the certain objective function achieves the minimum (or maximum) value. The most common performance indicators of the tuned FOPID controller are the transient response parameters of the closed-loop system: rise time, settling time, overshoot, and steady-state error. In order to present the results obtained by the recent studies that deal with FOPID tuning, Table 3 provides optimal values of the FOPID parameters and the corresponding transient response parameters of the acquired AVR system.

Table 3. FOPID parameters and transient response parameters from the literature.

Method Number	Reference	K_p	K_i	K_d	μ	λ	t_r (s)	t_s (s)	OS (%)	$|Ess|$ (pu)
1	[5]	0.408	0.374	0.1773	1.3336	0.6827	1.0083	1.512	0.0221	0.0155
2	[5]	0.9632	0.3599	0.2816	1.8307	0.5491	1.3008	1.6967	6.99	0.0677
3	[5]	1.0376	0.3657	0.6546	1.8716	0.5497	0.0104	1.8796	30.8479	0.0595
4	[6]	1.9605	0.4922	0.2355	1.4331	1.5508	0.1904	1.0259	4.8187	0.0102
5	[7]	1.0537	0.4418	0.251	1.1122	1.0624	0.2133	0.6145	5.2398	0.0153
6	[7]	0.9315	0.4776	0.2536	1.0838	1.0275	0.2259	0.564	3.7006	0.0098
7	[8]	0.9894	1.7628	0.3674	0.7051	0.9467	0.1823	1.8835	58.315	0.0409
8	[8]	0.8399	1.3359	0.3511	0.7107	0.9146	0.1998	1.8727	44.8059	0.0146
9	[8]	0.4667	0.9519	0.2967	0.2306	0.8872	0.3041	1.986	45.2452	0.1768
10	[9]	2.9737	0.9089	0.5383	1.3462	1.1446	0.0769	0.388	8.6266	0.0086
11	[10]	2.549	0.1759	0.3904	1.38	0.97	0.0963	0.9774	3.5604	0.0321
12	[10]	2.515	0.1629	0.3888	1.38	0.97	0.0967	0.9849	3.5141	0.033
13	[10]	2.4676	0.302	0.423	1.38	0.97	0.0902	0.9933	3.2504	0.0283
14	[11]	1.5338	0.6523	0.9722	1.209	0.9702	0.0614	1.3313	22.5865	0.0175
15	[12]	1.9982	1.1706	0.5749	1.1656	1.1395	0.1011	0.5633	13.2065	0.0068

It is important to mention that the transient response parameters presented in the table are the calculated values obtained by carrying out the simulations with the given FOPID parameters. In order to conduct the graphical comparison between given references in terms of the transient response parameters, Figures 6–9 present rise time, settling time, overshoot, and steady-state error, respectively, for each method from Table 3.

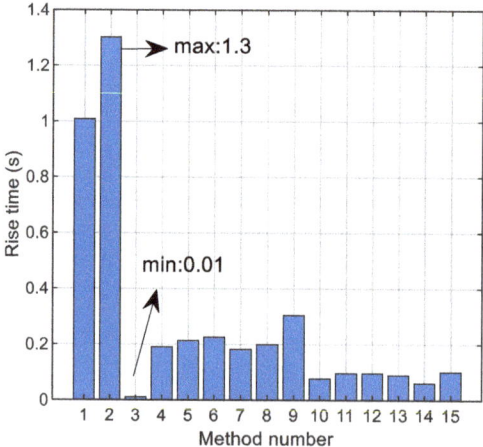

Figure 6. Rise time for each method from Table 3.

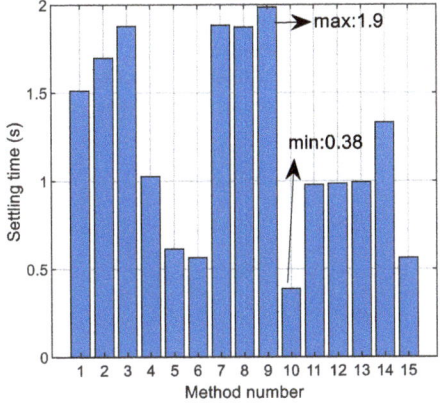

Figure 7. Settling time for each method from Table 3.

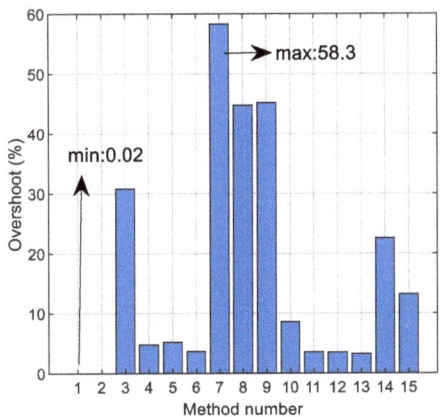

Figure 8. Overshoot for each method from Table 3.

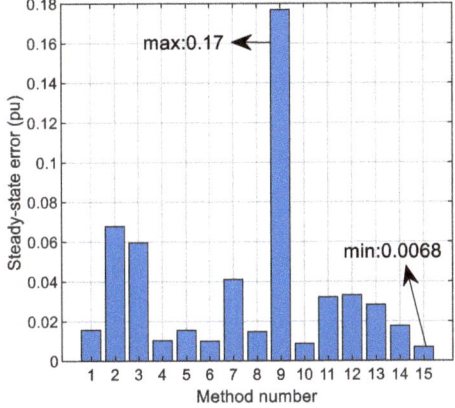

Figure 9. Steady-state error for each method from Table 3.

The process of the optimization of the FOPID parameters highly depends on the chosen objective function that has to be minimized (or maximized). Considering the importance of the objective function, the list of different used functions is given in Table 4. It can be observed that certain authors use a single objective function [4,6,7,10–12], while others perform multi-objective optimization [5,8,9].

Table 4. The list of the used objective functions.

Objective Function	Reference
$OF = w_1 \cdot OS + w_2 \cdot t_r + w_3 \cdot t_s + w_4 \cdot E_{ss} + \int \left(w_5 \cdot |e(t)| + w_6 \cdot V_f(t)^2\right)dt + \frac{w_7}{P_m} + \frac{w_8}{G_m}$	[4]
$J_1 = \omega_{gc}, J_2 = P_m$	[5]
$IAE = \int |e(t)|dt$	[6]
$ZLG = (1 - e^{-\beta}) \cdot (OS + E_{ss}) + e^{-\beta} \cdot (t_s - t_r)$	[7,10]
$J_1 = \int te^2(t)dt, J_2 = \int \Delta u^2(t)dt, J_3 = \int te_{load}^2(t)dt$	[8]
$J_1 = IAE, J_2 = 1000|E_{ss}|, J_3 = t_s$	[9]
$OF = w_1 \cdot OS + w_2 \cdot t_s + w_3 \cdot E_{ss} + w_4 \int |e(t)|dt + w_5 \int u^2(t)dt$	[11]
$OF = (w_1 \cdot OS)^2 + w_2 t_s^2 + \frac{w_3}{(max\_dv)^2}$	[11]
$ITAE = \int t|e(t)|dt$	[12]

In the previous table, e is the error signal (the difference between the reference voltage and the terminal voltage), V_f is the voltage of the generator field winding, ω_{gc} is the gain crossover frequency, u is the control signal (the output of the controller), e_{load} is the error signal when load disturbances are present, and max_dv is the maximum point of the voltage signal derivative. The weighting coefficients are marked as $w_1, w_2, w_3, ..., w_8$.

5. Proposed Chaotic-Yellow Saddle Goatfish Algorithm

The development of the Yellow Saddle Goatfish Algorithm (YSGA) is based on the model of the hunting process by a group of yellow saddle goatfishes, as proposed in [23]. According to this approach, the whole population of the fishes is split into sub-populations. Each sub-population has one fish that is called a chaser, while the others are called blockers. Also, the search space of the possible solutions for the optimization problem is represented by the hunting area of the goatfishes.

The first step of the YSGA is the initialization of the population. Assuming that a population P consists of m goatfishes ($P = \{p_1, p_2, ..., p_m\}$), each goatfish is initialized randomly between the low boundary (b^L) and the high boundary (b^H) of the search space [23]:

$$p_i = rand \cdot \left(b^H - b^L\right) + b^L, \; i = 1, 2, \ldots, m, \tag{11}$$

where $rand$ is a vector of random numbers between 0 and 1. It is very important to mention that p_i is a vector that consists of n decision variables (variables that are being optimized). Furthermore, b^L and b^H are also vectors that represent lower and upper boundaries for each decision variable.

According to (11), the initialization process of the original YSGA proposed in [23] is random, which does not ensure a good starting point in the optimization process. Namely, metaheuristic algorithms have extremely sensitive dependence on the initial conditions, so the improvements in this part may have a great effect on the overall performance of an algorithm. The idea of introducing Chaotic maps into the metaheuristic algorithms in order to replace the random parameters that appear in the algorithm is shown in [26–28]. Among the most interesting approaches are the ones presented in [29–31], where the random population is replaced with the population generated by Chaotic algorithm with different maps. There are many existing Chaotic maps, such as circle map, cubic map, Gauss map, ICMIC map, logistic map, sinusoidal map, and so on. In order to examine the performances of the mentioned maps, many authors provide a mutual comparison of the different maps [29–33]. Concretely, the comparison is carried out by solving concrete optimization problems employing the different Chaotic maps. The existing studies demonstrate that the logistic mapping is the most convenient to use, due to the better computational efficiency than other mentioned Chaotic maps [29–33].

Based on the previous analysis, this paper introduces the initialization of the population using Chaotic Logistic Mapping [24]. Thus, the random initialization is given by (11), which is proposed by the original YSGA algorithm, is replaced by the initialization provided as a result of Chaotic Logistic Mapping. The proposed model of the initialization of the population is described by (12) and (13). Firstly, vectors y_i that are the products of Logistic Mapping are introduced as follows:

$$y_1 = rand,$$
$$y_{i+1} = \mu \cdot y_i \cdot (1 - y_i), \ i = 1, 2, \ldots m, \tag{12}$$

where *rand* stands for a vector of random numbers in the interval [0,1], and μ is the coefficient that is chosen to be 4 in this study. In this manner, we gave the chaotic character of the basic Yellow Saddle Goatfish Algorithm.

Afterward, initialization in C-YSGA is realized according to the following Equation:

$$p_i = y_i \cdot \left(b^H - b^L\right) + b^L, \ i = 1, 2, \ldots, m. \tag{13}$$

Before starting the process of the hunt, the whole population must be divided into sub-populations or clusters. Each of the k clusters c_k has a chaser fish Φ_l and the blocker fish φ_g. Clustering can be made using any of the clustering algorithms. However, the YSGA algorithm uses the K-means clustering algorithm in order to divide the population, as it is described in [23] in detail. The cluster organization of the population is depicted in Figure 10.

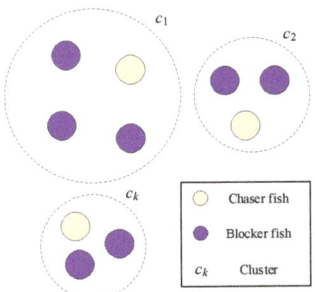

Figure 10. Graphical view of the goatfish population.

The chaser fish of each cluster is the one with the best fitness value. The first step of the hunting process is to update the position of the chaser fish. If the current position is denoted as Φ_l^t, the updated position is Φ_l^{t+1}, and the best chaser fish from all clusters is Φ_{best}^t, where t represents the number of the iteration; the updated law is given by the following Equation:

$$\Phi_l^{t+1} = \Phi_l^t + \alpha \left(\frac{u}{|v|^{1/\beta}}\right)\left(\Phi_l^t - \Phi_{best}^t\right), \tag{14}$$

where α defines the step size (it is set to 1 in this study), and β is Levy index that is calculated as follows (t_{max} stands for the maximum number of iterations):

$$\beta = 1.99 + 0.01t/t_{max}. \tag{15}$$

Parameters u and v from (14) are defined using the following equations:

$$u \sim N(0, \sigma_u^2), \ \sigma_u = \left(\frac{\Gamma(1+\beta) \cdot \sin \frac{\beta \pi}{2}}{\Gamma(1+\beta) \cdot \beta \cdot 2^{(\beta-1)/2}} \right)^{1/\beta}, \tag{16}$$

$$v \sim N(0, \sigma_v^2), \ \sigma_v = 1, \tag{17}$$

where Γ stands for Gamma function and N for normal distribution. In order to update the position of the best chaser fish from all clusters, it is necessary to use (18) instead of (14):

$$\Phi_{best}^{t+1} = \Phi_{best}^{t} + \alpha \left(\frac{u}{|v|^{1/\beta}} \right), \tag{18}$$

The next step in the optimization process is the update of the positions of the blocker fishes. The new position of the blocker fish φ_g^{t+1} can be determined based on the following Equation:

$$\varphi_g^{t+1} = |r \cdot \Phi_l - \varphi_g^t| \cdot e^{b\rho} \cdot \cos(2\pi p) + \Phi_l, \tag{19}$$

where ρ is a random number between a and 1, r is a random number between 0 and 1, and b is the constant that is set to be 1. The parameter a is called the exploitation factor and is linearly decreased from −1 to −2 during the iterations.

It is vital to keep in mind that during the optimization process, the exchange of roles may occur. Namely, if the blocker fish has a better fitness value than the chaser fish, they exchange the roles, and the blocker fish becomes a new chaser fish in the next iteration.

The YSGA model has the predefined parameter λ, which is called an overexploitation parameter. Precisely, if a solution is not improved in λ iterations, it is necessary to change an area of the hunt. Each goatfish, no matter if it is chaser or blocker, must change the hunting area according to the following Equation:

$$p_g^{t+1} = \frac{\Phi_{best} + p_g^t}{2}, \tag{20}$$

where p_g^t and p_g^{t+1} represent old and new positions of the goatfish, respectively. The whole described process is iteratively repeated until the maximum number of iterations is reached. The detailed description is provided with the pseudo-code presented in Table 5.

Table 5. Pseudo-code of the proposed C-YSGA.

Pseudo-Code of the C-YSGA
Enter the input data: m, k, t_{max}, λ
Initialize the population P using chaotic logistic mapping
According to the fitness values determine Φ_{best}
Split the population into k clusters and determine the chaser fish Φ_l for each cluster
while ($t < t_{max}$)
for each cluster
Update the position of the chaser fish and blocker fish
Calculate the fitness value of every fish
Exchange the roles if any blocker fish has better fitness value than the chaser fish
Update the Φ_{best} if the chaser fish has better fitness value
If the fitness value of the chaser fish has not improved, increase the counter q by 1
If the counter q is higher than λ then apply the formula for the change of the zone
end for
$t = t + 1$
endwhile
Φ_{best} is the output result of the algorithm

This section is not mandatory but can be added to the manuscript if the discussion is unusually long or complicated.

6. Simulation Results

This section presents the results that are obtained by applying the proposed C-YSGA method to optimize the FOPID parameters of the AVR system.

Firstly, the formulation of the optimization problem is provided, including the novel objective function presented in this paper. Afterward, the convergence characteristics of the different optimization algorithms used in the literature are compared to the one obtained by C-YSGA in order to demonstrate the convergence superiority of the proposed algorithm. Furthermore, the comparison is conducted in terms of the step response of the AVR system, as well as in the cases of different kinds of uncertainties and disturbances in the system.

6.1. Formulation of the Optimization Problem

From (2), it can be seen that the FOPID is defined with 5 parameters, K_p, K_i, K_d, λ, and μ, which need to be optimized so that the controller satisfies the desired performances. The optimization process is guided by the objective function that defines the performances of the AVR system.

In order to provide a good quality transient response (for the reference step signal), we tested all of the previously used objective functions. However, none of the mentioned functions provide the appropriate system responses as they do not take into account all of the essential characteristics (time-domain parameters or frequency parameters). Furthermore, they do not give an appropriate compromise between all the critical time-domain parameters. On the other side, the objective functions that are based on frequency parameters have higher execution time, which makes the optimization process slower. Observing the different mathematical formulations of the objective functions presented in [4,7,10,11], the authors in this paper propose a novel objective function (21) that contains a smaller number of weighting coefficients, but also outperforms other objective functions in the literature:

$$OF = w_1 \cdot \int t|e(t)|dt + w_2 OS + w_3|E_{ss}| + w_4 t_s. \tag{21}$$

Weighting coefficients are chosen carefully after many experiments, and the following values are considered in this paper: $w_1 = 1$, $w_2 = 0.02$, $w_3 = 1$, and $w_4 = 5$. The values of the coefficients are chosen after many experiments with different combinations. It can be seen that w_2 has a significantly lower value than the other three weighting coefficients. The reason for this is that the overshoot in (21) is given in percentage, and its value is always larger than the values of ITAE, settling time, and steady-state error. Concretely, from Table 3, it can be observed that the highest value of overshoot can go to 45%, while the settling time and the steady-state error reach maximum values of 1.9 s and 0.17 pu, respectively. However, it is very important to highlight that the presence of the FOPID controller can make the closed-loop system unstable. In order to surpass that, this paper uses optimization with constraints. In other words, each solution (each set of FOPID parameters) is first tested to examine if the obtained closed-loop system remains stable. If a certain solution makes the system unstable, it is automatically removed, ignoring its fitness value. The size of the population in the C-YSGA algorithm is selected to be 40, and the maximum number of iterations is 50. Also, the lower and upper boundary must be defined for each of the optimization variables. Taking into account previous studies related to this topic, the chosen boundaries that are used in this paper are presented in Table 6.

Table 6. Boundaries of the optimization variables.

Parameter	Lower Bound	Upper Bound
K_p	1	2
K_i	0.1	1
K_d	0.1	0.4
λ	1	2
μ	1	2

By using the proposed C-YSGA method and the novel objective function depicted above, the optimal FOPID parameters are: $K_p = 1.762$, $K_i = 0.897$, $K_d = 0.355$, $\mu = 1.26$, and $\lambda = 1.032$. The proposed method is compared with all methods presented in Table 3, and the results are provided in Table 7 where the best value is in bold. Note, the best solutions of each method, as it is shown in Table 3, are applied with the proposed fitness function given by (21). It is clear that the new fitness function proposed in this paper has the lowest value when the FOPID parameters obtained by C-YSGA algorithm are used.

Table 7. Comparison of the proposed method with other techniques from literature in terms of OF.

Method Number	Proposed	1	2	3	4	5	6	7
OF value	1.08	24.6	47.3	50	10.1	8.4	4.5	12.3
Method number	8	9	10	11	12	13	14	15
OF value	10.1	53.6	2.3	4.8	4.8	3	9.8	3.1

6.2. Convergence Characteristics

The main goal of hybridizing the concepts of two algorithms (classical YSGA and chaotic logistic mapping) is to accelerate the convergence speed of the original algorithm. Due to the fact that the initial population of the C-YSGA is not selected randomly, but it is the product of the chaotic logistic mapping, it is expected that the proposed algorithm will reach the optimal solution for the least number of iterations. In order to demonstrate that, the original YSGA algorithm, as well as PSO [11], CS [10], and GA [5] algorithms have been implemented to determine the optimal FOPID parameters using the proposed objective function (21). The convergence curves of all mentioned algorithms demonstrate that the C-YSGA algorithm converges in a minimum number of iterations (approximately 10) compared to the other algorithms, as it is depicted in Figure 11. In this figure, the convergence characteristics represent the mean value of convergence characteristics when we started all the algorithms multiple times. Therefore, the chaotic improvement of the standard YSGA algorithm enables obtaining better convergence characteristics. In that manner, it is demonstrated that Chaotic maps, in combination with the metaheuristic algorithm, improves the initial position, which is very important for the convergence speed of the algorithm.

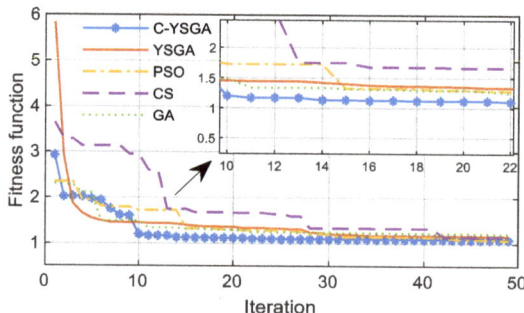

Figure 11. Convergence curves comparison for the different algorithms.

6.3. Step Response

Among all the presented results in Table 3, for the comparison with the proposed C-YSGA, the papers [5,10,11] are chosen. The main indicators of the step response quality-rise time, settling time, overshoot, and steady state-error, as well as the obtained FOPID parameters, are presented in Table 8, the best values are marked in bold.

Table 8. Comparison of the transient response parameters.

Algorithm	K_p	K_i	K_d	μ	λ	t_r (s)	t_s (s)	OS (%)	$|E_{ss}|$ (pu)
C-YSGA	1.7775	0.9463	0.3525	1.2606	1.1273	0.1347	0.2	1.89	0.0009
PSO [11]	1.5338	0.6523	0.9722	1.209	0.9702	0.0614	1.3313	22.58	0.0175
CS [10]	2.549	0.1759	0.3904	1.38	0.97	0.0963	0.9774	3.56	0.0321
GA [5]	0.9632	0.3599	0.2816	1.8307	0.5491	1.3008	1.6967	6.99	0.0677

Additionally, the step response of the AVR system with FOPID parameters from Table 4 is shown in Figure 12. Undoubtedly, it can be concluded that the FOPID controller tuned by the proposed C-YSGA method provides better transient response compared to the other considered algorithms. Precisely, the settling time, the overshoot, and the absolute value of the steady-state error have the least values when the C-YSGA is used, while the rise time has a very low value. Taking a look into the previous table, it can be seen that the overshoot with the PSO algorithm [11] is 22%, which is an unacceptably big value. Similarly, the rise time and the settling time with the GA algorithm are larger than 1 s, which makes the voltage response extremely slow.

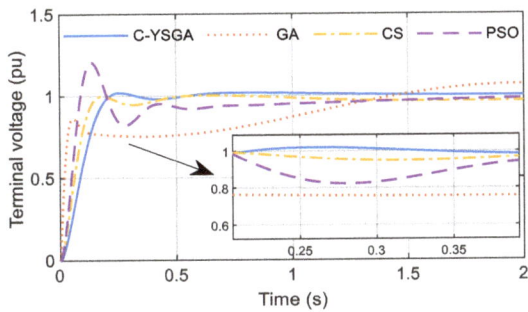

Figure 12. Step response of the AVR system using different algorithms.

6.4. Robustness Analysis

The analysis in the previous section is conducted under the nominal conditions. However, it may occur that the components of the AVR system change their parameters. One of the tasks of the FOPID controller is to ensure the stability of the system and the high quality of the step response in the case of the sudden change in the parameters' values. To that end, the robustness analysis of the AVR system with the C-YSGA FOPID controller is conducted, and the results are presented in Figures 13–16. Precisely, the study is carried out for the change of time constants T_A, T_E, T_G, and T_S from −50% to +50% of the nominal value, in steps of 25%. The step response of the AVR system is shown in Figures 13–16.

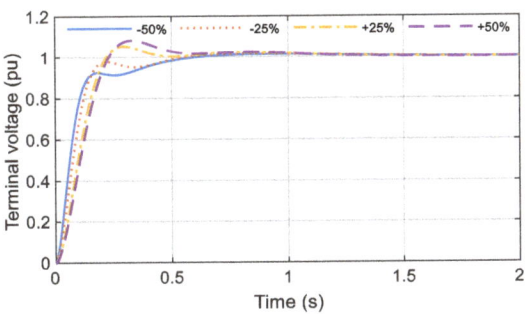

Figure 13. Step response under the variation of T_A.

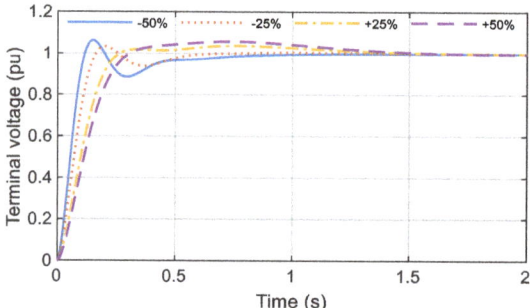

Figure 14. Step response under the variation of T_E.

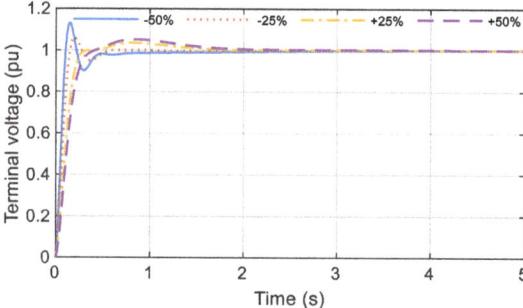

Figure 15. Step response under the variation of T_G.

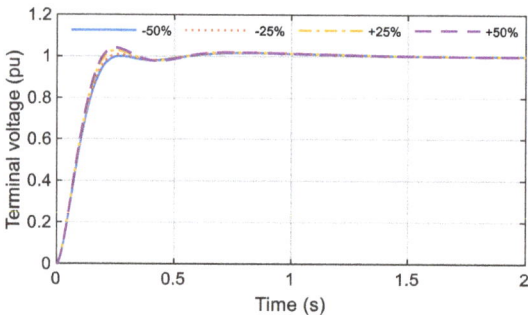

Figure 16. Step response under the variation of T_S.

The results of the previous analysis prove that the C-YSGA FOPID controller makes the AVR system very robust to the changes of each parameter. It is observed that the step response does not deviate a lot compared to the nominal conditions.

6.5. Rejection of the Disturbances

The ability of the FOPID controller to cope with the different disturbances is analyzed by introducing three kinds of disturbances into the AVR system: control signal disturbance, load disturbance, and measurement noise. The block diagram of the AVR with considered disturbances is depicted in Figure 17, while their detailed description is given as follows:

- One of the most common disturbances not only in the AVR system but generally in every control system is control signal disturbance. In this subsection, the obtained C-YSGA FOPID

controller is compared with FOPID controllers tuned by PSO [11], CS [10], and GA [5] algorithms. Control signal disturbance is presented as a constant step signal in the first case, and in the second case as a step signal that lasts from $t = 2$ s to $t = 8$ s. Step responses of the AVR system are shown in Figure 18 for both cases.
- Afterward, the load disturbance that is specific mainly for AVR systems is presented. Similarly to control signal disturbance, it is modeled as a step signal that lasts from $t = 2$ s to $t = 3.5$ s. The obtained step responses, in this case, are shown in Figure 19.
- The last type of disturbance is measurement noise, which is modeled as white Gaussian noise with the power 0.0001 dBW. Figure 20 presents the step responses of the AVR system when the measurement noise is present.

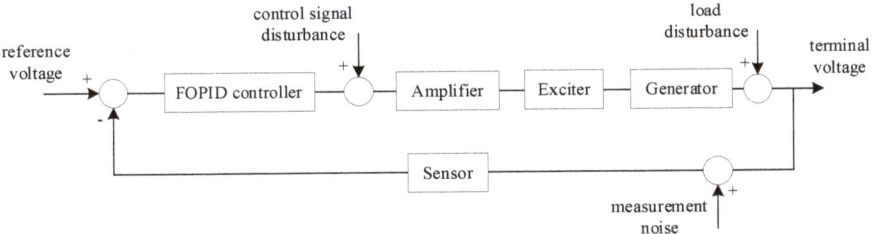

Figure 17. Block diagram of the AVR system considering different kinds of disturbances.

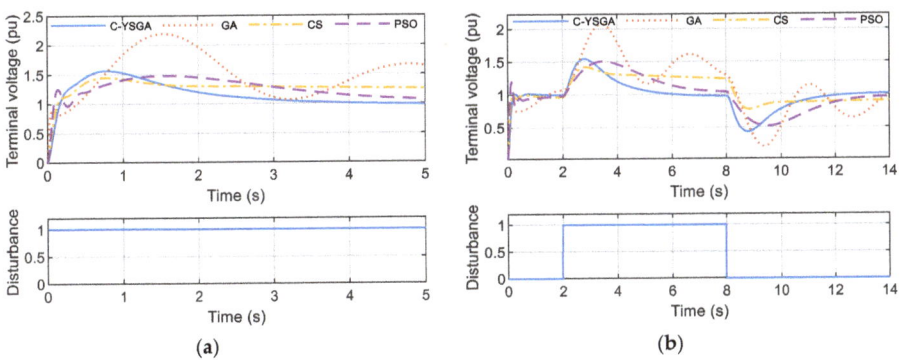

Figure 18. Step responses in the two different cases of the control signal disturbance. (**a**) constant signal, (**b**) step signal that lasts from $t = 2$ s to $t = 8$ s.

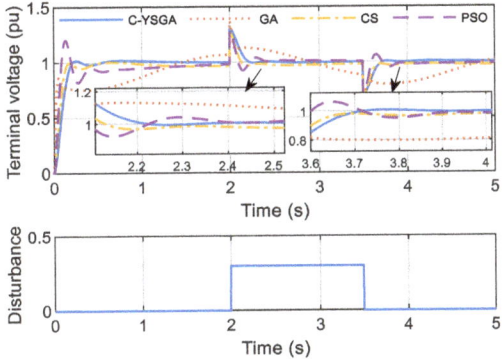

Figure 19. Step responses in the case of load disturbance.

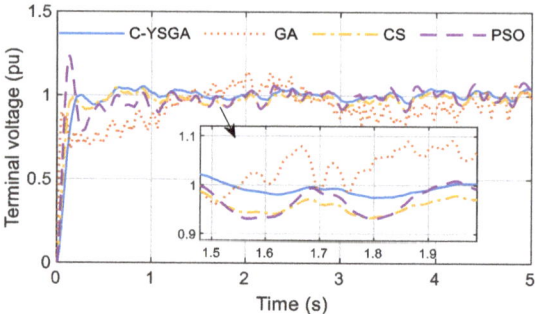

Figure 20. Step responses in the case of measurement noise.

Based on the previous figures, it is obvious that the FOPID controller whose parameters are optimized by using a novel C-YSGA algorithm provides a significantly better ability to reject different types of disturbances. Comparison is conducted with some of the most popular and most used algorithms, whose performances, in this case, are remarkably weaker than the proposed method. To be more precise, from Figures 18 and 19, it can be noted that the voltage does not reach its nominal value after the disturbance is introduced, when the controllers presented in [5,10,11] are used. Such fluctuations in the terminal voltage, caused by the inability of the controller to reject the disturbance, can present a major problem for the consumers of the electrical energy. Unlike them, the FOPID controller tuned by using C-YSGA provides a very stable level of the terminal voltage, whose value reaches the nominal value in a very short period after the disturbance in the system occurs.

7. Conclusions

This paper proposes the novel optimization algorithm in order to optimize the FOPID controller parameters in the AVR system. The proposed algorithm presents the compound of the Yellow Saddle Goatfish Algorithm (YSGA) and Chaotic Logistic mapping to obtain the innovative Chaotic-Yellow Saddle Goatfish Algorithm (C-YSGA). Instead of random initialization of the population, as in many existing metaheuristic algorithms, Chaotic Logistic mapping is used to determine the initial point in the optimization process. It is proved in the paper that such an approach significantly accelerates the convergence of the algorithm. Furthermore, to determine optimal FOPID controller parameters, a new objective function is presented. The results obtained by applying the proposed algorithm with the new objective function introduced in this paper provide significantly better voltage response of the AVR system compared to other considered algorithms. The robustness of the AVR system with such an obtained FOPID controller is tested by changing the AVR system's parameters. It is shown that in all examined cases, the step response of the AVR system has extremely small deviations compared to the nominal case, which means the system is robust to the uncertainties in the system. Moreover, three very often disturbances are introduced into the system, and the system's behavior with different FOPID controllers is analyzed. The mutual comparison shows that the C-YSGA FOPID controller is by far the best in rejecting all considered types of disturbances.

We think that in this way, the algorithm is improved, no matter what optimization problem is considered. In this paper, we tested its applicability and efficiency on the problem of optimal FOPID design. However, at the moment, we are working on proving its superiority over other literature known methods for solving the synchronous machine parameters estimation problem. To that goal, we consider field and armature current waveforms during the short circuit test.

Author Contributions: Conceptualization, M.M. and M.Ć.; methodology, M.M. and M.Ć.; software, M.M.; validation, M.Ć. and D.O.; formal analysis, M.M. and M.Ć.; investigation, M.M.; resources, M.M. and M.Ć.; data curation, M.M. and M.Ć.; writing—original draft preparation, M.M.; writing—review and editing, M.Ć. and D.O.; visualization, M.M. and M.Ć.; supervision, M.Ć. and D.O. All authors have read and agreed to the published version of the manuscript.

Funding: This research received no external funding.

Conflicts of Interest: The authors declare no conflict of interest.

Abbreviations

AVR	Automatic Voltage Regulation
SVC	Static Var Compensator
SG	Synchronous generator
PID	Proportional-Integral-Derivative
FOPID	Fractional Order Proportional-Integral Derivative
YSGA	Yellow Saddle Goatfish Algorithm
C-YSGA	Chaotic Yellow Saddle Goatfish Algorithm
PSO	Particle Swarm Optimization
GA	Genetic Algorithm
CNC-ABC	Improved Artificial Bee Colony
CAS	Chaotic Ant Swarm
MOEO	Multi-Objective Extremal Optimization
CS	Cuckoo Search
SSO	Salp Swarm Optimization
IAE	Integrated Absolute Error
ITSE	Integrated Time Squared Error

Nomenclature

K_A	amplifier gain
K_E	exciter gain
K_G	generator gain
K_S	sensor gain
T_A	amplifier time constant
T_E	exciter time constant
T_G	generator time constant
T_S	sensor time constant
t_r	rise time
t_s	settling time
OS	overshoot
E_{ss}	steady-state error
G_m	gain margin
P_m	phase margin
K_p	proportional gain
K_i	integral gain
K_d	derivative gain
λ	order of the integral
μ	order of the derivative
e	error signal
V_f	voltage of the generator field winding
ω_{gc}	gain crossover frequency
u	control signal
e_{load}	error signal when load disturbances are present
max_dv	maximum point of the voltage signal derivative
$w_1, w_2, w_3, ..., w_8$	weighting coefficients

P	population
m	number of goatfishes
b^L	low boundary
b^H	high boundary
rand	vector of random numbers between 0 and 1
y_i	product vector of Logistic Mapping
k	number of clusters
c_k	cluster
Φ_l	chaser fish
φ_g	blocker fish
Φ_{best}	best chaser fish
t	number of the current iteration
t_{max}	maximum number of iterations
α	step size
β	Levy index
Γ	gamma function
N	normal distribution
a	exploitation factor
r	random number between 0 and 1
ρ	random number between a and

References

1. Lipo, T.A. *Analysis of Synchronous Machines*, 2nd ed.; CRC Press: Boca Raton, FL, USA, 2012.
2. Boldea, I. *Synchronous Generators*, 2nd ed.; CRC Press: Boca Raton, FL, USA, 2016.
3. Shah, P.; Agashe, S. Review of fractional PID controller. *Mechatronics* **2016**, *38*, 29–41. [CrossRef]
4. Zamani, M.; Karimi-Ghartemani, M.; Sadati, N.; Parniani, M. Design of a fractional order PID controller for an AVR using particle swarm optimization. *Control Eng. Pract.* **2009**, *17*, 1380–1387. [CrossRef]
5. Pan, I.; Das, S. Frequency domain design of fractional order PID controller for AVR system using chaotic multi-objective optimization. *Int. J. Electr. Power Energy Syst.* **2013**, *51*, 106–118. [CrossRef]
6. Zhang, D.-L.; Tang, Y.-G.; Guan, X.-P. Optimum Design of Fractional Order PID Controller for an AVR System Using an Improved Artificial Bee Colony Algorithm. *Acta Autom. Sin.* **2014**, *40*, 973–979. [CrossRef]
7. Tang, Y.; Cui, M.; Hua, C.; Li, L.; Yang, Y. Optimum design of fractional order PI λD μ controller for AVR system using chaotic ant swarm. *Expert Syst. Appl.* **2012**, *39*, 6887–6896. [CrossRef]
8. Pan, I.; Das, S. Chaotic multi-objective optimization based design of fractional order PI λD μ controller in AVR system. *Int. J. Electr. Power Energy Syst.* **2012**, *43*, 393–407. [CrossRef]
9. Zeng, G.Q.; Chen, J.; Dai, Y.X.; Li, L.M.; Zheng, C.W.; Chen, M.R. Design of fractional order PID controller for automatic regulator voltage system based on multi-objective extremal optimization. *Neurocomputing* **2015**, *160*, 173–184. [CrossRef]
10. Sikander, A.; Thakur, P.; Bansal, R.C.; Rajasekar, S. A novel technique to design cuckoo search based FOPID controller for AVR in power systems. *Comput. Electr. Eng.* **2018**, *70*, 261–274. [CrossRef]
11. Ortiz-Quisbert, M.E.; Duarte-Mermoud, M.A.; Milla, F.; Castro-Linares, R.; Lefranc, G. Optimal fractional order adaptive controllers for AVR applications. *Electr. Eng.* **2018**, *100*, 267–283. [CrossRef]
12. Khan, I.A.; Alghamdi, A.S.; Jumani, T.A.; Alamgir, A.; Awan, A.B.; Khidrani, A. Salp Swarm Optimization Algorithm-Based Fractional Order PID Controller for Dynamic Response and Stability Enhancement of an Automatic Voltage Regulator System. *Electronics* **2019**, *8*, 1472. [CrossRef]
13. Gaing, Z.L. A particle swarm optimization approach for optimum design of PID controller in AVR system. *IEEE Trans. Energy Convers.* **2004**, *19*, 384–391. [CrossRef]
14. Blondin, M.J.; Sicard, P.; Pardalos, P.M. Controller Tuning Approach with robustness, stability and dynamic criteria for the original AVR System. *Math. Comput. Simul.* **2019**, *163*, 168–182. [CrossRef]
15. Ekinci, S.; Hekimoglu, B. Improved Kidney-Inspired Algorithm Approach for Tuning of PID Controller in AVR System. *IEEE Access* **2019**, *7*, 39935–39947. [CrossRef]
16. Mosaad, A.M.; Attia, M.A.; Abdelaziz, A.Y. Whale optimization algorithm to tune PID and PIDA controllers on AVR system. *Ain Shams Eng. J.* **2019**, *10*, 755–767. [CrossRef]

17. Blondin, M.J.; Sanchis, J.; Sicard, P.; Herrero, J.M. New optimal controller tuning method for an AVR system using a simplified Ant Colony Optimization with a new constrained Nelder–Mead algorithm. *Appl. Soft Comput. J.* **2018**, *62*, 216–229. [CrossRef]
18. Calasan, M.; Micev, M.; Djurovic, Z.; Mageed, H.M.A. Artificial ecosystem-based optimization for optimal tuning of robust PID controllers in AVR systems with limited value of excitation voltage. *Int. J. El. Eng. Educ.* **2020**, *1*, 1–25. [CrossRef]
19. Mosaad, A.M.; Attia, M.A.; Abdelaziz, A.Y. Comparative Performance Analysis of AVR Controllers Using Modern Optimization Techniques. *Electr. Power Components Syst.* **2018**, *46*, 2117–2130. [CrossRef]
20. Bingul, Z.; Karahan, O. A novel performance criterion approach to optimum design of PID controller using cuckoo search algorithm for AVR system. *J. Frankl. Inst.* **2018**, *355*, 5534–5559. [CrossRef]
21. Al Gizi, A.J.H.; Mustafa, M.W.; Al-geelani, N.A.; Alsaedi, M.A. Sugeno fuzzy PID tuning, by genetic-neutral for AVR in electrical power generation. *Appl. Soft Comput. J.* **2015**, *28*, 226–236. [CrossRef]
22. Mohanty, P.K.; Sahu, B.K.; Panda, S. Tuning and assessment of proportional-integral-derivative controller for an automatic voltage regulator system employing local unimodal sampling algorithm. *Electr. Power Compon. Syst.* **2014**, *42*, 959–969. [CrossRef]
23. Zaldívar, D.; Morales, B.; Rodríguez, A.; Valdivia-G, A.; Cuevas, E.; Pérez-Cisneros, M. A novel bio-inspired optimization model based on Yellow Saddle Goatfish behavior. *BioSystems* **2018**, *174*, 1–21. [CrossRef] [PubMed]
24. Ausloos, M.; Dirickx, M. *The Logistic Map and the Route to Chaos*; Springer: Berlin, Germany, 2006.
25. Oustaloup, A.; Levron, F.; Mathieu, B.; Nanot, F.M. Frequency-band complex noninteger differentiator: Characterization and synthesis. *IEEE Trans. Circuits Syst. I. Fundam. Theory Appl.* **2000**, *47*, 25–39. [CrossRef]
26. Zhang, H.; Zhou, J.; Zhang, Y.; Fang, N.; Zhang, R. Short term hydrothermal scheduling using multi-objective differential evolution with three chaotic sequences. *Int. J. Electr. Power Energy Syst.* **2013**, *47*, 85–99. [CrossRef]
27. Dos Coelho, L.S.; Alotto, P. Multi-objective electromagnetic optimization based on a nondominated sorting genetic approach with a chaotic crossover operator. *IEEE Trans. Magn.* **2008**, *44*, 1078–1081. [CrossRef]
28. Coelho, L.d.S. A quantum particle swarm optimizer with chaotic mutation operator. *Chaos Solitons Fractals* **2008**, *37*, 1409–1418. [CrossRef]
29. Alatas, B. Chaotic harmony search algorithms. *Appl. Math. Comput.* **2010**, *216*, 2687–2699. [CrossRef]
30. Ahmadi, M.; Mojallali, H. Chaotic invasive weed optimization algorithm with application to parameter estimation of chaotic systems. *Chaos Solitons Fractals* **2012**, *45*, 1108–1120. [CrossRef]
31. Ma, Z. Chaotic populations in genetic algorithms. *Appl. Soft Comput. J.* **2012**, *12*, 2409–2424. [CrossRef]
32. Talatahari, S.; Farahmand Azar, B.; Sheikholeslami, R.; Gandomi, A.H. Imperialist competitive algorithm combined with chaos for global optimization. *Commun. Nonlinear Sci. Numer. Simul.* **2012**, *17*, 1312–1319. [CrossRef]
33. Zilong, G.; Sun'an, W.; Jian, Z. A novel immune evolutionary algorithm incorporating chaos optimization. *Pattern Recognit. Lett.* **2006**, *27*, 2–8. [CrossRef]

© 2020 by the authors. Licensee MDPI, Basel, Switzerland. This article is an open access article distributed under the terms and conditions of the Creative Commons Attribution (CC BY) license (http://creativecommons.org/licenses/by/4.0/).

Article

Genetic Programming Guidance Control System for a Reentry Vehicle under Uncertainties

Francesco Marchetti * and Edmondo Minisci

Intelligent Computational Engineering Laboratory (ICE-Lab), University of Strathclyde, Glasgow G11XJ, UK; edmondo.minisci@strath.ac.uk
* Correspondence: francesco.marchetti@strath.ac.uk

Abstract: As technology improves, the complexity of controlled systems increases as well. Alongside it, these systems need to face new challenges, which are made available by this technology advancement. To overcome these challenges, the incorporation of AI into control systems is changing its status, from being just an experiment made in academia, towards a necessity. Several methods to perform this integration of AI into control systems have been considered in the past. In this work, an approach involving GP to produce, offline, a control law for a reentry vehicle in the presence of uncertainties on the environment and plant models is studied, implemented and tested. The results show the robustness of the proposed approach, which is capable of producing a control law of a complex nonlinear system in the presence of big uncertainties. This research aims to describe and analyze the effectiveness of a control approach to generate a nonlinear control law for a highly nonlinear system in an automated way. Such an approach would benefit the control practitioners by providing an alternative to classical control approaches, without having to rely on linearization techniques.

Keywords: evolutionary optimization; genetic programming; control; differential evolution; reusable launch vehicle

1. Introduction

Several decades after its formulation, and despite the great abundance of different control schemes and paradigms in the academic literature, the PID controller is still the most used control approach in industrial and real world applications. This is due to its simplicity, effectiveness and to the decades long active research. Nonetheless, it needs to be applied to linear or linearized models to work efficiently. This is particularly difficult in the aerospace domain, where the physics of the controlled plant and of the environment are defined by nonlinear models. To overcome the limitations of the PID controller and improve the performances and robustness of a control system, several control system design approaches have been proposed in the past, such as optimal control, adaptive control and robust control. However, all these different control system design approaches require accurate mathematical models and they are challenged by the nonlinear character of these models [1]. Moreover, as discussed by Haibin et al. [2] and Xu et al. [3], these complex systems have to meet stringent constraints on reliability, performances and robustness and the current control methods can only satisfy these constraints partially, usually by relying on simplified physical models. An in depth review of guidance control algorithms is presented by Chai et al. [4]. In their work, several guidance control algorithms are analyzed and divided into three main categories: Stability-theory based, optimisation-based and AI-based. Each of these three approaches present pros and cons which can be summarized as follows: Stability-theory based methods are characterized by a well defined mathematical formulation and proof of their stability; nonetheless, they present issues when dealing with uncertainties and when the controlled system models are not well defined. As for stability-theory based methods, also the optimisation-based methods' robustness and stability are proved mathematically and they are flexible in the sense that they can be easily combined

with other tools. However, they are computationally expensive, they have issues when dealing with constraints and nonlinearities and the reliability of the approaches devised to overcome these limitations, e.g., convexification, can be questioned. To overcome these issues and to fully exploit the available nonlinear models, AI could be used and integrated in classical control schemes. Despite the application of AI for control purposes is still in an early stage, it could greatly enhance the control capabilities and the robustness of classical control schemes. However, as pointed out in [4], they lack of mathematical proofs of reliability. A more in depth survey on some of the latest applications of AI in spacecraft guidance dynamics and control was produced by Izzo et al. [5] where the potential benefits of this research direction were assessed. A promising AI technique which can be used for control purposes is GP. GP is a CI technique belonging to the class of EAs. It was made known by Koza in 1992 [6] and it is capable of autonomously finding a mathematical input-output model from scratch to minimize/maximize a certain fitness function defined by the user. As other EAs, it starts from an initial population of random individuals and then evolves them using evolutionary operators such as crossover, mutation and selection until a termination condition is met. The evolution is guided by the principle of the evolution of the fittest and it proceeds by finding increasingly better individuals to solve the desired problem. In contrast to other EAs, in GP an individual is structured as a tree which represents a mathematical function. This clear representation of the models produced by GP is an advantage when compared to other ML techniques, like NNs, especially in a control context. In fact, to assess the reliability and robustness of a control scheme it is desirable to have the complete mathematical representation of it. This can be naturally achieved by GP while a NN produces a so called black-box model, where inputs are provided and outputs are produced but what is inside the model is not known to the user. The effectiveness of a GP based controller was discussed in [7], where it was shown that GP is capable of producing human-competitive results. In general, in a control context GP can be used to autonomously find a control law, starting from little or no knowledge of the control system, capable of performing the guidance or attitude control of a desired plant. It is not limited by the aforementioned nonlinearities and by being autonomous, it permits to avoid the cumbersome mathematical formulation of other control approaches. Hence, the aim of this work is to present a controller design approach based on GP applied to the reentry guidance control problem of the FESTIP-FSS5 RLV, considering the presence of uncertainties in the environment and plant models. The purpose of the presented control approach is to overcome the challenges related to the control of highly nonlinear systems. In particular, the proposed approach allows avoiding the use of linearization techniques that introduce modelization discrepancies and over simplified controllers that can fail in real conditions. This is particularly true for reentry applications, where the environment and aerodynamic models vary rapidly and modelization and control simplifications can result in system failures. To the best of the authors knowledge no application of GP to the guidance control of a reentry vehicle was found in the literature. The learning process is performed offline due to hardware limitations, but, as the hardware technologies improve, there will be the possibility to perform the process online in the future. A recent example of progress in this direction is represented by the work of Baeta et al. [8], where they proposed a framework to deploy GP using TensorFlow. If such an approach proves to be easily applicable to other GP applications, with improvements in the computational time of the same order of magnitude of those found in [8], it would represent a big step forward in the online usage of GP for control purposes. In fact, what was done so far in the literature consists in applying the GP offline to find a control law. On the other hand, if GP is used online, it is possible to define that control approach, IC [9].

The reminder of the paper is organized as follows: In Section 2 an overview of the different control approaches for reentry applications is provided; in Section 3 the theory behind GP and the particular GP algorithm employed in this work, the IGP, are presented along with a description of the IGP settings for this work. In Section 4 the chosen test case is described, while in Section 5 the obtained results are presented. Finally, in Section 6 the found conclusions and future work directions are discussed.

2. Related Work

Among the different control approaches employed for reentry applications, the most commonly used are MPC, Optimal Control and Sliding Mode Control. MPC is a control strategy akin to optimal control, in the sense that a finite-horizon optimization of the control variables is performed online while satisfying the imposed constraints. Luo et al. [10] used this technique to solve the attitude control problem of a reentry vehicle while considering the actuator dynamics and failure scenarios. A more recent approach involving MPC is proposed by Wang et al. [11], where they used a convex optimization with pseudospectral optimization inside an MPC framework, to solve a constrained rocket landing problem. Both these approaches proved to be successful, but in order to be applied, they must rely on simplified aerodynamic models or on the constraints convexification as done for the convex optimization approach. These restrictions are common to Optimal Control. In fact, the use of an optimizer, especially gradient based, inside a control framework is an issue if the models involved are nonlinear. Nonetheless, Optimal Control is widely used, both for ascent and reentry trajectory optimization. An example of that is represented by [12], where a pseudospectral optimal control approach is used to generate online the optimal guidance controls for a reentry vehicle in the presence of disturbances and constraints, but simplified aerodynamic models are used. A similar approach is used in [13], where a pseudospectral optimal control algorithm is combined with a sliding mode controller to take into account uncertainties, but also here a simplified aerodynamic model is used. Other usage of a sliding mode controller can be found in [14], where a sliding mode controller is used for the approach and landing phase of a reentry vehicle but simplified aerodynamic models are used and constraints are imposed only on the states variables. While in [15], an adaptive twisting sliding mode controller is used for the attitude tracking of an hypersonic reentry vehicle. Such an approach is not tested on a constrained problem and involves a cumbersome mathematical formulation.

All the aforementioned techniques can be considered "classical" approaches, in contrast to those based on AI. AI control approaches for reentry vehicles are more rare in the literature, but, as said above, the incorporation of AI in control systems is becoming more and more frequent. Few examples of these approaches are represented by the work done by Wu et al. [16] and Gao et al. [17]. In the former, a fuzzy logic-based controller is developed for the attitude control of the X-38 reentry vehicle; while on the latter, a Deep Reinforcement Learning approach is used for the reentry trajectory optimization of a reusable launch vehicle.

Among the various AI techniques employed, EAs are among the less employed. In particular to the best of the authors knowledge no control approach for a reentry vehicle involving GP was studied in the literature. For reentry applications, EAs are mostly used to find an optimal reentry trajectory, as it is done in [18–20]. Other uses of EAs in reentry applications can be found in [21], where GA is used to optimize the parameters of a sliding mode controller applied to the attitude control of a reentry vehicle. While, in [22] the evolutionary method Pigeon Inspired Optimization is combined with a Gauss Newton gradient based method to form a predictor-corrector guidance algorithm. The goal of this guidance scheme is to use the EA to generate an initial condition for the gradient-based method to solve the entry guidance problem.

3. Genetic Programming for Control

Genetic Programming (GP) is a Computational Intelligence (CI) technique pertaining to the class of Evolutionary Algorithms (EAs) made known by Koza in 1992 [6]. As other EAs, GP consists in evolving a program from an initial population of randomly generated programs, to minimize/maximize a defined fitness function. Such evolution is performed according to the defined crossover, mutation and selection operators, in order to steer the evolution of the population in the desired direction and avoid local minima. Compared to other EAs, in GP the individuals are shaped as trees which corresponds to symbolic mathematical equations as depicted in Figure 1, where a, b and 3 are called terminal nodes and a and b are the input variables. The other nodes in the GP tree corresponds to the primitive functions provided by the user and employed by the GP algorithm to build programs autonomously.

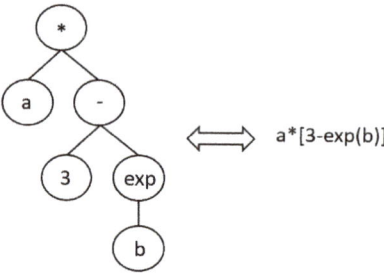

Figure 1. Structure of an individual in GP. The tree can be read as the equation on the right.

About the evolutionary operators, crossover consists in an exchange of genes between two parents individuals to produce other two individuals as offspring. Looking at Figure 2, two genes highlighted by the green and orange boxes are exchanged between the parents to form the offspring. Regarding the mutation operation, a randomly chosen gene of a chosen individual is randomly mutated to generate a new offspring, as depicted in Figure 3. Several different types of the crossover and mutation operators were devised in the past decades and which one to choose is highly dependant on the particular problem to solve. Since it is not the aim of this work to provide a comprehensive description of the peculiarities of the GP algorithm, the reader is referred to [6] for more details.

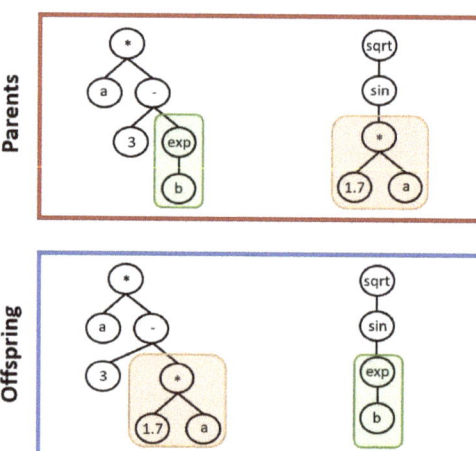

Figure 2. Schematic of crossover operation. From two parents, two offspring are generated.

Regarding the control applications of GP, several examples can be found in the literature. In [23] a multi-objective GP is used to evolve controllers for a UAV allowing it to navigate towards a radar source. In [24] an adapted GP approach is used to generate the control law for a UAV which provides recovery onto a ship considering real world disturbances and uncertainties. Other control approaches involving GP are represented by [25] where GP is used to generate a control Lyapunov function and the modes of a switched state feedback controller; while in [26] GP is used to generate a PID based controller. All these approaches involving GP for control applications cannot be considered IC. In fact, the GP evaluation is done offline and no online learning is present. An example of GP used in an IC framework is represented by the work of Chiang [27] where the control law for a small robot is produced online using GP, with the aim of moving the robot in an environment filled with obstacles. A similar approach was used in [28], where a guidance controller for a Goddard rocket is designed online to cope with different kinds of uncertainties. In contrast to this last example, the work presented here aims to generate a control law for a much more complex and nonlinear system, also considering more severe uncertainties. This results in a longer computational time, hence the impossibility to use it online with the current hardware technology. In fact, to generate the control law online successfully, the considered plant must be very simple in order to perform the evolutionary process in a useful time interval.

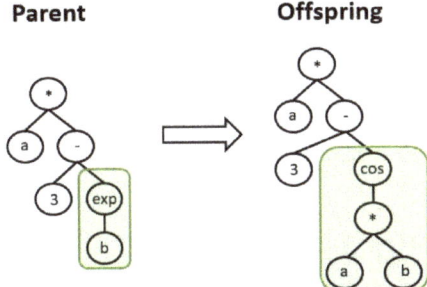

Figure 3. Schematic of mutation operation. A randomly selected gene of an individual is mutated randomly.

As opposed to all these examples, the approach proposed in this work is developed considering a more complex plant model and uncertainties are also taken into account. In fact, the proposed control approach takes into account the nonlinearities present in the models of the vehicle dynamics, the aerodynamics and the environment while uncertainties are applied to the environment and aerodynamic models. Moreover, to the best of the authors' knowledge this is the first application of GP to perform guidance control of the reentry trajectory of an RLV.

The work produced in [28] was later used as a foundation to design the Hybrid GP-NN controller presented in [29]. In this control system, the GP was used offline to generate a control law that was later optimized online by a NN. The approach to generate offline the GP control law is the same used in this work, with the following differences: (1) A different more challenging application is considered, (2) more severe uncertainties are applied and (3) a more thorough analysis of the GP performances is conducted than the one performed in [29].

3.1. Inclusive Genetic Programming

The particular GP algorithm used in this work is the Inclusive Genetic Programming (IGP), which was originally presented in [29] and later analyzed in [30]. IGP was developed in Python 3 relying on the open source library DEAP [31]. IGP was designed to promote an maintain the population's diversity throughout the evolutionary process, so to avoid losing big individuals due to bloat control operators. In fact, it was observed in [29] that bigger individuals are capable of handling the nonlinearities of the treated problem better than smaller individuals. Therefore, to maintain the population's diversity and hence also preserve big individuals, three main features were inserted in the GP algorithm so to make what is now called the IGP. These features are: (1) A niches creation mechanism; (2) the Inclusive Reproduction and (3) the Inclusive Tournament. The creation of the niches implies that IGP belongs to the class of niching method. Niching methods are among the various methods used to tackle the diversity issue in GP. In contrast to standard niches approaches, the IGP makes use of the niches in a different manner. In fact, in the IGP the niches are used to make sure that the majority of the individuals are considered during the evolutionary process and that a flow of genes is established between the different niches. While in standard niching methods, the niches are separated and are used to find different local optima in a multimodal optimization, by parallel evolving the different niches towards different optima [32].

The whole Inclusive Evolutionary Process is based on the $\mu + \lambda$ evolutionary strategy [33] and it is schematized in Figure 4. The process starts with an initial population which is subdivided into niches. Then the Inclusive Reproduction is performed, λ offspring is produced and the total of the individuals (parents+offspring) is subdivided into niches. These new niches are used to select μ individuals from the new population, which is again subdivided into niches and the process starts again.

Regarding the niches creation mechanism, the individuals in the IGP populations are divided into niches according to their genotypic diversity, which is the number of nodes inside each individual, i.e., their length. Each niche will cover a lengths interval, defined by the maximum and minimum lengths of the individuals in the population, and the number of niches. This way, each niche will be placed between the extremes of each interval and these extremes are evenly spaced between the maximum and minimum length of the individuals in the population. For example, if the maximum length is 30, the minimum is 5 and 5 niches are created, the lengths intervals that the niches will cover will be [5, 11.25], [11.25, 17.5], [17.5, 23.75], [23.75, 30]. So the first niche will contain those individuals with a length between 5 and 11.25 and so on. The number of niches to create is decided by the user and this number is kept constant throughout the evolutionary process. Nonetheless, the lengths intervals that they cover will change during the evolution since the maximum and minimum lengths of the individuals in the population will change. This allows for a shifting of the individuals between contiguous niches, helping to maintain diversity.

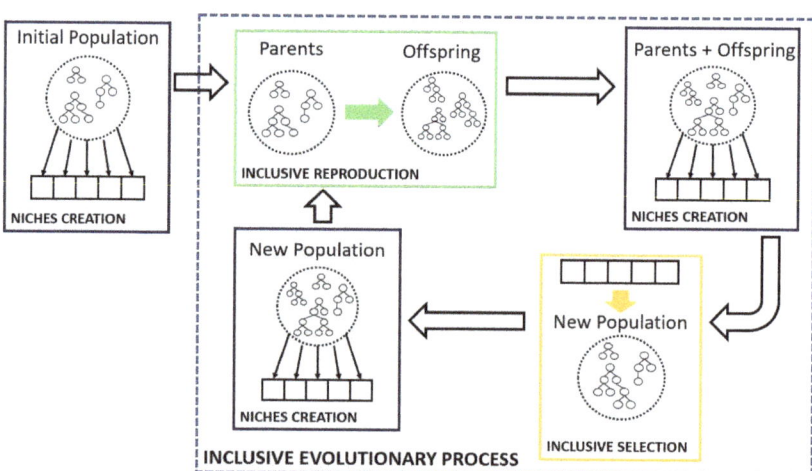

Figure 4. Schematic representation of the Inclusive Evolutionary Process.

About the Inclusive Reproduction, it is designed to consider all the niches during the mating process. To perform crossover, two niches are selected from a list of available niches which is updated as the process goes on. Such an update consists in removing the selected niches and when all of them are selected, the list of available niches is reset to its initial state. Regarding the two selected niches, the best performing individual is selected from the first niche and a random one is selected from the second, this to consider the best performing individuals but also to avoid losing diversity. When mutation is selected, a niche is selected from the list of the available niches and such list is updated as said above. Then a random individual is chosen from the list. The individual is chosen randomly since applying mutation to the best performing ones does not guarantee an improvement. While, when 1:1 reproduction is selected, the best individuals is passed unaltered to the offspring. These selection criteria are summarized in Table 1.

Finally, the Inclusive Tournament consists in applying a Double Tournament [34] to all the niches sequentially, where each niche can be considered at most t times, where t is the number of individuals inside them. This is done to reduce the probability of having clones in the population. More on the Inclusive Reproduction and Selection can be found in [30]. Finally, other two peculiarities of the IGP in comparison to other GP formulations are its ability to handle constraints and the fact that is set to evolve more than one GP tree simultaneously.

3.2. Inclusive Genetic Programming Settings

The ability of the IGP to handle constraints and to evolve more than one GP tree at the same time, makes it particularly suited to solve control problems. In fact, the possibility to evolve more than one GP tree at the same time, allows to solve a control problem with more than one control parameter, which is often the case for most applications.

To produce the results presented in Section 5, the IGP was set as in Table 1, and two GP trees are evolved simultaneously, one for each control parameter of the considered problem. These will referred as α-individual and σ-individual from this point onward.

Table 1. Settings of IGP algorithm. The percentages near the mutation mechanisms refers to the probability of that mutation mechanism to be chosen when the mutation is performed. The selection criteria refers to the criteria used to select the individuals from the niches when performing crossover, mutation or 1:1 reproduction.

Population Size	300 individuals
Maximum Generations	150
Stopping criteria	Reaching maximum number of generations or successful trajectory found
Number of niches	10
Crossover probability	0.2 (+0.01 at every generation if $Fit_2 = 0$) $\rightarrow$ 0.65
Mutation probability	0.7 (−0.01 at every generation if $Fit_2 = 0$) $\rightarrow$ 0.35
1:1 Reproduction probability	0.1
Crossover selection criteria	Best, random
Mutation selection criteria	Random
1:1 Reproduction selection criteria	Best
Evolutionary strategy	$\mu + \lambda$
μ	Population size
λ	Population Size $\times$ 1.2
Number of Ephemeral constants	2
Limit Height	30
Limit Size	50
Selection Mechanism	Inclusive Tournament
Double Tournament fitness size	2
Double Tournament parsimony size	1.6
Tree creation mechanism	Ramped half and half
Mutation mechanisms	Uniform (50%), Shrink (5%), Insertion (25%), Mutate Ephemeral (20%)
Crossover mechanism	One point crossover
Primitives Set	$+, -, *, add3, tanh, psqrt$ $plog, pexp, sin, cos$
Fitness measures	Fit_1, Fit_2

Regarding the fitness functions, they are designed to solve the considered problem. The vehicle's final position must be inside the FAC box, as described in Section 4, while satisfying the imposed constraints on normal acceleration, dynamic pressure and heat rate. To reach this goal, the fitness functions were implemented as in Equation (1). The evolutionary strategy on which the IGP was built, is such that first it favors those individuals that minimize and satisfy the constraints violation. So it first tries to get $Fit_2 = 0$, then it favors those individuals with a better, i.e., lower, value of the fitness Fit_1, which tells how close the final position is to the desired one. RMSE in Equation (1) is the Root Mean Square Error.

$$Fit_1 = RMSE([T_e/100, FAC_R \cdot 1000])$$
$$Fit_2 = RMSE(CV) \qquad (1)$$

In Equation (1), T_e is an array composed as in Equation (2) by the values of the IAE between the reference states and the actual ones evaluated on the last 50 points of the trajectory, as in Equation (3), where e_x is evaluated as in Equation (4). For more details on the states and the whole test case, see Section 4.

$$T_e = [IAE_{50}(V), IAE_{50}(\chi), IAE_{50}(\gamma), IAE_{50}(\theta), IAE_{50}(\lambda), IAE_{50}(h)] \qquad (2)$$

$$IAE_{50} = \int_{t_f}^{t_f-50} |e_x| dt \qquad (3)$$

$$e_x = \begin{cases} \frac{x_{ref}-x}{x_{max}} & \text{if } x \text{ is } V, \chi, \gamma, h \\ x_{ref} - x & \text{if } x \text{ is } \theta, \lambda \end{cases} \qquad (4)$$

The integrals inside T_e were not evaluated on the whole trajectory but only on its last 50 points. This was done to encode in the fitness function the instruction for the GP to find a control law that produces a trajectory similar to the reference one in the last part, but it leaves it freedom on the rest of it, except for the constraints satisfaction. This way, the GP has more freedom to find a control law that satisfies the constraints but does not necessarily needs to track exactly the reference trajectory, which might be impossible due to the presence of uncertainties. In fact, the trajectory is considered successful if the constraints are satisfied and the vehicle's final position is inside the FAC box. This last information is encoded in the FAC_R array of Equation (1) which is composed as in Equation (5) and where the errors are evaluated as in Equation (4).

$$FAC_R = [e_{\theta_f}, e_{\lambda_f}, e_{h_f}/10] \qquad (5)$$

The FAC_R array contains the errors between the final value of h, λ and θ and their final value of reference. The different way to evaluate the errors on the states, as shown in Equation (4), is done to balance their values since the errors on λ and θ tends to be much smaller that those of the other states. For the same reason, e_{h_f} was divided by 10 in Equation (5). Moreover, to make the FAC requirement more important than the minimization of the tracking errors, in Equation (1) FAC_R was multiplied by 1000 and T_e was divided by 100. This combination of scaling factors is an heuristic that was defined experimentally, so it could be improved. Nonetheless, it is important since the magnitude of the different errors and IAEs varies of several orders of magnitude.

Fit_2 is implemented in a more straightforward way. CV in Equation (1) represents an array composed by the constraints violations measured during the evaluation of the trajectory. The constraints violation is evaluated as the error between the constrained quantity and its maximum or minimum value. If such an array is empty, meaning that the trajectory satisfies all the constraints, Fit_2 is set to 0.

About the other features listed in Table 1, the mutation and crossover rates are changed dynamically during the evolutionary process. The mutation rate starts at 0.7 and it is decreased by 0.01 at every generation while the crossover rate starts at 0.2 and it is increased by 0.01 at every generation, until the mutation rate becomes 0.35 and the crossover rate 0.65. This change is done if at least one individual with $Fit_2 = 0$ is found at the current generation. The rationale behind this approach is that exploration is favored at the beginning of the evolutionary process and until at least one individual that satisfies the imposed constraints is found. While, exploitation is favored towards the end of the evolutionary process to find better individuals. Then, limit height and size are two parameters used by the bloat control operator, which is the same implemented in the DEAP library but modified to take into account individuals composed by multiple GP trees. Moreover, the crossover and mutation operations, which are the same of the DEAP library, are modified to take into account individuals composed by multiple GP trees. Four different mutation mechanisms were used as listed in Table 1 and the probability of them being selected when mutation is applied is reported in the brackets near them. Finally, about the operations listed in the primitive set, $add3$ is a ternary addition, $psqrt$, $plog$ and $pexp$ are protected square root, logarithm and exponential to avoid numerical errors.

4. Test Case

To test the proposed control approach, a reentry mission of the FESTIP-FSS5 RLV is simulated. This vehicle was chosen since by being a spaceplane, it possess more control capabilities than a classical rocket and its guidance control during the reentry phase is particularly challenging due to the rapidly varying aerodynamic forces. The model of the vehicle, as well with the aerodynamic and atmospheric models, were taken from [35] and it is characterized by a lifting body and an aerospike engine. For the considered mission, two control parameters were considered, the angle of attack α and the bank angle σ. The aerodynamics models are composed by two lookup tables that give the values of c_l and c_d as a function of the $Mach$ and angle of attack. These data were obtained from experimental data, hence they form a nonlinear model. The aerodynamic data were smoothed with the CSG approach described in [36]. The atmospheric model is the USSA-1962 model. Regarding the mission, its details were taken from [37] and they are summarized in Table 2.

Table 2. Summary of the reentry mission.

States	$V, \chi, \gamma, \theta, \lambda, h$
Controls	α, σ
Initial	$V_0 = 2600$ m/s,
Conditions	$\chi_0 = 0$ deg,
	$\gamma_0 = -1.3$ deg
	$\theta_0 = -85$ deg,
	$\lambda_0 = 30$ deg,
	$h_0 = 51$ km,
Final	$V_f = 91.44$ m/s,
Conditions	$\chi_f = -60$ deg,
	$\gamma_f = -6$ deg,
	$\theta_f = -80.7112 \pm 0.0014$ deg,
	$\lambda_f = 28.6439 \pm 0.0014$ deg,
	$h_f = 609.6 \pm 121.92$ m
Controls	-2 deg $\leq \alpha \leq 40$ deg
Bounds	-90 deg $\leq \sigma \leq 90$ deg,
Constraints	-25 m/s$^2 \leq a_z \leq 25$ m/s^2,
	$q \leq 40$ kPa,
	$\dot{Q} \leq 4$ MW/m^2

As described in [37], the final conditions corresponds to the target FAC box so to have an horizontal landing at the Space Shuttle Landing Facility at the NASA Kennedy Space Center. The FAC is defined by the final values of h, λ, θ and their respective tolerances. A trajectory was considered successful if the final position of the vehicle was inside the FAC box.

Regarding the constraints in Table 2, the one on the dynamic pressure q comes from the original formulation of the problem from [35]. The constraint on the normal acceleration a_z was increased from 15 to 25 m/s^2, to correspond to a load factor of about 2.5 g's. This choice was made to be more in line with the load factor constraint imposed in [37], since the aim was to fly a similar trajectory. The heat rate constraint was taken from [38] and was modeled as in Equation (6)

$$\dot{Q} = C\sqrt{\rho}V^3(1 - 0.18\sin^2 \Lambda)\cos\Lambda \quad (6)$$

where $C = 9.12 \times 10^{-4}$ kg$^{0.5}$m$^{1.5}$s^{-3} and $\Lambda = 45$ deg. This heat rate model was chosen since no data of the FESTIP-FSS5 vehicle were available, and as reported in [38], the chosen heat rate model is a conservative approach. For more details on the heat rate model, please refer to [38].

4.1. Reference Trajectory

The trajectory used as reference for the controller was obtained by solving an optimal control problem using a Multiple-Shooting transcription. The initial guess research for the Multiple-Shooting was performed with the approach presented in [36]. The objective function was set to minimize the difference between the actual final position and the desired one while satisfying the constraints in Table 2. The obtained reference trajectory is shown in Figure 5.

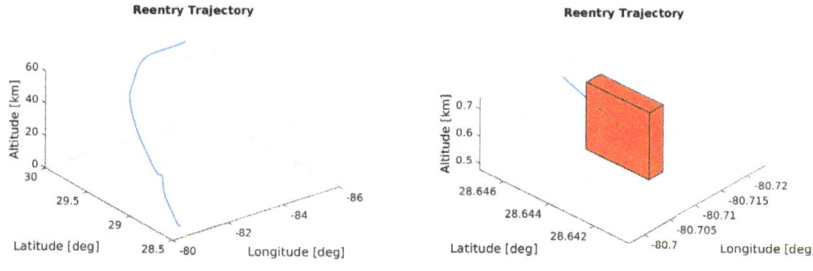

(a) Reference trajectory (b) Close up on FAC box

Figure 5. Reference trajectory and close up on its final part. The red box highlights the FAC box.

4.2. Uncertainty Model

The control mission is simulated by inserting uncertainties into the aerodynamic and atmospheric models. The formulation of the uncertainties was taken from [39]. According to these models, the uncertainties were formulated by creating different sets of interpolating surfaces which were applied to the atmospheric and aerodynamic models to make them vary within a certain range from their unperturbed values, as in Equation (7).

$$x_{unc}(t) = x_{nom}(t)[r(t)((1+\epsilon) - (1-\epsilon)) + (1-\epsilon)] \quad (7)$$

In Equation (7), x_{nom} is the nominal value of the quantity on which the uncertainty is applied, ϵ is the uncertainty bounding parameter and r is a random parameter varying in the interval [0, 1]. Using Equation (7), the quantity nominal value is randomly varied in the interval $[1-\epsilon, 1+\epsilon]$. The random value r is picked from an uncertainty profile obtained by interpolating 50 points randomly generated and uniformly distributed as in Figure 6.

Here, the time starts at 100 s since the uncertainties were inserted at 100 s, as explained in Section 5.

In this work, three interpolating surfaces, for the pressure, temperature and aerodynamics, were created for each different uncertainty scenario in the uncertainty set. Each uncertainty set is composed of 20 uncertainty scenarios and 11 uncertainty sets were created in this work, for a total of 220 uncertainty scenarios. According to what was done in [39], three ϵ parameters were evaluated, one for each perturbed quantity. The reference values of the upper and lower bounds, and the other quantities used to evaluate the ϵ parameters are the same as in [39], and are summarized in Table 3, with the exception of h_c, M_c and α_c which were changed according to the considered problem.

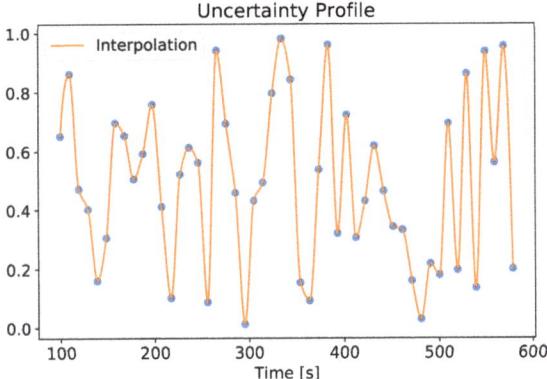

Figure 6. Example of an uncertainty profile.

Table 3. Reference values used to evaluate ϵ as in [39].

h_c	121,920 m
M_c	10
α_c	40 deg
$l_{b,T}$	0.1
$u_{b,T}$	0.5
$l_{b,P}$	0.01
$u_{b,P}$	0.5
$l_{b,h}$	0.1
$u_{b,h}$	0.2
$l_{b,M}$	0.1
$u_{b,M}$	0.2
$l_{b,\alpha}$	0.1
$u_{b,\alpha}$	0.2

5. Results

All the results in this work were obtained on a laptop with 16 GB of RAM and an Intel® Core™ i7-8750H CPU @ 2.20GHz × 12 threads and multiprocessing was used. The code was developed in Python 3 and it is open source and can be found at https://github.com/strath-ace/smart-ml (Last accessed on 4 April 2021).

The methodology to test the proposed control approach was structured so to prove the ability of GP to find an adequate control law in the presence of various levels of uncertainties. These uncertainties were applied after 100 s from the start of the mission. After the applications of the uncertainties, the trajectory was flown using the reference control values until one of the states went outside the range of 1% from its reference value. When that happened, the GP evaluation was started. This was done to simulate the scenario where the system has to adapt during flight to changes in the environment and plant models.

To define the level of uncertainty, the ϵ parameter in Equation (7) was varied. As explained in Section 4, three ϵ parameters were used, one for each perturbed quantity. These parameters were varied by increasing the boundaries used to evaluate them, by 10% in each different simulation. The starting values are those listed in Table 3 and they correspond to those used in Simulation 0 in Table 4. Then they were increased by 10% in Simulation 1, 20% in Simulation 2 and so on. These settings are summarized in Table 4. For each simulation a different uncertainty set was produced which was composed by 20 different uncertainty profiles, where an example of uncertainty profile is shown in Figure 6. Therefore, 20 GP runs were performed for each different simulation and a total of 11 different simulations was performed, so up to 100% increment in the bounds values. This lead to a total of 220 GP runs performed. Twenty uncertainty profiles were produced for each uncertainty set so to avoid having too big computational times and a total of 11 simulations was performed, since it was physically meaningless to have too big uncertainties.

Table 4. Bounds values for each simulation.

	$l_{b,T}$	$u_{b,T}$	$l_{b,P}$	$u_{b,P}$	$l_{b,h}$	$u_{b,h}$	$l_{b,M}$	$u_{b,M}$	$l_{b,\alpha}$	$u_{b,\alpha}$
Simulation 0	0.1	0.5	0.01	0.5	0.1	0.2	0.1	0.2	0.1	0.2
Simulation 1	0.11	0.55	0.011	0.55	0.11	0.22	0.11	0.22	0.11	0.22
Simulation 2	0.12	0.6	0.012	0.6	0.12	0.24	0.12	0.24	0.12	0.24
Simulation 3	0.13	0.65	0.013	0.65	0.13	0.26	0.13	0.26	0.13	0.26
Simulation 4	0.14	0.7	0.014	0.7	0.14	0.28	0.14	0.28	0.14	0.28
Simulation 5	0.15	0.75	0.015	0.75	0.15	0.3	0.15	0.3	0.15	0.3
Simulation 6	0.16	0.8	0.016	0.8	0.16	0.32	0.16	0.32	0.16	0.32
Simulation 7	0.17	0.85	0.017	0.85	0.17	0.34	0.17	0.34	0.17	0.34
Simulation 8	0.18	0.9	0.018	0.9	0.18	0.36	0.18	0.36	0.18	0.36
Simulation 9	0.19	0.95	0.019	0.95	0.19	0.38	0.19	0.38	0.19	0.38
Simulation 10	0.2	1.0	0.02	1.0	0.2	0.4	0.2	0.4	0.2	0.4

Figures 7 and 8 show a qualitative depiction of some of the profiles of the perturbed quantities. The profiles on these pictures come from the first GP run of each simulation. The continuous lines represent the successful cases while the dashed lines the unsuccessful ones. The shaded ares represent the regions contained between the uncertainty bounds and in which the perturbed quantities could have varied when applying the disturbance profiles like the one in Figure 6. As already said, Figures 7 and 8 are purely qualitative but they give an idea of the magnitude of the applied uncertainties. In particular, in Figure 7b the progressive increase of the uncertainty bounds from one simulation to the next is clearly visible.

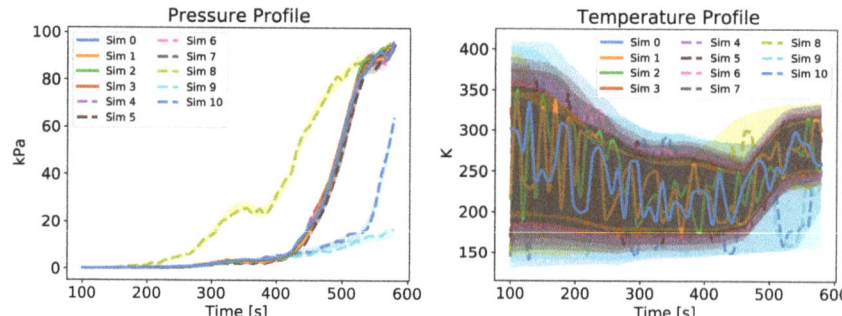

(a) Perturbed pressure profiles   (b) Perturbed temperature profiles

Figure 7. Perturbed pressure and temperature profiles of one GP run for each different simulation. The continuous lines represent the successful cases and the dashed one the unsuccessful ones. The time starts at 100 s since the uncertainties were applied at 100 s.

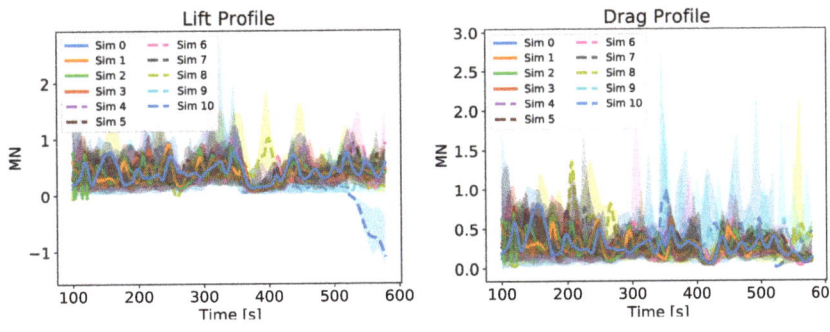

(a) Perturbed lift profiles (b) Perturbed drag profiles

Figure 8. Perturbed lift and drag profiles of one GP run for each different simulation. The continuous lines represent the successful cases and the dashed one the unsuccessful ones. The time starts at 100 s since the uncertainties were applied at 100 s.

As shown in Table 1, each GP simulation was run until a successful trajectory was found or it was stopped after 150 generations. As stated above, a trajectory was considered successful if the final position of the vehicle was within the FAC box and the constraints were satisfied for the whole trajectory. A qualitative example of the performed trajectories is depicted in Figures 9–11. Here the performed trajectories of the 20 GP runs for Simulations 0, 5 and 10 are shown. In plots Figures 9a–11a the full reentry trajectories are plotted in terms of altitude h, longitude θ and latitude λ, where the red dashed line is the reference trajectory, the green lines are the successful trajectories and the black lines are the failed trajectories. These trajectories show the three-dimensional path of the vehicle during the reentry phase and it is clear how, when increasing the magnitude of the applied uncertainties, the vehicle tends to stray from the reference trajectory. The plots Figures 9b–11b show a closeup on the final part of the trajectory to highlight the FAC box and how close the obtained trajectories got to it. The FAC box is projected onto its components θ, λ and h and it is represented by the black horizontal lines. Despite being a qualitative depiction, it can be seen from Figures 9–11 how the performed trajectories deviates more and more from the reference as the magnitude of the applied uncertainties increases.

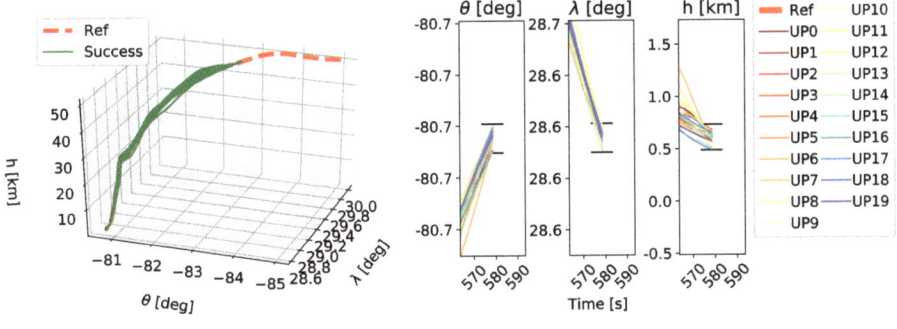

(a) Simulation 0 results (b) Simulation 0 results-closeup

Figure 9. Trajectories obtained as a results of the 20 GP runs in Simulation 0. Plot *a* shows an overview of the performed trajectories while plot *b* depicts a closeup on the final part of the trajectory to show which trajectory ended inside the FAC box and which did not. The black horizontal lines in plot *b* show the FAC bounds for θ, λ and h. UP refers to the different uncertainty profiles.

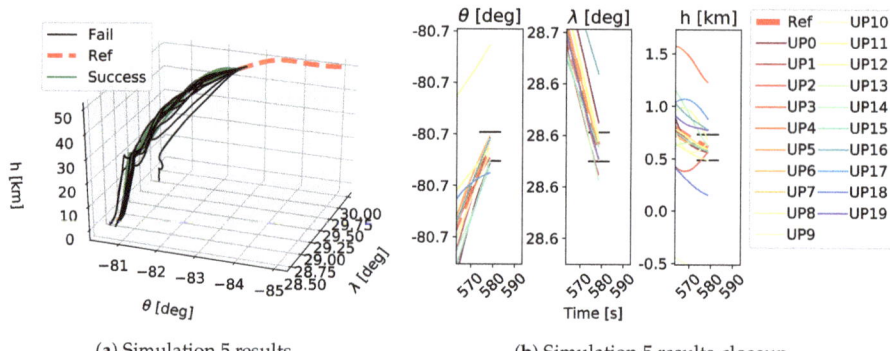

(a) Simulation 5 results

(b) Simulation 5 results-closeup

Figure 10. Trajectories obtained as a results of the 20 GP runs in Simulation 5. Plot *a* shows an overview of the performed trajectories while plot *b* depicts a closeup on the final part of the trajectory to show which trajectory ended inside the FAC box and which did not. The black horizontal lines in plot *b* show the FAC bounds for θ, λ and h. UP refers to the different uncertainty profiles.

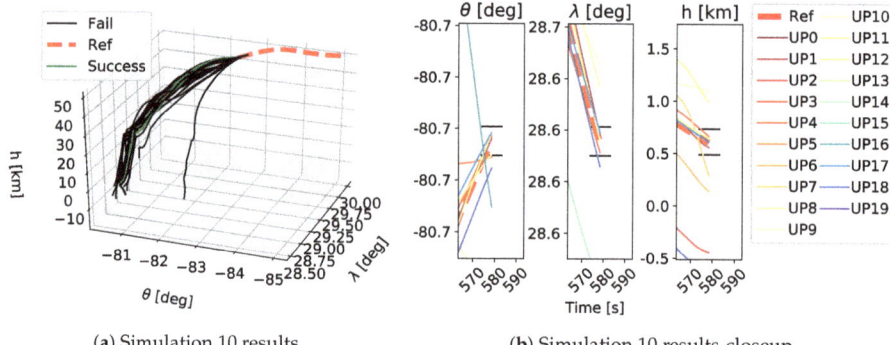

(a) Simulation 10 results

(b) Simulation 10 results-closeup

Figure 11. Trajectories obtained as a results of the 20 GP runs in Simulation 10. Plot *a* shows an overview of the performed trajectories while plot *b* depicts a closeup on the final part of the trajectory to show which trajectory ended inside the FAC box and which did not. The black horizontal lines in plot *b* show the FAC bounds for θ, λ and h. UP refers to the different uncertainty profiles.

For a more quantitative point of view, the number of successes for each simulation is shown in Figure 12. Here, it is clear that with the reference values of the uncertainty bounds, the GP is able to find a successful control law 100% of the times and this success rate tends to decrease as the uncertainty bounds increase, as expected. Three cases go against this trend, namely Simulations 1, 6 and 10. This is probably due to the fact that 20 GP runs for each simulation are not enough to constitute a statistically significant sample. Therefore by increasing the number of GP runs for each simulation, a more accurate trend could be achieved. Nonetheless, the obtained results are useful to understand the extent of the capabilities of the GP. In fact, even with an increase of 100% of the uncertainty bounds, the GP is capable of finding four good control laws.

This success trend is also highlighted in Figure 13, where the median and standard deviation values of the first fitness function Fit_1 are shown. Here the median fitness value increases as the uncertainty bounds increase and the smaller value is found at Simulation 0. Moreover, the standard deviation also increases as the uncertainty bounds increase, as expected. A different perspective of the same phenomenon is observable in Figure 14. Here the number and magnitude of the constraints violation for each simulation are shown. The bar plots show the number of GP runs whose best individual violated the constraints, for each simulation. For example, in Simulation 3, two individuals out of 20 violated the constraints. The line plot refers to the mean value of the constraints violation for each simulation. These results show an increase in the GP runs that violated the constraints as the uncertainty bounds increase. Concurrently, the magnitude of constraints violation also increases as the uncertainty bounds increases. Nonetheless, as said before, the number of GP runs for each simulation was not enough to have a proper statistical sample.

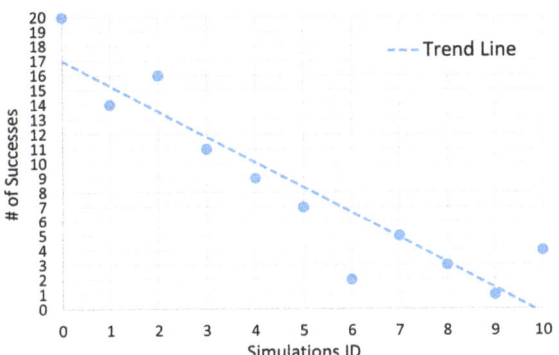

Figure 12. Number of successful GP evaluations on each simulation.

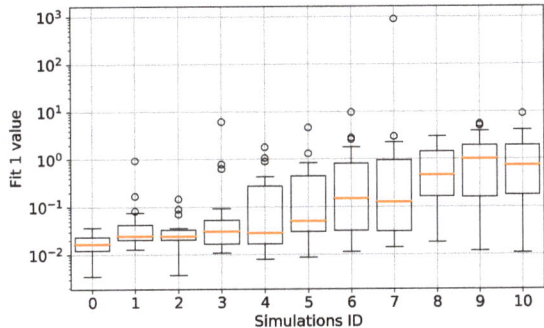

Figure 13. Median an standard deviations of the Fit_1 values evaluated on the 20 different GP runs for each simulation.

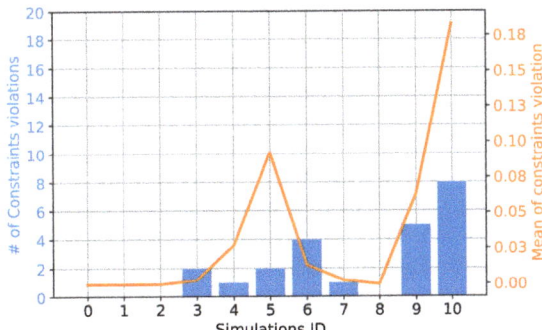

Figure 14. Number of solutions that violated the constraints for each simulation and amount of the constraints violation. The left ordinate axis refers to the bar plots which shows the number of GP runs that violated the constraints. The right ordinate axis refers to the line plot and shows the mean constraints violation among the 20 GP runs for each simulation.

To have a greater insight into the produced individuals from a genotypic perspective the lengths of the individuals were analyzed. In Figure 15, the lengths of the successful and unsuccessful individuals for each simulation are plotted. As said in Section 3, two individuals were simultaneously produced in each GP run, since two control parameters had to be found. The lengths of the α-individuals are plotted in Figure 15a and those of the σ-individuals are depicted in Figure 15b. These lengths are represented by the green dots for the successful individuals and the red crosses for the unsuccessful ones. On top of these, the mean length of the successful α and σ individuals are plotted to observe how they evolved as the magnitude of the applied uncertainties increased. From the lengths of each particular individual found in the various simulations, it can be observed that they vary greatly but the majority of the successful individuals are below 100 nodes. Moreover, small successful individuals are still found also when the magnitude of the uncertainties increases greatly. Considering the mean length of the α and σ individuals, it can be observed that they oscillate around 50 nodes but no particular increase in the length of the individuals is observed as the magnitude of the uncertainties increases. It was expected to observe an increase of the length of the individuals as the uncertainties become more severe. With GP usually the complexity of the produced models increases if the data to fit are distributed in a more complex manner, also causing problems of overfitting. In contrast, what was produced highlights how the IGP is capable of maintaining the mean length of the successful individuals approximately constant even when the magnitude of the uncertainties increases.

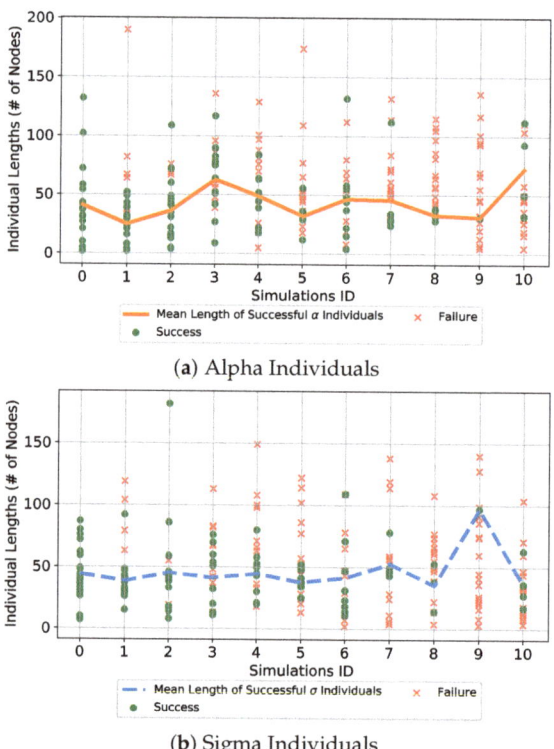

(a) Alpha Individuals

(b) Sigma Individuals

Figure 15. Lengths of successful and unsuccessful individuals for each simulation. The green dots represent the successful individuals while the red crosses the unsuccessful ones. (**a**) depicts the first control parameter α, and the second control parameter σ is in (**b**). In (**a**) the continuous orange line represents the mean values of the lengths of the successful α individuals for each simulation, while in (**b**) the dashed blue line represents the mean values of the lengths of the successful σ individuals.

The obtained results show how GP is capable of generating a good guidance control law even when severe uncertainties in the environment and aerodynamic models are considered. Current research directions focus more on classical approaches to generate the guidance control law rather than using an AI technique, and not all of them are tested considering uncertainties and/or disturbances. As an example, Zang et al. [40] propose a guidance algorithm based on the height-range (H-R) and height-velocity (H-V) joint design method that is able to generate online a reference trajectory. Unfortunately, the robustness of such an approach is not tested against disturbances and/or uncertainties. Hence the approach using GP appears to be more robust.

Another guidance control approach which does not involve AI is the one presented by Wang et al. [41]. Their approach consists in performing an online generation and tracking of an optimal reentry trajectory using convex optimization. Such an approach proves to be successful in generating and tracking an optimal trajectory online even considering disturbances. It would be interesting to compare the latter approach with the GP guidance controller presented in this work, to understand which solution is the more robust and versatile. Moreover, it would also be interesting to compare the effort needed to apply both algorithms to different systems, so to understand their degree of generalizability.

6. Conclusions

In this work, a guidance control algorithm based on GP for the reentry phase of the FESTIP-FSS5 RLV was developed. The reentry trajectory was flown in the presence of uncertainties in the atmospheric and aerodynamic models. Eleven simulations were performed where, starting from Simulation 1, the bounds of the uncertainty ranges were increased progressively by 10% in each subsequent simulation, until an increase of 100% was reached in Simulation 10. For each simulation, a total of 20 GP runs was performed considering 20 different uncertainty profiles, for a total of 220 GP runs. The particular GP algorithm employed is the Inclusive Genetic Programming algorithm, whose characteristic is the ability to maintain the population's diversity throughout the evolutionary process. Moreover, it is also set to handle constraints and to evolve more than one GP tree simultaneously, one for each control law of the considered problem. The results show that the IGP is capable of always producing at least one successful individual even when the uncertainty bounds were increased 100% from their reference value. Moreover, by analyzing the size of the produced individuals, the IGP is capable of keeping the length of the successful individuals approximately constant as the magnitude of the uncertainties increases. Meaning that the size of the successful individuals does not become too big in order to better fit a more complex distribution of data. Overall, the shown results demonstrate that GP can be used successfully to generate the guidance control law of a reentry vehicle which can deal with nonlinearities and big uncertainties in the considered models while satisfying a set of imposed constraints.

The GP evaluation was done offline, since it is a computationally expensive process but, as the hardware technologies improve, it will be possible to perform such operations online so as to constitute what is defined as Intelligent Control. Such a controller would be able to deal in real time with unforeseen disturbances and to take into account the uncertainties in the models considered at the design phase. Future work directions would comprehend both a development of computationally more efficient GP algorithms and different ways to apply them. In fact, in this work the guidance control was performed but GP could also be employed for an attitude control task.

Author Contributions: Conceptualization, F.M. and E.M.; methodology, F.M. and E.M.; software, F.M.; validation, F.M.; formal analysis, F.M.; investigation, F.M. and E.M.; resources, F.M.; writing—original draft preparation, F.M.; writing—review and editing, F.M. and E.M.; visualization, F.M.; supervision, E.M. Both authors have read and agreed to the published version of the manuscript.

Funding: This research received no external funding.

Institutional Review Board Statement: Not applicable.

Informed Consent Statement: Not applicable.

Conflicts of Interest: The authors declare no conflict of interest.

Abbreviations

The following abbreviations are used in this manuscript:

AI	Artificial Intelligence
CI	Computational Intelligence
CSG	Cubic Spline Generalization
EA	Evolutionary Algorithm
FAC	Final Approach Corridor
GA	Genetic Algorithm
GP	Genetic Programming
IAE	Integral of Absolute Error
IC	Intelligent Control
IGP	Inclusive Genetic Programming
ML	Machine Learning
MPC	Model Predictive Control
NN	Neural Network
RLV	Reusable Launch Vehicle

References

1. Xie, Y.C.; Huang, H.; Hu, Y.; Zhang, G.Q. Applications of advanced control methods in spacecrafts: Progress, challenges, and future prospects. *Front. Inf. Technol. Electron. Eng.* **2016**, *17*, 841–861. [CrossRef]
2. Haibin, D.; Pei, L. Progress in control approaches for hypersonic vehicle. *Sci. China Technol. Sci.* **2012**, *55*, 2965–2970. [CrossRef]
3. Xu, B. Robust adaptive neural control of flexible hypersonic flight vehicle with dead-zone input nonlinearity. *Nonlinear Dyn.* **2015**, *80*, 1509–1520. [CrossRef]
4. Chai, R.; Tsourdos, A.; Savvaris, A.; Chai, S.; Xia, Y.; Philip Chen, C.L. Review of advanced guidance and control algorithms for space/aerospace vehicles. *Prog. Aerosp. Sci.* **2021**, *122*, 100696. [CrossRef]
5. Izzo, D.; Märtens, M.; Pan, B. A survey on artificial intelligence trends in spacecraft guidance dynamics and control. *Astrodynamics* **2019**, *3*, 287–299. [CrossRef]
6. Koza, J.R. *Genetic Programming: On the Programming of Computers by Means of Natural Selection*; MIT Press: Cambridge, MA, USA, 1992.
7. Koza, J.; Keane, M.; Yu, J.; Bennett, F., III; Mydlowec, W. Automatic Creation of Human-Competitive Programs and Controllers by Means of Genetic Programming. *Genet. Program. Evolvable Mach.* **2000**, *1*, 121–164. [CrossRef]
8. Baeta, F.; Correia, J.; Martins, T.; Machado, P. TensorGP—Genetic Programming Engine in TensorFlow. In *Applications of Evolutionary Computation. EvoApplications 2021*; Springer: Cham, Switzerland, 2021.
9. Wilson, C.; Marchetti, F.; Di Carlo, M.; Riccardi, A.; Minisci, E. Classifying Intelligence in Machines: A Taxonomy of Intelligent Control. *Robotics* **2020**, *9*, 64. [CrossRef]
10. Luo, Y.; Serrani, A.; Yurkovich, S.; Oppenheimer, M.W.; Doman, D.B. Model-predictive dynamic control allocation scheme for reentry vehicles. *J. Guid. Control Dyn.* **2007**, *30*, 100–113. [CrossRef]
11. Wang, J.; Cui, N.; Wei, C. Optimal Rocket Landing Guidance Using Convex Optimization and Model Predictive Control. *J. Guid. Control Dyn.* **2019**, *42*, 1078–1092. [CrossRef]
12. Bollino, K.P.; Ross, I.M. A pseudospectral feedback method for real-time optimal guidance of reentry vehicles. In Proceedings of the American Control Conference, New York, NY, USA, 9–13 July 2007; pp. 3861–3867. [CrossRef]
13. Tian, B.; Fan, W.; Zong, Q. Real-Time Trajectory and Attitude Coordination Control for Reusable Launch Vehicle in Reentry Phase. *IEEE Trans. Ind. Electron.* **2015**, *3*, 1639–1650. [CrossRef]
14. Harl, N.; Balakrishnan, S.N. Reentry terminal guidance through sliding mode control. *J. Guid. Control Dyn.* **2010**, *33*, 186–199. [CrossRef]
15. Guo, Z.; Chang, J.; Guo, J.; Zhou, J. Adaptive twisting sliding mode algorithm for hypersonic reentry vehicle attitude control based on finite-time observer. *ISA Trans.* **2018**, *77*, 20–29. [CrossRef]
16. Wu, S.F.; Engelen, C.J.H.; Chu, Q.P.; Bubuška, R.; Mulder, J.A.; Ortega, G. Fuzzy logic based attitude control of the spacecraft X-38 along a nominal re-entry trajectory. *Control Eng. Pract.* **2001**, *9*, 699–707. [CrossRef]
17. Gao, J.; Shi, X.; Cheng, Z.; Xiong, J.; Liu, L.; Wang, Y.; Yang, Y. Reentry trajectory optimization based on Deep Reinforcement Learning. In Proceedings of the 31st Chinese Control and Decision Conference, CCDC 2019, Nanchang, China, 3–5 June 2019; pp. 2588–2592. [CrossRef]
18. Kumar, G.N.; Sarkar, A.K.; Ahmed, M.S.; Talole, S.E. Reentry Trajectory Optimization using Gradient Free Algorithms. *IFAC-PapersOnLine* **2018**, *51*, 650–655. [CrossRef]

29. Duan, H.; Li, S. Artificial bee colony-based direct collocation for reentry trajectory optimization of hypersonic vehicle. *IEEE Trans. Aerosp. Electron. Syst.* **2015**, *51*, 615–626. [CrossRef]
30. Wu, Y.; Yan, B.; Qu, X. Improved Chicken Swarm Optimization Method for Reentry Trajectory Optimization. *Math. Probl. Eng.* **2018**, *2018*, 8135274. [CrossRef]
31. Vijay, D.; Bhanu, U.S.; Boopathy, K. Multi-objective genetic algorithm-based sliding mode control for assured crew reentry vehicle. *Adv. Intell. Syst. Comput.* **2017**, *517*, 465–477. [CrossRef]
32. Sushnigdha, G.; Joshi, A. Evolutionary method based integrated guidance strategy for reentry vehicles. *Eng. Appl. Artif. Intell.* **2018**, *69*, 168–177. [CrossRef]
33. Oh, C.K.; Barlow, G.J. Autonomous controller design for unmanned aerial vehicles using multi-objective genetic programming. In Proceedings of the 2004 Congress on Evolutionary Computation (IEEE Cat. No.04TH8753), Portland, OR, USA, 19–23 June 2004; Volume 2, pp. 1538–1545. [CrossRef]
34. Bourmistrova, A.; Khantsis, S. Genetic Programming in Application to Flight Control System Design Optimisation. In *New Achievements in Evolutionary Computation*; Books on Demand: Norderstedt, Germany, 2010.
35. Verdier, C.F.; Mazo, M., Jr. Formal Controller Synthesis via Genetic Programming. *IFAC-PapersOnLine* **2017**, *50*, 7205–7210. [CrossRef]
36. Łapa, K.; Cpałka, K.; Przybył, A. Genetic programming algorithm for designing of control systems. *Inf. Technol. Control* **2018**, *47*, 668–683. [CrossRef]
37. Chiang, C.H. A genetic programming based rule generation approach for intelligent control systems. In Proceedings of the 3CA 2010—2010 International Symposium on Computer, Communication, Control and Automation, Tainan, Taiwan, 5–7 May 2010; Volume 1, pp. 104–107. [CrossRef]
38. Marchetti, F.; Minisci, E.; Riccardi, A. Towards Intelligent Control via Genetic Programming. In Proceedings of the International Joint Conference on Neural Networks (IJCNN), Glasgow, UK, 19–24 July 2020.
39. Marchetti, F.; Minisci, E. A Hybrid Neural Network-Genetic Programming Intelligent Control Approach. In *International Conference on Bioinspired Methods and Their Applications, Proceedings of the 9th International Conference, BIOMA 2020, Brussels, Belgium, 19–20 November 2020*; Filipič, B., Minisci, E., Vasile, M., Eds.; Springer: Cham, Switzerland, 2020.
40. Marchetti, F.; Minisci, E. Inclusive Genetic Programming. In *Proceedings of the 24th European Conference on Genetic Programming, EuroGP 2021, Virtual Event, 7–9 April 2021*; Lecture Notes in Computer Science; Springer: Cham, Switzerland, 2021.
41. Fortin, F.A.; De Rainville, F.M.; Gardner, M.A.; Parizeau, M.; Gagńe, C. DEAP: Evolutionary algorithms made easy. *J. Mach. Learn. Res.* **2012**, *13*, 2171–2175.
42. Shir, O.M. Niching in Evolutionary Algorithms. In *Handbook of Natural Computing*; Rozenberg, G., Bäck, T., Kok, J.N., Eds.; Springer: Berlin/Heidelberg, Germany, 2012; pp. 1035–1069. [CrossRef]
43. Beyer, H.G.; Schwefel, H.P. Evolution strategies—A comprehensive introduction. *Nat. Comput.* **2002**, *1*, 3–52. [CrossRef]
44. Luke, S.; Panait, L. Fighting bloat with nonparametric parsimony pressure. In *Parallel Problem Solving from Nature—PPSN VII*; Guervós, J.J.M., Adamidis, P., Beyer, H.G., Schwefel, H.P., Fernández-Villacañas, J.L., Eds.; Springer: Berlin/Heidelberg, Germany, 2002; pp. 411–421.
45. D'Angelo, S.; Minisci, E.; Di Bona, D.; Guerra, L. Optimization Methodology for Ascent Trajectories of Lifting-Body Reusable Launchers. *J. Spacecr. Rocket.* **2000**, *37*, 761–767. [CrossRef]
46. Marchetti, F.; Minisci, E.; Riccardi, A. Single-Stage to Orbit Ascent Trajectory Optimisation with Reliable Evolutionary Initial Guess. In *Optimization and Engineering*; Springer: Berlin, Germany, 2021. Available online: https://www.springer.com/journal/11081 (accessed on 2 August 2021).
47. Bollino, K.P.; Ross, I.M.; Doman, D.D. Optimal nonlinear feedback guidance for reentry vehicles. In Proceedings of the Collection of Technical Papers—AIAA Guidance, Navigation, and Control Conference and Exhibit, Keystone, CO, USA, 21–24 August 2006; Volume 1, pp. 563–582. [CrossRef]
48. ul Islam Rizvi, S.T.; He, L.; Xu, D. Optimal trajectory and heat load analysis of different shape lifting reentry vehicles for medium range application. *Def. Technol.* **2015**, *11*, 350–361. [CrossRef]
49. Pescetelli, F.; Minisci, E.; Maddock, C.; Taylor, I.; Brown, R.E. Ascent trajectory optimisation for a single-stage-to-orbit vehicle with hybrid propulsion. In Proceedings of the 18th AIAA/3AF International Space Planes and Hypersonic Systems and Technologies Conference, Tours, France, 24–28 September 2012; pp. 1–18.
50. Zang, L.; Lin, D.; Chen, S.; Wang, H.; Ji, Y. An on-line guidance algorithm for high L/D hypersonic reentry vehicles. *Aerosp. Sci. Technol.* **2019**, *89*, 150–162. [CrossRef]
51. Wang, Z.; Grant, M.J. Autonomous entry guidance for hypersonic vehicles by convex optimization. *J. Spacecr. Rocket.* **2018**, *55*, 993–1006. [CrossRef]

article

Multi-Objective Optimisation under Uncertainty with Unscented Temporal Finite Elements

Lorenzo A. Ricciardi, Christie Alisa Maddock * and Massimiliano Vasile

Aerospace Centre of Excellence, University of Strathclyde, Glasgow G1 1XJ, UK; lorenzo.a.ricciardi@gmail.com (L.A.R.); massimiliano.vasile@strath.ac.uk (M.V.)
* Correspondence: christie.maddock@strath.ac.uk

Abstract: This paper presents a novel method for multi-objective optimisation under uncertainty developed to study a range of mission trade-offs, and the impact of uncertainties on the evaluation of launch system mission designs. A memetic multi-objective optimisation algorithm, named MOD-HOC, which combines the Direct Finite Elements in Time transcription method with Multi Agent Collaborative Search, is extended to account for model uncertainties. An Unscented Transformation is used to capture the first two statistical moments of the quantities of interest. A quantification model of the uncertainty was developed for the atmospheric model parameters. An optimisation under uncertainty was run for the design of descent trajectories for a spaceplane-based two-stage launch system.

Keywords: optimal control; multi-objective optimisation; robust design; trajectory optimisation; uncertainty quantification; unscented transformation; spaceplanes; space systems; launchers

1. Introduction

This paper presents a novel method for multi-objective optimisation under uncertainty, developed to study a range of mission trade-offs and the impact of uncertainties on system models for space launch systems. This is applied to the analysis and design of descent trajectories for a two-stage, partially re-usable launch system based on the Orbital-500R, a commercial system developed by Orbital Access Ltd. (Prestwick, UK) [1]. The set of Pareto-optimal solutions show the trade-off between minimising the induced acceleration limits and maximising the robustness of the solutions by minimising the sensitivity to uncertainties.

Uncertainty quantification (UQ), the science of quantifying the uncertainty in the desired performance of a system, can be a key step in analysing the robustness of a control solution and of the whole guidance, navigation and control chain. Common approaches to UQ use extensive Monte Carlo simulations to account for errors, unmodelled components and disturbances. At a system level, UQ analysis can translate into the assessment of the reliability of the system as a whole, or only of one or more components. An uncertainty quantification analysis is, therefore, a fundamental step towards de-risking any technological solution as it provides a quantification of the variation in performance and probability of recoverable or unrecoverable system failures, given existing information.

The goal here is, therefore, to design a robust guidance trajectory considering the uncertainty related to the atmospheric model which in turn affects the aero-, aerothermal and flight dynamics.

The trajectories are designed using a MODHOC (Multi-Objective Direct Hybrid Optimal Control) solver [2,3]. MODHOC, first developed in collaboration with ESA, is based on a transcription using temporal finite elements (DFET, Direct Finite Elements in Time) of the optimal control problem and a solution of the transcribed problem with a multi-agent multi-objective optimisation algorithm (MACS, Multi-Agent Collaborative Search).

This paper presents an extension of the multi-objective optimal control to account for uncertainties. The extension is based on an unscented transformation to capture the first two statistical moments of the quantities of interest. The result is an unscented multi-objective optimal control approach that can efficiently handle the level of uncertainty in model parameters.

Ross et al. [4] introduced unscented optimal control as the combination of the unscented transform by Julier and Uhlmann [5] combined with deterministic optimal control theory to directly manage uncertainties within an open-loop control framework. This has been applied to single objective optimisation problems mainly in the field of guidance and attitude control [6,7], using pseudospectral optimal control methods [8] with the common Legendre and Chebyhsev polynomials as the bases. Ross et al. [9] later extended the work to account for path constraints on the states and controls. This paper builds on this work by adapting and applying it to a multi-objective optimal control problem, subject to nonlinear boundary and path constraints. The transcription method here uses finite elements on a temporal basis, using Bernstein polynomials, which have been mathematically shown to ensure both the states and controls representations remains feasible over the entire time domain, not just at collocation nodes [10].

The methods presented will be applied to trade-off studies on the first stage of a multi-stage horizontal take-off and landing launch system. The analysis focuses on the unpowered descent trajectory of a spaceplane, starting from the stage separation point at 100 km altitude. Uncertainties are introduced on the atmospheric parameters, which in turn strongly affect the aerodynamics and flight performance of the vehicle. The optimisation will trade-off robustness, by minimising the effect of these uncertainties, against the structural loads induced by the flight dynamics.

The paper is structured first introducing the mathematical method and implementation, followed by the description of the applied test case and results. Specifically, Section 2 describes the integration of an unscented transform into a multi-objective optimal control problem, including the direct transcription method using temporal finite elements with Bernstein bases. Section 3 details the solution of the multi-objective nonlinear programming problem through an adaption of a global evolutionary Multi-Agent Collaborative Search (MACS) algorithm tailored to optimal control problems. Section 4 presents the quantification of the uncertainty on the atmospheric parameters predicted by the International and US-76 Standard Atmospheric models. Section 5 describes the vehicle and environment models for the launch vehicle test case, with Section 6 presenting and discussing the results, including a validation of the methods and results against a Monte Carlo analysis.

2. Unscented Multi-Objective Optimal Control

In order to perform robust optimisation of the trajectory, an Unscented Transformation [5] was included in the formulation of the optimal control problem. An unscented transformation is defined as "the application of a given nonlinear transformation to a discrete distribution of points, computed so as to capture a set of known statistics of an unknown distribution, is referred to as an unscented transformation" [11]. These points are referred to as sigma points.

Unscented transformations capture the first statistical moments, mean and covariance, of the distributions of the states of a system subject to uncertainty and undergoing arbitrary nonlinear transformations by propagating a small number of sigma points. If the system depends on N_{uq} uncertain variables, whose mean and covariances are known, the unscented transformation requires the propagation of $(2N_{uq} + 1)$ samples. The first sigma point takes the mean value for all the uncertain variables, while the others assume the mean plus (or minus) the square root of the matrix of the covariances of the uncertain variables. All the sigma points are propagated simultaneously with the mean and covariance of the final states computed as a weighted combination of the final states of each sigma point.

Let the dynamics of the system be given by

$$\dot{\mathbf{x}} = f(\mathbf{x}(\mathbf{u}, t), \mathbf{u}(t), \mathbf{b}, t) \qquad (1)$$

where **x** is the system state vector, **u** are the controls, t is time, and **b** are additional static (time independent) parameters. Similar to (1), the dynamics of each sigma point χ_i are given by

$$\dot{\chi}_i = f(\chi_i(\mathbf{u},t), \mathbf{u}(t), \mathbf{b}_i, t) \tag{2}$$

where $i = [1, \ldots, (2N_{uq}+1)]$. Each sigma point has a different value for the static variables, its dynamics evolve independently of the other sigma points, but all sigma points are controlled by the same control law **u**. The goal is to find a single control law that, when applied to all sigma points, allows the system to reach a desired final condition and to be optimal, in some sense. The particular values for each static variable $\mathbf{b}_i$ is decided by the application of the Unscented Transformation.

A known problem of the unscented transformation is that it can generate covariance matrices that are not semidefinite positive. To avoid this problem, the Square Root Unscented Transformation [12] was implemented. Algorithmically, it is very similar to the standard Unscented Transformation but differs in the way the samples are generated and has the advantage that the resulting covariance matrix is guaranteed to be semidefinite positive (up to machine precision). The sigma points are computed from the Cholesky factorisation of the covariance matrix, which decomposes the matrix into lower triangular matrix with real and positive diagonal entries, and its conjugate transpose [12].

The problem can be described as follows. Let **X** be a state vector of length $N_\sigma N_x$ defined as

$$\mathbf{X} = [\chi^0, \chi^1, \cdots, \chi^{N_\sigma}]^T \tag{3}$$

where N_σ is the number of sigma points and N_x is the number of states of the system. The state dynamics are then defined as

$$\dot{\mathbf{X}} = \begin{bmatrix} F(\chi_1, \mathbf{u}, \mathbf{b}_0, t) \\ F(\chi_2, \mathbf{u}, \mathbf{b}_1, t) \\ \vdots \\ F(\chi_{N_\sigma}, \mathbf{u}, \mathbf{b}_{N_\sigma}, t) \end{bmatrix} = \mathbf{F}(\mathbf{X}, \mathbf{u}, \mathbf{B}, t) \tag{4}$$

The multi-objective unscented optimal control problem is then formulated as

$$\min_{\mathbf{u} \in U} \mathbf{J}(\mathbf{X}, \mathbf{u}, \mathbf{B}, t) \tag{5}$$

$$\dot{\mathbf{X}} = \mathbf{F}(\mathbf{X}, \mathbf{u}, \mathbf{B}, t) \tag{6}$$

$$\mathbf{g}(\mathbf{X}, \mathbf{u}, \mathbf{B}, t) \geq 0 \tag{7}$$

$$\boldsymbol{\psi}\left(\mathbf{X}(t_0), \mathbf{X}(t_f), \mathbf{u}(t_0), \mathbf{u}(t_f), \mathbf{b}, t_0, t_f\right) \geq 0 \tag{8}$$

$$t \in [t_0, t_f]$$

where $\mathbf{J} = [J_1, \ldots, J_m]^T$ is, in general, a vector function of the state variables $\chi_i : [t_0, t_f] \to \mathbb{R}^{N_x}$, control variables $\mathbf{u} \in L^\infty(U \subset \mathbb{R}^{N_u})$ and time t. Functions **X** belong to the Sobolev space $W^{1,\infty}$, objective functions are $J_i : \mathbb{R}^{2N_x+N_x} \times \mathbb{R}^{N_u} \times [t_0, t_f] \to \mathbb{R}$, $\mathbf{F} : \mathbb{R}^{N_x N_\sigma} \times \mathbb{R}^{N_u} \times [t_0, t_f] \to \mathbb{R}^{N_x N_\sigma}$, algebraic constraint function $\mathbf{g} : \mathbb{R}^{N_x N_\sigma} \times \mathbb{R}^{N_u} \times [t_0, t_f] \to \mathbb{R}^{N_g}$, and boundary condition functions $\boldsymbol{\psi} : \mathbb{R}^{2N_x N_\sigma + 2} \to \mathbb{R}^{N_\psi}$.

Direct Transcription with Temporal Finite Elements

The optimal control problem in (5)–(8) is transcribed into a many-objective, nonlinear programming problem via Direct Transcription with Finite Elements in Time (DFET) [13]. DFET was first proposed by Vasile [14] in 2000, and uses finite elements in time on spectral bases to transcribe the differential and algebraic constraints, and objective function into a set of algebraic equations. The formation allows different bases to be selected for both the states and controls, and for different segments. As a scheme, DFET has been proven to be robust, accurate and flexible [13].

For the continuous optimal control problem in (5), the time domain $\mathcal{T} = [t_0, t_f]$ is decomposed into N finite elements $\mathcal{T}_j(\tau_{j-1}, \tau_j)$, with each element normalised to the interval $[-1, 1]$ through the transformation

$$\tau = 2\frac{t - \frac{1}{2}(t_j - t_{j-1})}{t_j - t_{j-1}} \qquad t_{j-1} \leq t \leq t_j \qquad \text{for } j = 1, \ldots, N. \tag{9}$$

This ensures the domain of the basis functions are consistent irrespective of the element size. The differential constraints in (6) are first recast in weak variation form as

$$\int_{\mathcal{T}_j} \dot{\mathbf{w}}^T \mathbf{X} + \mathbf{w}^T \mathbf{F}(\mathbf{X}, \mathbf{u}, \mathbf{b}, t) \, dt - \mathbf{w}^T(t_j) \mathbf{X}_j^b + \mathbf{w}^T(t_{j-1}) \mathbf{X}_{j-1}^b = 0 \tag{10}$$

where $\mathbf{w}$ are generalised weight functions, and $\mathbf{X}_j^b$ and $\mathbf{X}_{j-1}^b$ are the values of the states at the boundaries of each element. For each element $\mathcal{T}_j$, the states, controls, and weight functions can be parameterised according to the basis functions $\mathbf{f}_{s,j}$ such that:

$$\mathbf{X}_j = \sum_{s=0}^{l_x} f_{s,j}^{X_j}(\tau) \mathbf{X}_{s,j} \tag{11}$$

$$\mathbf{u}_j = \sum_{s=0}^{l_u} f_{s,j}^{u_j}(\tau) \mathbf{u}_{s,j} \tag{12}$$

$$\mathbf{w}_j = \sum_{s=0}^{l_x+1} f_{s,j}^{w_j}(\tau) \mathbf{w}_{s,j} \tag{13}$$

For this paper, Bernstein polynomials are used as the bases functions for all the elements, of order l_x for the states, l_u for the controls and $(l_x + 1)$ for the weights. Bernstein basis polynomials are defined generally as

$$b_{\nu,n}(t) = \binom{n}{\nu} t^\nu (1-t)^{n-\nu} \qquad \text{for } \nu = 0, \ldots, n \text{ and } 0 \leq t \leq 1 \tag{14}$$

where n is the order. Bernstein bases have the advantage of smooth control profiles with no oscillations near discontinuities or step changes, meaning the polynomial representation of both states and controls remains within the feasible set [10,15].

Recasting (10) into Gauss quadrature using the polynomials in (11)–(13) gives

$$\sum_{k=0}^{l_u} \beta_k \left[\dot{\mathbf{w}}_j(\tau_k)^T \mathbf{X}_j(\tau_k) + \mathbf{w}_j(\tau_k)^T \mathbf{F}_j(\tau_k) \frac{\Delta t_j}{2} \right] - \mathbf{w}^T(1) \mathbf{X}_j^b + \mathbf{w}^T(-1) \mathbf{X}_{j-1}^b = 0 \tag{15}$$

where τ_k and β_k are Gauss nodes and weights, respectively, $\Delta t_j = (t_j - t_{j-1})$ and $\mathbf{F}_j(\tau_k)$ is the shorthand notation for $\mathbf{F}(\mathbf{X}_j(\tau_k), \mathbf{u}_j(\tau_k), \mathbf{b}, t(\tau_k))$. Since (15) must be valid for every arbitrary $\mathbf{w}_{s,j}$, this can be written as a system of equations for each element:

$$\sum_{k=0}^{l_u} \beta_k \left[\dot{f}_{0,j}(\tau_k) \mathbf{X}_j(\tau_k) + f_{1,j}(\tau_k) \mathbf{F}_j(\tau_k) \frac{\Delta t_j}{2} \right] + \mathbf{X}_{j-1}^b = 0 \qquad \text{for } k = 0 \tag{16}$$

$$\sum_{k=0}^{l_u} \beta_k \left[\dot{f}_{s,j}(\tau_k) \mathbf{X}_j(\tau_k) + f_{s,j}(\tau_k) \mathbf{F}_j(\tau_k) \frac{\Delta t_j}{2} \right] = 0 \qquad \text{for } k = 1, \ldots, l_x \tag{17}$$

$$\sum_{k=0}^{l_u} \beta_k \left[\dot{f}_{l_x+1,j}(\tau_k) \mathbf{X}_j(\tau_k) + f_{l_x+1,j}(\tau_k) \mathbf{F}_j(\tau_k) \frac{\Delta t_j}{2} \right] - \mathbf{X}_j^b = 0 \qquad \text{for } k = (l_x + 1) \tag{18}$$

The path constraints from (7) are directly collocated at the Gauss nodes, generating a set of constraint equations for each element given by

$$\mathbf{g}(\mathbf{X}_j(\tau_k), \mathbf{u}_j(\tau_k), \mathbf{b}, t(\tau_k)) \geq 0 \tag{19}$$

Constraints are also imposed on the boundary states of all adjacent elements to ensure continuity.

The transcribed objective functions from (5) are therefore

$$\tilde{J}_i = \begin{cases} \sum_{k=0}^{l_u} \phi_k(\mathbf{X}_k, \mathbf{u}_k, \tau_k, \mathbf{b}) \\ \phi_i(\mathbf{X}_0^b, \mathbf{X}_f^b, t_0, t_f, \mathbf{b}) \end{cases} \tag{20}$$

which equate to the two terms in a Bolza optimisation problem [16].

The time domain $\mathcal{T}$ corresponds to a single time period $[t_0, t_f]$. For launch systems, however, trajectories often have multiple phases either in series, or in parallel. For example, a multi-stage vehicle can have one phase per vehicle stage with all phases connected in series for the ascent, and/or branching parallel phases for the upper stage ascent, and first stage descent and landing. For a problem with N_p distinct phases, the dynamic constraints (15), path constraints (19), boundary constraints (8) and objective functions (20) are defined per phase. An additional set of N_p boundary constraints are introduced to manage the connections between phases defined by

$$\psi_{s_p}\left(\mathbf{X}_{0,\mathbf{I}_{s,p}}^b, \mathbf{X}_{f,\mathbf{I}_{s_p}}^b, t_{0,\mathbf{I}_{s_p}}, t_{f,\mathbf{I}_{s_p}}\right) \geq 0 \quad s_p = 1, \ldots, N_p \tag{21}$$

where the index vector $\mathbf{I}_{s_p}$ collects all the indexes of the phases that are connected by the constraint ψ_{s_p}. Note that, while the number of phases N_p is fixed, their temporal order is defined by the phase boundary constraints (21).

The resulting multi-objective nonlinear programming (MONLP) problem coming from the transcription of (5)–(8), with the inclusion of phase constraints (21), is given by

$$\min_{\mathbf{y} \in Y, \mathbf{p} \in \Pi} \tilde{\mathbf{J}}(\mathbf{y}, \mathbf{p}) \tag{22}$$

$$\mathbf{C}(\mathbf{y}, \mathbf{p}) \geq 0$$

where $\mathbf{y} = [\mathbf{X}_{0,1}, \ldots, \mathbf{X}_{s,j}, \ldots, \mathbf{X}_{l_x,N}]^T$, Y is a box in $\mathbb{R}^{n_Y}$ with $n_Y = n(l_x + 1)N$, $\mathbf{p} = [\mathbf{u}_{0,1}, \ldots, \mathbf{u}_{s,j}, \ldots, \mathbf{u}_{l_u,N}, \mathbf{b}^*]^T$ is a solution, or decision, vector that collects all the static and discretised control variables with $\mathbf{b}^* = [\mathbf{b}, \mathbf{x}_0^b, \mathbf{x}_f^b, t_0, t_f]^T$, $\Pi \subseteq \mathbb{R}^{n_s} \times \mathbb{R}^{n_b^*}$ with $n_s = n_u(l_u + 1)N$ (assuming that each element has the same number of control parameters) and $n_{b^*} = n_b + 2n + 2$, and $\mathbf{C}$ collects all path and boundary constraints.

Similar to (5), the solution of (22) is a subset $\Omega_\Pi \subset \Pi$ that satisfies the constraints $\mathbf{C}$ and contains solution vectors that are Pareto efficient. Given the subset Ω_Π of feasible solution vectors, a solution vector $\mathbf{p}^* \in \Omega_\Pi$ is said to be Pareto efficient if $\mathbf{p}^* \not\succ \mathbf{p}, \forall \mathbf{p} \in \Omega_\Pi$. The symbol of dominance $\succ$ indicates that, if $\mathbf{p}_1 \succ \mathbf{p}_2$, then $\tilde{J}_i(\mathbf{p}_2) \leq \tilde{J}_i(\mathbf{p}_1)$ for $i = 1, \ldots, m$ and $\exists j$ such that $\tilde{J}_j(\mathbf{p}_2) < \tilde{J}_j(\mathbf{p}_1)$. In other words, a solution is non-dominated if the values of any of the objective functions, using that solution, cannot be improved without sacrificing at least one of the other objectives [17]. For continuous functions, the subset Ω_Π is a manifold in $\mathbb{R}^{n_s + n_{b^*}}$ with dimension $(n_s + n_{b^*}) \leq (m - 1)$ [18]. In the following, the goal is to identify a pre-defined countable number of Pareto-efficient solutions contained in Ω_Π.

3. Solution of the Transcribed Problem

The MONLP problem in (22) is solved with an adaption of the Multi-Agent Collaborative Search (MACS) tailored to optimal control problems [10]. MACSoc combines the stochastic agent-based global search in MACS [19,20] with a local refinement of the solutions [21,22] (see Algorithm 1).

Algorithm 1 MACS optimal control (MACSoc)

1: Initialise population $\mathcal{P}_0$ and global archive $\mathcal{A}_0$, $k = 0$, $\rho_B = 1$
2: Initialise weight vectors ω
3: **while** $n\_fun\_eval < max\_fun\_eval$ **do**
4:    Run individualistic heuristics on $\mathcal{P}_k$ using bi-level formulation
5:    $\mathcal{P}_k \to \mathcal{P}_k^+$
6:    Update archive $\mathcal{A}_k$ with potential field filter
7:    Run social heuristics combining $\mathcal{P}_k^+$ and $\mathcal{A}_k$ using bilevel formulation
8:    Update archive $\mathcal{A}_k$ with potential field filter
9:    $\mathcal{P}_k^+ \to \mathcal{P}_k^\dagger$
10:    **if** local search triggered **then**
11:      Run gradient based refinement using single level formulation
12:      $\mathcal{P}_k^\dagger \to \mathcal{P}_k^*$
13:      Update archive $\mathcal{A}_k$ with potential field filter
14:      $\mathcal{P}_k^* \to \mathcal{P}_{k+1}$
15:    **else**
16:      $\mathcal{P}_k^\dagger \to \mathcal{P}_{k+1}$
17:    **end if**
18:    $k = k + 1$
19:    Update ρ_B
20: **end while**

At the start of MACSoc, an initial population $\mathcal{P}_0$ is generated with N_a agents representing feasible candidate solutions. Next, a set of N_w uniformly spread weight vectors ω are generated. Each agent is associated with a different weight vector, allowing the agent to converge to a different part of the Pareto-optimal set (set of non-dominated solutions for the multi-objective optimisation problem).

The global search generates candidate solutions for the decision vector using a combination of social and individualistic actions (lines 4 and 7 in Algorithm 1). Each action generates a candidate decision vector, starting from the current solution allocated to a given agent j, and submits it to a bi-level optimisation problem, where the inner level makes the candidate decision vector feasible with respect to constraints, and the outer level assesses whether the solution of the inner level represents an improvement with respect to the current solution allocated to agent j. All feasible and non-dominated solutions are added to the current population $\mathcal{P}_k$, and saved in an archive $\mathcal{A}_k$ (lines 5, 6, 8, 9 and 13 in Algorithm 1). A local refinement is triggered periodically after a user-defined number of iterations, and at the end of the algorithm, which update the current population and archive (lines 10–17 in Algorithm 1). The local refinement solves a single level scalarised version of (22).

The entire process alternates between social and individualistic actions, with periodic local refinement, until a maximum number of calls to the objective vector max_fun_eval is reached.

3.1. Bi-Level Global Optimisation Problem

The NLP problem for the global optimisation is defined by

$$\min_{\mathbf{p}^*} \tilde{\mathbf{J}}(\mathbf{y}^*, \mathbf{p}^*) \tag{23}$$

$$(\mathbf{y}^*, \mathbf{p}^*) = \underset{\mathbf{y}, \mathbf{p}}{\mathrm{argmin}}\{\delta_p(\mathbf{y}, \mathbf{p}) \mid \mathbf{C}(\mathbf{y}, \mathbf{p}) \geq 0\}$$

and represents two optimisation sub-problems at two different levels. The outer level minimises the objective function vector $\tilde{\mathbf{J}}$ and generates a first set of candidate solutions

p. The inner level looks for state $\mathbf{y}^*$ and control $\mathbf{p}^*$ vectors that satisfy the constraint functions $\mathbf{C}$, and minimise a cost function based on the candidate solutions of the outer loop $\delta_p = \|\mathbf{p}^* - \mathbf{p}\|$ that look the closest feasible solution to the candidate solution provided by the outer loop. The feasible solution is then passed back to the outer loop to evaluate the objective functions $\tilde{\mathbf{J}}$ with $(\mathbf{y}^*, \mathbf{p}^*)$. The inner level problem is solved with a local, gradient-based optimiser such as SQP or interior point.

In order to reduce the number of iterations required by the inner level to converge, the outer level stores the feasible states $\mathbf{y}^*$ from one iteration to be used as a first guess for the inner level at the next iteration. As shown in Figure 1, the feasible states $\mathbf{y}_k^*$ are preserved from iteration k to iteration $(k+1)$; therefore, the outer level only generates candidate solutions for $\mathbf{p}_{k+1}$. Thus, for iteration $(k+1)$, the inner level is given $(\mathbf{y}_k^*, \mathbf{p}_{k+1})$ as initial guesses for states and controls. Despite $\mathbf{y}_k^*$ being associated with $\mathbf{p}_k^*$, it has been shown to work well as an initial guess also when associated with $\mathbf{p}_{k+1}$.

When individualistic actions are applied, each agent generates one or more candidate solution vectors through three mechanisms that are triggered sequentially in the order: Inertia → Pattern Search → Differential Evolution. If any of these mechanisms produces an improved solution, the process is stopped and proceeds to update the population and archive (line 5 and 6 of Algorithm 1).

Inertia is triggered by agent j only if, in the previous iteration, agent j generated an improved solution. In this case, a step with random length is taken in the direction defined by $(\mathbf{p}_k^* - \mathbf{p}_{k-1}^*)$.

Pattern Search will change one optimisation parameter at a time, by a random amount in each direction, within a given neighbourhood $\mathcal{B}_j$ of agent j. The order by which the parameters are changed is a random permutation of the number of decision parameters. The process is repeated until either an improvement is registered or the maximum number of trials has been reached. As in Ricciardi and Vasile [20], the maximum number of trials is dynamically adjusted during the optimisation process: when the archive is empty, the maximum number of parameters scanned is equal to the total number of optimisation parameters. This maximum value is decreased linearly as the archive fills up, until only one optimisation parameter is changed when the archive is full. The neighbourhood $\mathcal{B}_j$ is a box centred in the position of the agent in parameter space and with the edges equal to the edges of the search space Π multiplied by the scaling parameter $\rho_{\mathcal{B}_j}$.

Differential Evolution generates a sample with the simple heuristic:

$$\mathbf{p}_{trial,j} = \mathbf{p}_j + \xi_1 \mathbf{e}\big((\mathbf{p}_j - \mathbf{p}_{j_1}) + c_F(\mathbf{p}_{j_2} - \mathbf{p}_{j_3})\big) \tag{24}$$

where $\mathbf{p}_j$ is the current candidate solution, $\mathbf{p}_{j_1}, \mathbf{p}_{j_2}, \mathbf{p}_{j_3}$ are three randomly chosen solutions from the current population $\mathcal{P}_k$, ξ_1 is a uniformly distributed random number in the unit interval, c_F is a user-defined constant and $\mathbf{e}$ is a mask vector defined as

$$e_j = \begin{cases} 1, & \text{if } \xi_2 < CR \\ 0, & \text{otherwise} \end{cases} \tag{25}$$

where ξ_2 is another uniformly distributed random number in the unit interval, and CR is the crossover rate. For the following test cases, $c_F = 0.9$ and $CR = 1$.

If no improvement is made after trying all the three heuristics, $\rho_B \to 0.5\rho_B$; if instead an improvement is made, $\rho_B \to 2\rho_B$ until the initial value $\rho_B = 1$ is reached again.

When social actions are applied, the outer level uses the population to generate a candidate solution using the same heuristics of Differential Evolution (24), with the parent solutions $\mathbf{p}_{j_1}, \mathbf{p}_{j_2}, \mathbf{p}_{j_3}$ chosen from the union of the current population $\mathcal{P}_k$ and the current archive $\mathcal{A}_k$.

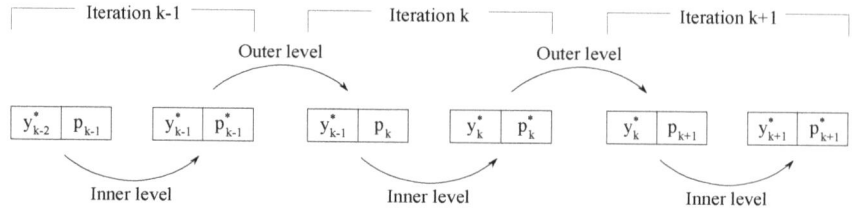

Figure 1. Schematic representation of the bilevel approach acting on a single solution.

A candidate solution $(\mathbf{y}^*, \mathbf{p}^*)$ generated by the inner level is evaluated in the outer level by computing the weighted Chebychev norm

$$\Phi_i = \max_i \omega_i \left(\tilde{J}_i(\mathbf{y}^*, \mathbf{p}^*) - z_i \right) \qquad (26)$$

$$z_i = \min_{\mathcal{P}_k \cup \mathcal{A}_k} \tilde{J}_i$$

where ω is the weight vector in objective space and $\mathbf{z}$ is the current utopia point.

Given the value of Φ_i at step k, or Φ_i^k, an improvement corresponds to $\Phi_i^{k+1} < \Phi_i^k$. This improvement criterion has two very important properties: first, it allows the search to reach even non-convex parts of the Pareto front, and, second, if the weights are chosen appropriately, it enables an efficient convergence to the global minimum for each objective function.

Note that, if the inner level does not converge to the required tolerance, the objective functions of the outer level are recalculated to be the infinity norm of the constraint violation plus the maximum values of each objective functions in the archive and population. This creates an adaptive rejection mechanism: if none of the agents are feasible, the ones that best satisfy the feasibility are entered in the archive, with the next iterations trying to improve their feasibility. Once an agent finds a feasible solution, it will explore the search space through the global bi-level approach, generating several feasible and non-dominated solutions. These solutions will enter in the archive as they will dominate many of the existing infeasible solutions, and due to the social actions, some agents will be directly moved onto those solutions, allowing the entire population to converge to feasible solutions in a handful of iterations.

Finally, if any candidate solutions for $\mathbf{y}$ and $\mathbf{p}$ fall outside the boundaries of the search space $Y \times \Pi$, the solution vector is reduced until it is back within the search space.

3.2. Single Level Local Search

The local refinement solves the following scalarised problem for each agent j:

$$\min_{\epsilon \geq 0} \epsilon \qquad (27)$$

$$\omega_{i,j} \vartheta_{i,j}(\mathbf{y}, \mathbf{p}) \leq \epsilon \qquad \text{for } i = 1, \ldots, m$$

$$C(\mathbf{y}, \mathbf{p}) \geq 0$$

where $\omega_{i,j}$ is the ith component of the weights for the jth agent, $\vartheta_{i,j}$ is the ith component of the rescaled objective vector of the jth agent, and ϵ is a slack variable.

This reformulation of the problem, which uses Pascoletti–Serafini scalarisation [23], constrains the movement of the agent to within a descent cone defined by the point $(\epsilon \mathbf{d}_j + \mathbf{1}_j)$ and along the direction $\mathbf{d}_j = (1/\omega_{1,j}, \ldots, 1/\omega_{i,j}, \ldots, 1/\omega_{m,j})$. The rescaled vector of the objective functions is therefore

$$\vartheta_j(\mathbf{y}, \mathbf{p}) = \frac{\tilde{J}_{i,j}(\mathbf{y}, \mathbf{p}) - \tilde{z}_i}{z_{i,j}^* - \tilde{z}_i} \qquad \text{for } i = 1, \ldots, m \qquad (28)$$

where $\mathbf{z}_j^*$ is equal to $\tilde{\mathbf{J}}_j(\bar{\mathbf{y}}, \bar{\mathbf{p}})$, $(\bar{\mathbf{y}}, \bar{\mathbf{p}})$ is the initial guess for the solution of (27), and $\tilde{\mathbf{z}} = (\mathbf{z} - \mathbf{z}_A)$ where $\mathbf{z}_A$ is the nadir of the archive. The components of the vector ζ_j are derived from the normalisation

$$\zeta_{i,j} = \frac{z_i}{z_{i,j}^* - \tilde{z}_i} \quad \text{for } i = 1, \ldots, m \tag{29}$$

This allows the components of $\vartheta_j(\mathbf{y}, \mathbf{p})$ to have values of 1 at the beginning of the local search, and 0 if the agent converges to the target point $\tilde{\mathbf{z}}$. Thus, the single level approach avoids biases when the objectives have significantly different scales.

The weighted Chebychev norm in (26) and (27) are equivalent and lead to the same optimal solution if the target point for the Pascoletti–Serafini scalarisation coincides with the utopia point, and the weight vectors are the same [24]. By combining (26) in the global search phase with (27) in the refinement phase, the algorithm ensures a smooth transition from global exploration of the Pareto set, to local convergence.

3.3. Archiving Strategy

MACSoc, through MACS, employs an archiving strategy described in Ricciardi and Vasile [20]. When the elements in the archive $\mathcal{A}$ are less than the maximum allowed cardinality of $\mathcal{A}$, every new feasible and non-dominated solution is recorded in the archive. Once the defined maximum size for the archive is reached, new elements are added to $\mathcal{A}$ only if they minimise the potential function,

$$E(\tilde{\mathbf{J}}_1, \cdots, \tilde{\mathbf{J}}_{N_\mathcal{A}}) = \sum_{i=1}^{N_\mathcal{A}} \sum_{j=i+1}^{N_\mathcal{A}} \frac{1}{(\tilde{\mathbf{J}}_i - \tilde{\mathbf{J}}_j)^T (\tilde{\mathbf{J}}_i - \tilde{\mathbf{J}}_j)} \tag{30}$$

where $N_\mathcal{A}$ is the number of elements in the archive $\mathcal{A}$.

To avoid biasing in the rejection–retention process when the objectives have different scales, the objective values of the set of non-dominated solutions are all normalised between 0 and 1. This leads to a combinatorial problem that can be solved approximately but efficiently and returns a uniformly spread set of points [20].

3.4. Generation of the Initial Feasible Population

Before the optimisation starts, MACSoc generates an initial population of agents $\mathcal{P}_0$ representing feasible candidate solutions.

A first guess for the candidate solutions is generated using Latin Hypercube sampling within the given boundaries, which gives a near-random sample of parameter values from a multidimensional distribution [18]. State variables for each phase are initialised with a linear interpolation between initial and final conditions. For each phase, each equation in (15) is optimised using the inner level subproblem (23) to ensure feasibility through a local gradient-based optimiser. Additional constraints within the phase are then added in, and the problem is re-optimised. Lastly, the linking constraints between phases are then included, and the full resulting problem is optimised for a final time (always using a local optimiser with the inner level objective function in (23)). If at the end of the initialisation phase, an agent is associated with a solution that is not feasible within the prescribed tolerance, that solution is still included in the initial population $\mathcal{P}_0$ and submitted to the subsequent optimisation cycle.

For the test case, the tolerance on feasibility is set to 10^{-6}, both for the initial population generation, and for determining feasibility within loop. By default, the maximum number of calls to the constraint function is set equal to $10(n_b^* + n_s + n_Y)$. An interior point NLP solver was used as it delivered a more robust and consistent convergence to feasible solutions.

3.5. Definition of the Descent Directions and Target Points

The weight vectors for the bi-level global search are generated as follows: first, a simplex in objective space is generated through simplex lattice design [25]. Then, the points of this simplex lattice are projected on the unit sphere by dividing their position vectors by their distance to the origin. This gives a fairly uniform distribution of weight vectors (and thus descent directions) in any N_w dimensional space.

In order to generate a more uniform distribution, however, these weight vectors are refined using a local optimisation of the same potential function E given in (30),

$$\min E(\omega_1, \ldots, \omega_{N_w}) \qquad (31)$$
$$\omega_i^T \omega_i = 1$$

While this approach is valid for general m-objective problems, for two objectives, it is simpler and faster to generate uniform angularly spaced weight vectors. In the following, $N_a = N_w$ and each agent is associated with the closest descent direction in criteria space, at the initialisation stage, with the constraint that no two agents can have the same descent direction.

For the single level approach, the weight vectors $\omega_j = [\sqrt{2}, \ldots, \sqrt{2}]^T$ are allocated to all agents except to those m agents that minimise each individual objective function. For these m agents, the weight vectors are $\omega_j = [0, \ldots, j, \ldots, 0]^T$ with $j = 1, \ldots, m$. These weights are orthogonal because they correspond to the m orthogonal directions in criteria space. If agent j associated with weight ω_j does not generate any improvement after two iterations, a new random orthogonal weight is associated with j and (27) is solved with the added constraints,

$$\tilde{J}_i \leq z_i \ \forall i \neq j \qquad (32)$$

The reason for the different choice of weight vectors between the bi-level and the single level formulation can be explained as follows: the bi-level formulation explores globally the search space with a population of agents, thus there is the need to maximise the spreading of the solutions; on the contrary, the single-level is used to improve the local convergence of each agent in a normalised criteria space. Thus, the goal of the single level is to return dominating solutions without altering too much their spreading in criteria space.

4. Uncertainty Model for Atmospheric Parameters

Preliminary design and trade-off studies, in particular those employing computationally intensive multidisciplinary design and multi-objective optimisations, typically use global static atmospheric models such as US-76 Standard Atmospheric model, or the International Standard Atmospheric (ISA) model. These models predict the atmospheric pressure p, temperature T and density ρ as function solely of altitude, and employ simplified algebraic expressions for the different atmospheric layers. The models are based on average values for year-round, mid-latitude conditions with moderate solar activity. The ISA is valid up to an altitude above mean sea level of 86 km, while US-76 has an extension up to 1000 km. Due to this, US-76 is often preferred in the field of space launchers over ISA. These two models rely on similar assumptions and methodologies, differing only in the prediction of the temperature in the upper atmosphere (above 32 km).

In order to assess the robustness of the mission design against uncertainties in the atmospheric model, a model of the uncertainty was developed using higher fidelity atmospheric models to create a data set. A number of empirical, global reference atmospheric models exist, for example, to analyse the effect of atmospheric drag on satellites. The Committee on Space Research (COSPAR) in their International Reference Atmosphere reports details three of these models [26]: NRLMSISE-00 [27], Jacchia Bowman reference atmospheric model JB2008 [28], and DTM2013 [29].

The NRLMSISE-00 is a model developed by the US Naval Research Laboratory and accounts for geographic, temporal, solar and magnetic effects through inputs for: date, time of day, geodetic altitude from 0 to 1000 km, geodetic latitude and longitude, local apparent solar time, 81-day average of F10.7 solar flux, daily F10.7 solar flux for previous day, and daily magnetic index.

A statistical analysis of the difference in the models was performed treating those all the input parameters save altitude as uncertain. A set of 10^5 quasi-random samples were generated for the altitude using a low discrepancy Halton sequence [30], with the corresponding values for T, P and ρ computed using the NRLMSISE-00 model for all altitudes in the range between 0 and 100 km. The altitude limit of 100 km was driven by the test case of the descent phase of a first stage, reusable spaceplane which is designed to operate within this range.

Every sample of the NRLMSISE-00 model was treated as an equivalent static global atmospheric model, and used to compare against the US-76 model. These differences were treated as random fluctuations in order to quantify the uncertainty.

As shown in Figure 2, the means and standard deviations σ were determined for each of the three atmospheric parameters. It can be observed that, for the temperature, the mean differences, or relative errors, are very low, with a 1σ relative error around 5% for altitudes below 80 km. The mean relative errors for pressure and density are also low for relatively low altitudes, but increase at higher altitudes, with the largest 1σ bands around 60 km. Above 40 km, the pressure and density have a very low absolute value, so a large relative error still means a low absolute error.

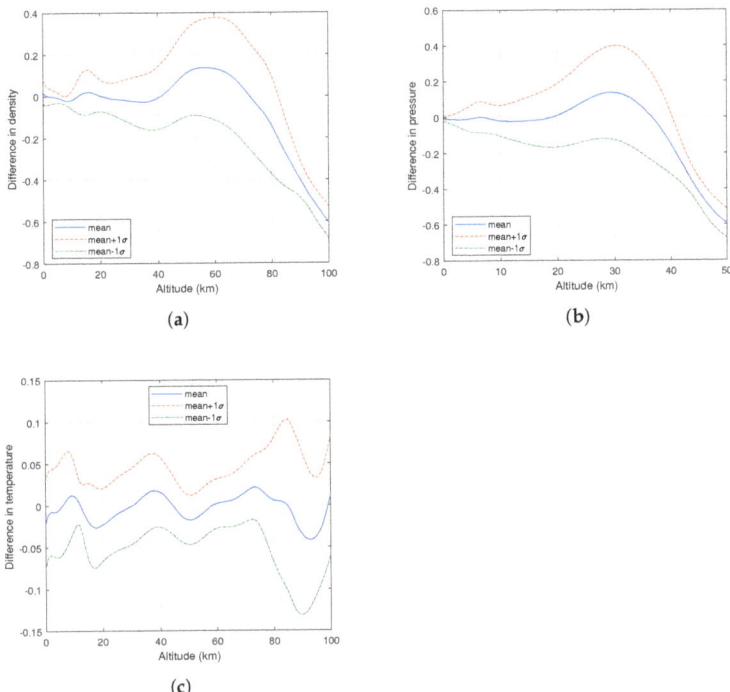

Figure 2. Mean and standard deviation ($\pm 1\sigma$) of the differences between every sample (representing an equivalent global static atmospheric model) and the US-76 model for each of the atmospheric parameters as a function of altitude. (**a**) Atmospheric density (kg/m^3); (**b**) Atmospheric pressure (kPa); (**c**) Atmospheric temperature (K).

Since the approach employed is that of the Square Root Unscented Transformation, different models were generated: one employing the mean relative error, and the others adding to the mean the Cholesky factorisation of the covariance matrix of the uncertain quantities at each altitude. Figure 3 shows the five generated temperature profiles, and their comparison with respect to the US-76 model. A similar approach was followed for the density.

As it can be seen, the Sigma Point 0 model is quite close to the US-76 model. Sigma points 1 and 2 add the standard deviation to this mean, while Sigma points 3 and 4 also include the correlation between variations in temperature and variations in density. The correlation between temperature and density changes sign repeatedly as the altitude changes, thus the profiles assume values lower than one standard deviation only to cross the mean and assume values higher than one standard deviation elsewhere.

Among the 10^5 samples generated by the Halton sequence on the NRLMSISE-00 model, several profiles do indeed have this kind of shape, which is significantly different to the US-76 model. The temperature affects the computation of the Mach number, on which the aerodynamic coefficients depend. The dependence of the aerodynamic coefficients on the Mach number is stronger around Mach 1 and weaker for high Mach numbers, thus it is not easy to foresee the effect of these variations.

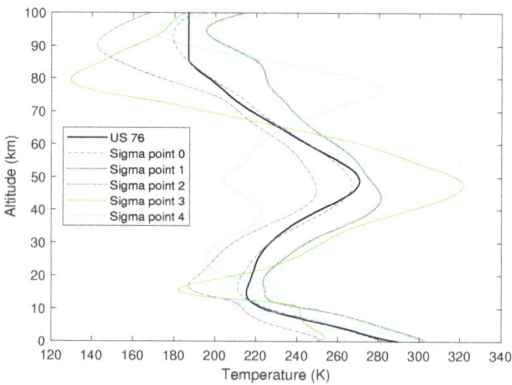

Figure 3. Temperature profiles for the models of each sigma point, and comparison with the US76 model.

5. Vehicle System Models

The Orbital-500R launch system is composed of a first stage reusable spaceplane, capable of rocket-powered ascent and an unpowered, glided descent, and an expendable, rocket-based upper stage. The system is air-launched from a carrier aircraft. The system was designed to launch from the UK and carry small payloads (500 kg) to Low Earth Orbits (e.g., up to 1500 km). The dry mass of the spaceplane is 20 tonnes.

The flight dynamics are modelled as a variable-mass point with three degrees of freedom in the Earth-centered Earth-fixed reference frame, subject to gravitational, aerodynamic lift and drag forces. The state vector contains the translational position and velocity components, $\mathbf{x} = [h, \lambda, \theta, v, \gamma, \chi]$, where $h = r_E - r$ is the altitude given radial distance r and the Earth's radius $r_E(\lambda, \theta)$, (λ, θ) are the geodetic latitude and longitude, v is the magnitude of the relative velocity vector directed by the flight path angle γ and the flight heading angle χ. The vehicle is controlled through the angle of attack α, and the bank angle μ of the vehicle. The dynamic model is therefore [31]

$$\dot{r} = \dot{h} = v \sin \gamma \tag{33}$$

$$\dot{\lambda} = \frac{v \cos \gamma \cos \chi}{r} \tag{34}$$

$$\dot{\theta} = \frac{v \cos \gamma \sin \chi}{r \cos \lambda} \tag{35}$$

$$\dot{v} = \frac{-D}{m} - g_r \sin \gamma + g_\phi \cos \gamma \cos \chi + \omega_E^2 r \cos \lambda (\sin \gamma \cos \lambda - \cos \gamma \cos \chi \sin \lambda) \tag{36}$$

$$\dot{\gamma} = \frac{1}{v}\left(\frac{L \cos \mu}{m} - g_r \cos \gamma - g_\phi \sin \gamma \cos \chi\right) + \frac{v}{r} \cos \gamma$$
$$+ \frac{\omega_E^2 r}{v} \cos \lambda (\sin \gamma \cos \chi \sin \lambda + \cos \gamma \cos \lambda) + 2\omega_E \sin \chi \cos \lambda \tag{37}$$

$$\dot{\chi} = \frac{1}{v \cos \gamma}\left(\frac{L \sin \mu}{m} - g_\phi \sin \chi\right) + \frac{v}{r} \cos \gamma \sin \chi \tan \lambda$$
$$+ \omega_E^2 r \frac{\sin \chi \sin \lambda \cos \lambda}{v \cos \gamma} + 2\omega_E(\sin \lambda - \tan \gamma \cos \chi \cos \lambda) \tag{38}$$

The Earth is modelled using WSG-84, with gravitational acceleration expressed in radial g_r and tangential g_ϕ components.

The accelerations induced on the vehicle are determined, using the relative vehicle body reference frame $\mathcal{B}$.

$$a_x^\mathcal{B} = \frac{-D \cos \alpha + L \sin \alpha}{m} - g_r(\sin \gamma \cos \alpha + \cos \gamma \cos \mu \sin \alpha)$$
$$+ g_\phi(\cos \gamma \cos \chi \cos \alpha - \cos \chi \sin \gamma \cos \mu \sin \alpha - \sin \chi \sin \mu \sin \alpha)$$
$$- \frac{v}{r} \cos \gamma \sin \alpha(\cos \mu - \sin \chi \tan \lambda \sin \mu) \tag{39}$$

$$a_y^\mathcal{B} = g_r \cos \gamma \sin \mu + g_\phi(\cos \chi \sin \gamma \sin \mu - \cos \mu \sin \chi)$$
$$+ \frac{v}{r} \cos \gamma (\sin \mu + \cos \mu \sin \chi \tan \lambda) \tag{40}$$

$$a_z^\mathcal{B} = \frac{-D \sin \alpha - L \cos \alpha}{m} + g_r(\cos \gamma \cos \mu \cos \alpha - \sin \gamma \sin \alpha)$$
$$+ g_\phi(\cos \chi \sin \gamma \cos \mu \cos \alpha + \sin \chi \sin \mu \cos \alpha + \cos \gamma \cos \chi \sin \alpha)$$
$$+ \frac{v}{r} \cos \alpha \cos \gamma (\cos \mu - \sin \chi \tan \lambda \sin \mu) \tag{41}$$

Surrogate models were used for the lift c_L and drag c_D coefficients as a function of Mach number M, angle of attack α and altitude h. An artificial neural network was employed, using an aerodynamic database for the training data coming from a mix of panel methods and CFD simulations (see Stindt et al. [32] for details on the generation of the aerodynamic database). The lift and drag forces were computed assuming no wind, with the relative velocity $\mathbf{v}_{rel} = \mathbf{v} - \omega_E \mathbf{r}$.

$$L = \frac{1}{2} C_L(M, \alpha, h) \rho S_{ref} v_{rel}^2 \tag{42}$$

$$D = \frac{1}{2} C_D(M, \alpha, h) \rho S_{ref} v_{rel}^2 \tag{43}$$

where S_{ref} is the reference area of the vehicle, and ρ is the atmospheric density.

Maddock et al. [1], Ricciardi et al. [10] performed a number trade-off studies through a multi-disciplinary design optimisation on design of the Orbital-500R launch system and on various missions, analysing the full trajectories for both stages. Extrapolating from these results, a nominal starting point for the descent was set. Table 1 lists the initial and final conditions of the state variables. A final condition was imposed on the expected value of the altitude, with a value of 10 km. In addition, the trajectories of all sigma points were required to have a flight path angle greater or equal to −20°.

Table 1. Values and constraints on the boundary values of the state variables.

	Initial Condition	Final Condition (Requirement)
Altitude h	90 km	$h = 10$ km
Latitude λ	60° N	
Longitude θ	−12° E	
Velocity v	3 km/s	
Flight path angle γ	−5°	
Heading angle χ	20°	$\chi \geq -20°$

6. Application and Results

6.1. Problem Formulation and Set-Up

For the applied test case here, the initial conditions are fixed. The final conditions are affected by the uncertainty, and are expressed as difference between the mean and target values. A boundary constraint is added as follows:

$$\psi(\mathbf{X}(t_f)) = \mu_\chi - \bar{x}(t_f) = 0 \tag{44}$$

where μ_χ is the mean of the final states of the sigma points, and $\bar{x}(t_f)$ is the target value.

The first objective uses a metric to minimise the average induced acceleration on the vehicle. The general formulation is

$$J_1 = \int_{t_0}^{t_f} E\left(\|\mathbf{a}\|^2\right) dt \tag{45}$$

where the transcribed version using DFET is

$$J_1 = \sum_j^N \frac{\delta t_j}{2} \sum_k^{l_u} \beta_k \sum_i^{N_\sigma} a^2(\chi_i, \tau_k) W_{\sigma,i} \tag{46}$$

where $a = \|\mathbf{a}\| = \sqrt{a_x^2 + a_y^2 + a_z^2}$, given in (39)–(41). As part of the generation of the sigma points, N_σ weights W_σ are generated, associated with each sigma point. The computation of the mean of a quantity, within the transcribed NLP, is the weighted sum of the quantities associated with each sigma point [12].

The second objective aims to reduce the uncertainty of the final state by minimising the sum of the square of all the entries of the covariance matrix:

$$J_2 = \sum_{i,j} \left(\text{Cov}_{i,j}(\mathbf{X}(t_f))\right)^2 \tag{47}$$

where $\text{Cov}_{i,j}$ is computed using the standard algebraic manipulations employed for the Square Root Unscented transformation, with the additional consideration that no update of the Cholesky factorisation is needed since no measurement is here performed and thus no error is present. This formulation has the advantage that the quantity to compute is smooth and differentiable, it involves all components of the covariance matrix, and does not require iterative procedures like decomposition in eigenvalues to compute the principal axes of the ellipsoid of the uncertainty. In order to give each element of the covariance matrix the same weight even if the quantities of interest have different scales, the state variables were scaled by the same factors internally employed by MODHOC to ensure that all variables assume values between 0 and 1.

The equations were transcribed through DFET, using 6 elements of order 7 for all the states and the controls. Since all five sigma points are propagated simultaneously, there are 30 elements for the states and 12 elements for the controls. MODHOC was run for a total of 30,000 function evaluations, keeping 10 solutions in the archive.

6.2. Trajectory Results

The computed Pareto front is shown in Figure 4, which confirms a trade-off between the two objectives corresponding to metrics to minimising the induced acceleration load on the vehicle (J_1), and maximising the robustness of the trajectory by minimising the covariance of the final values of the state variables (J_2).

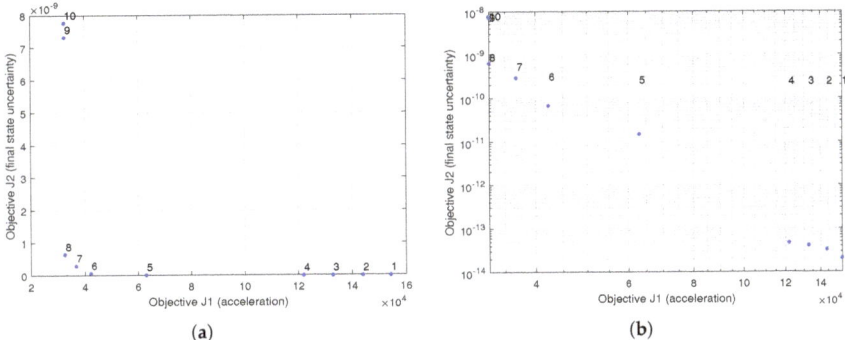

Figure 4. Pareto Front for the two-objective problem. (**a**) Linear scale; (**b**) Log-log scale.

Figure 5 shows the time histories of the state variables for altitude, velocity vector components given by the magnitude, flight path and heading angles. The solutions are shown for a set of Pareto-optimal solutions from Figure 4, along with the $\pm 1\sigma$ uncertainty indicated by the dashed and dotted lines. It can be seen that Solutions 1 to 4 (in greens and blues) have lower uncertainty for the final state, as expected as Solution 1 represents the extrema for min(J_2). This can also be seen from Figure 6, which show the standard deviation of the same four state variables in Figure 5. As previously stated, even if the uncertainty on atmospheric density is relatively high at high altitudes, its effect is quite limited for the first part of the trajectory as the absolute value is very small (e.g., 10^{-6} at 85 km, 10^{-3} Pa at 40 km). The uncertainty in the density starts to have a noticeable effect as the altitudes get to approximately 40 km and below.

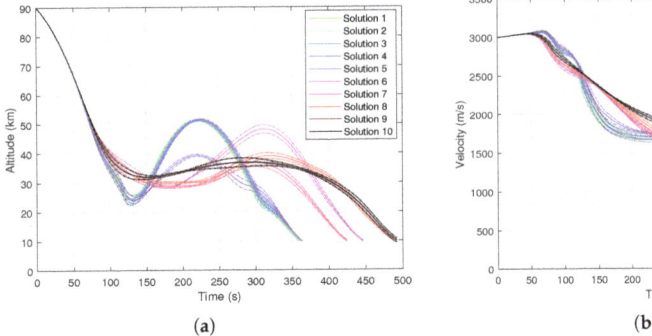

Figure 5. *Cont.*

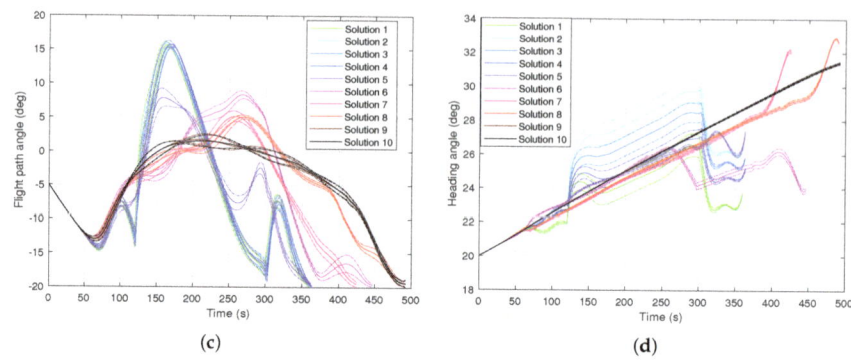

(c) (d)

Figure 5. State variables for set of 10 Pareto-optimal solutions, with dashed and dotted lines showing $\pm 1\sigma$ uncertainty. (**a**) Altitude; (**b**) Magnitude of the velocity; (**c**) Flight path angle; (**d**) Heading angle.

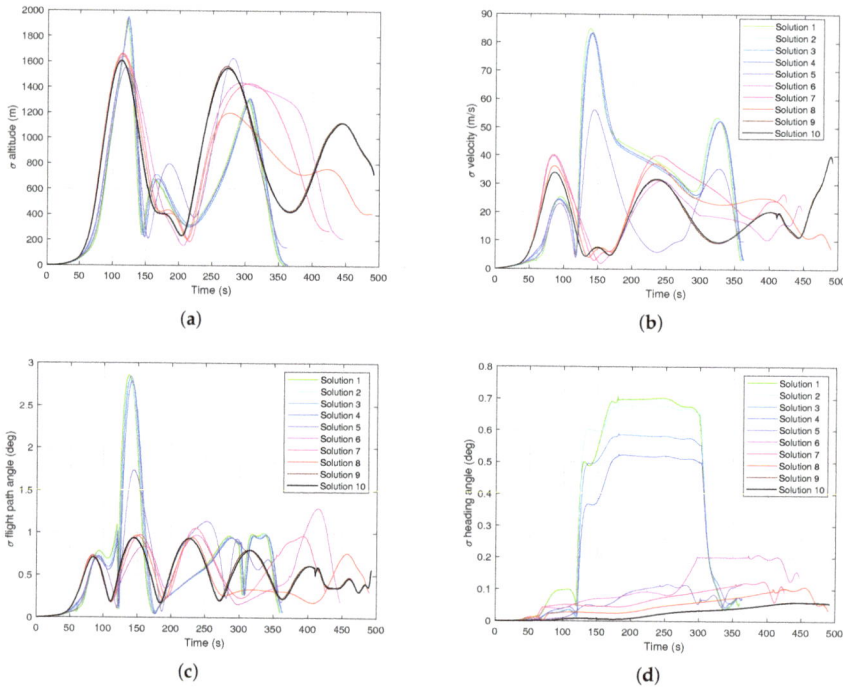

Figure 6. Standard deviations of state variables. (**a**) Altitude; (**b**) Magnitude of velocity; (**c**) Flight path angle; (**d**) Heading angle.

As the Pareto front (Figure 4) indicated, the solutions with a lower uncertainty of final states have higher acceleration loads, as shown in Figure 7. This further shows the breakdown between the accelerations in the three vehicle (or body) axes, as well as the magnitude of the acceleration vector.

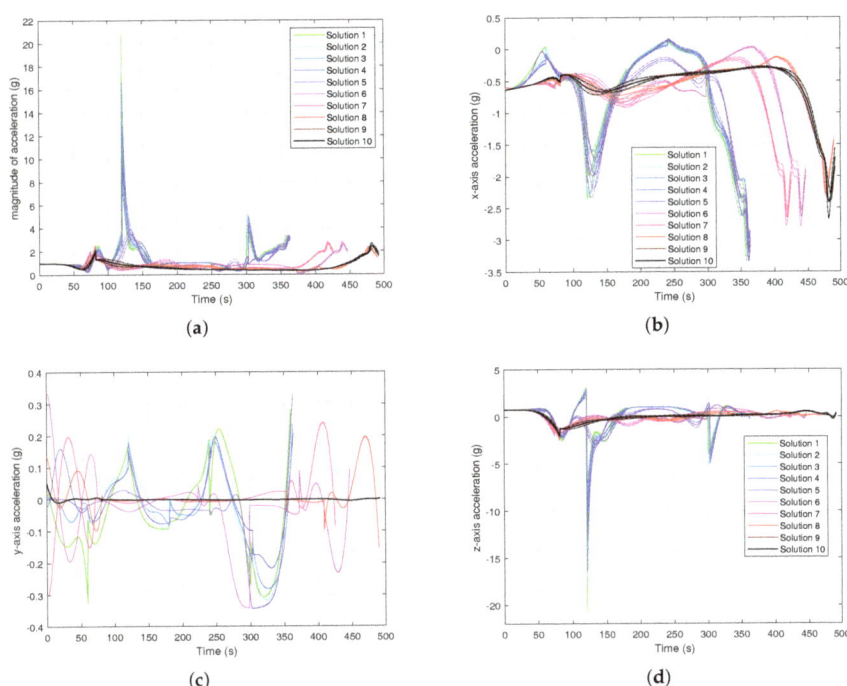

Figure 7. Magnitude and components (in vehicle body reference frame) of the induced accelerations, normalised against sea level gravity $g = g_0$. (**a**) Magnitude of acceleration; (**b**) Component in x-axis; (**c**) Component in y-axis; (**d**) Component in z-axis.

Figure 8 shows the time history for the two control variables: angle of attack α and bank angle μ. In all cases, the angle of attack starts with the maximum possible value of 45 deg, and then progressively decreases to a more moderate value around 5–10 deg. Solutions with lower accelerations stay in this regime for a while, and finally conclude with a value around 0 deg. Solutions with lower final uncertainty instead have a progressive decrease in α until small negative values, then a sharp increase to values around 10–15 deg, and finally stabilise around 0 deg.

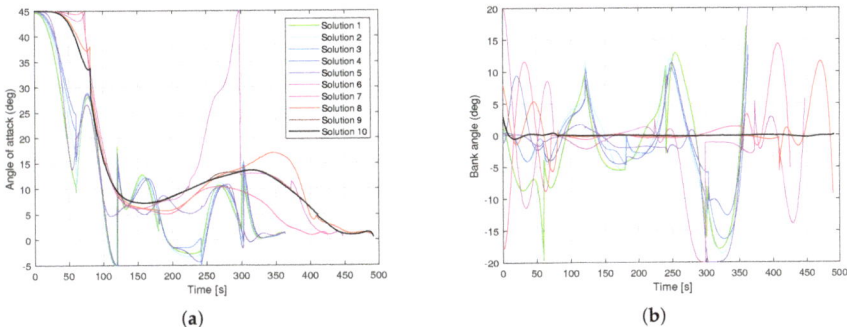

Figure 8. Control variables, which are consistent across all solutions in the Pareto-optimal set. (**a**) Angle of attack; (**b**) Bank attack.

Finally, Figures 9 show the time history of the states for all sigma points for extrema Solutions 1 and 10. As it is evident, the green lines have a much lower scattering at the final time than the black lines, indicating that Solution 1 (green) is subject to less uncertainty than solution 10 (black). These figures also give an idea of the complexity of the problem tackled by this approach, where the same control law is applied to multiple independent sigma points (lines with the same colour) and is able to steer the system to a given expected final state while also reducing the uncertainty associated with the final state, or reducing the expected acceleration load.

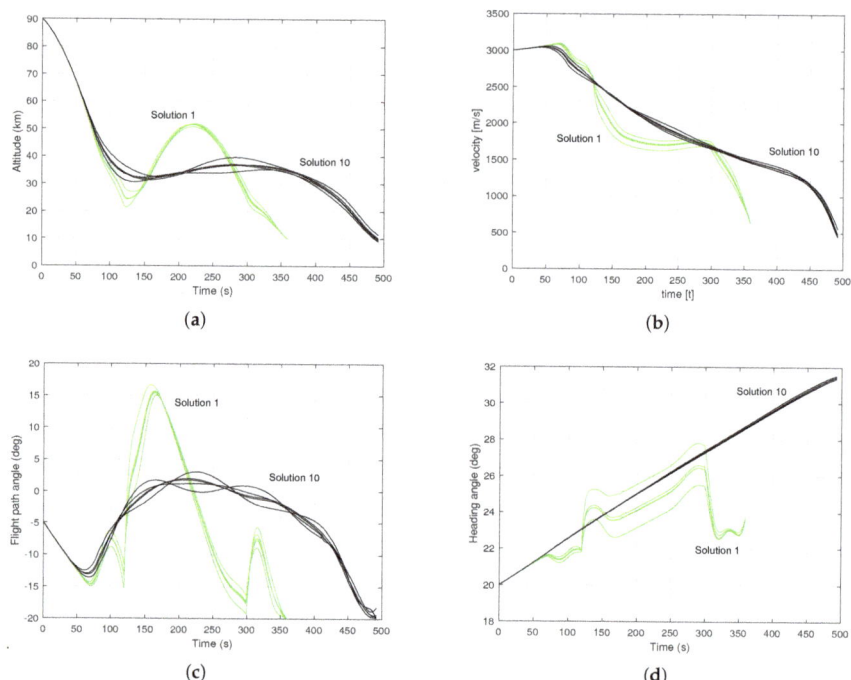

Figure 9. State variables of sigma points for the two extrema of the Pareto-optimal set (Solutions 1 and 10). (**a**) Altitude; (**b**) Velocity; (**c**) Flight path angle; (**d**) Heading angle.

6.3. Validation

A Monte Carlo analysis was run to validate the method and results. Starting from the Pareto-optimal set of 10 solutions, the solution was re-integrated using a different sample for the atmospheric model. A 100 different samples (for the atmospheric model) were taken for each Pareto-optimal solution, using the same sampling bounds as used to generate the sigma points (i.e., 1 standard deviation for each of the two atmospheric parameters, density and temperature).

Figure 10 shows the results of the Monte Carlo analysis for a middle set solutions (Solution 7, Figure 10a,b), and the two extrema (Solutions 1 and 10, Figure 10c,d). In each plot, the five sigma points are shown (Point 0 is green, and Points 2–4 in blue), with each of the 100 Monte Carlo runs shown as a grey line.

The values of the objective functions are compared in Table 2, looking at the optimised values against the mean values for J_1 and J_2 across the Monte Carlo runs.

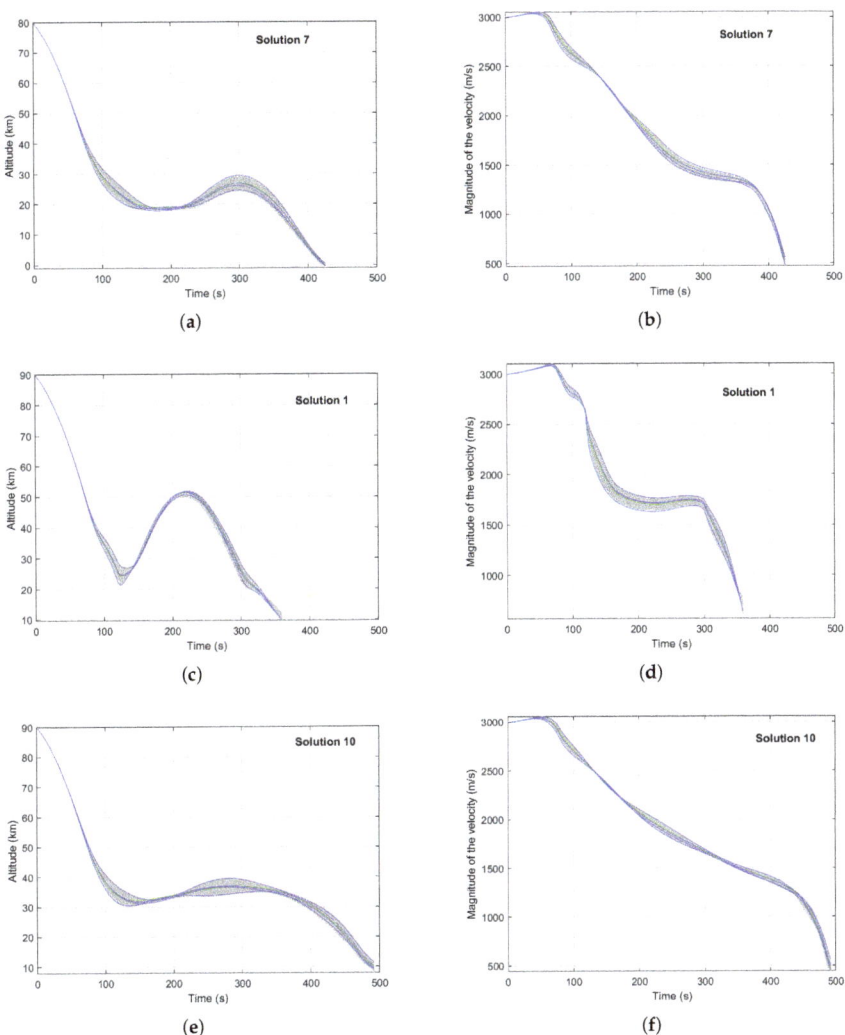

Figure 10. Monte Carlo analysis for Solution 7 (**a**,**b**), Solution 1 (**c**,**d**) and Solution 10 (**e**,**f**). Green is the central Sigma point 0, blue are Sigma points 1–4, and grey lines are the 100 Monte Carlo samples. (**a**) Altitude, Solution 7; (**b**) Velocity, Solution 7; (**c**) Altitude, Solution 1; (**d**) Velocity, Solution 1; (**e**) Altitude, Solution 10; (**f**) Velocity, Solution 10.

Table 2. Comparison of the mean value of the objective functions across the 100 runs of the Monte Carlo analysis with the optimised values.

		Optimisation	Monte Carlo
Solution 1	Objective function J_1	1.5375×10^5	1.4696×10^5
	Objective function J_2	2.1029×10^{-14}	1.0542×10^{-8}
	Mean final altitude	10.00 km (target)	10.354 km
Solution 7	Objective function J_1	3.6761×10^4	3.5939×10^4
	Objective function J_2	2.9916×10^{-10}	5.7082×10^{-10}
	Mean final altitude	10.00 km (target)	9.9277 km
Solution 10	Objective function J_1	3.2629×10^4	3.2396×10^4
	Objective function J_2	7.7731×10^{-9}	1.8189×10^{-8}
	Mean final altitude	10.00 km (target)	9.9014 km

The results from Monte Carlo analysis are consistent with the results found using the unscented transform integrated into the multi-objective optimal control solver MODHOC. Small deviations can be seen towards the middle-end of the trajectory at altitudes below 50 km. The five sigma points chosen are all orthogonal, however the sample points for the Monte Carlo run are not so constrained (though the boundaries are the same). The absolute values for atmospheric parameters, in particular density ρ and pressure p, have higher values in lower atmosphere, so it is consistent with the expectation that this area would experience an increased impact, coupled with the larger standard deviation between 40–60 km.

7. Conclusions

This paper presented an extension of a memetic multi-objective optimisation algorithm MODHOC to perform optimisation under uncertainty. This was applied to a test case to determine a set of Pareto-optimal solutions for the descent trajectory of a first-stage spaceplane, based on the conceptual commercial launch vehicle Orbital 500R. The robust multi-objective optimisation traded off a performance objective, the induced accelerations on the vehicle corresponding to the dynamics loads, with a metric on the robustness of the trajectory to uncertainties on the atmospheric temperature, pressure and density.

Through a square root unscented transformation, different atmospheric models were generated for five sigma points. All the sigma points share the same control law, thus making the trajectory robust against model uncertainty. While only the first two statistical moments of the uncertain values were considered for this work, future work will account for higher order moments, making the resulting trajectory even more robust. A larger number of sigma points will also be required, however, and the resulting optimal control problem becomes progressively larger, requiring the use of large scale optimisation code.

Author Contributions: Conceptualization, L.A.R., C.A.M. and M.V.; Formal analysis, L.A.R., C.A.M. and M.V.; Funding acquisition, C.A.M. and M.V.; Investigation, L.A.R. and C.A.M.; Methodology, M.V.; Project administration, C.A.M., M.V.; Software, L.A.R.; Supervision, C.A.M. and M.V.; Writing—original draft, L.A.R. and C.A.M.; Writing—review & editing, C.A.M. and M.V. All authors have read and agreed to the published version of the manuscript.

Funding: This work has been partially funded by the UK Space Agency and European Space Agency General Support Technology Programme (GSTP) 4000126322/18/NL/LvH/zk on De-risking of the GNC of the Orbital 500-R, and ESA Network Partnership Initiative (NPI) 4000112763/14/NL/GLC/ats on Multi-Objective Hybrid Optimal Control Problems.

Data Availability Statement: The software tools DFET and MACS are available open source under the Strathclyde Mechanical and Aerospace Research Toolbox (SMART) for Optimisation and Optimal Control (o2c) hosted at https://github.com/strath-ace/smart-o2c, accessed on 15 November 2021.

Acknowledgments: The work here uses as a test case a variant of the Orbital-500R launch system that was created and developed by Orbital Access Ltd., and analysed through the Future Space Payload Launcher UK programme which included University of Strathclyde, BAE Systems, Fluid Gravity Engineering Ltd, University of Glasgow and others and was led by Orbital Access Ltd.

Conflicts of Interest: The authors declare no conflict of interest. The funders had no role in the design of the study; in the collection, analyses, or interpretation of data; in the writing of the manuscript, or in the decision to publish the results.

References

1. Maddock, C.A.; Ricciardi, L.; West, M.; West, J.; Kontis, K.; Rengarajan, S.; Evans, D.; Milne, A.; McIntyre, S. Conceptual design analysis for a two-stage-to-orbit semi-reusable launch system for small satellites. *Acta Astronaut.* **2018**, *152*, 782–792. [CrossRef]
2. Ricciardi, L.A.; Vasile, M. MODHOC: Multi Objective Direct Hybrid Optimal Control. In Proceedings of the International Conference on Astrodynamics Tools and Techniques, Oberpfaffenhofen, Germany, 6–9 November 2018.
3. Vasile, M. Multi-objective optimal control: A direct approach. In *Satellite Dynamics and Space Missions*; Springer: Berlin/Heidelberg, Germany, 2019; pp. 257–289. [CrossRef]
4. Ross, I.M.; Proulx, R.J.; Karpenko, M. Unscented optimal control for space flight. In Proceedings of the International Symposium on Space Flight Dynamics, Laurel, MD, USA, 5–9 May 2014.
5. Julier, S.J.; Uhlmann, J.K. *A General Method for Approximating Nonlinear Transformations of Probability Distributions*; Technical Report; Robotics Research Group, Department of Engineering Science, University of Oxford: Oxford, UK, 1996.
6. Ross, I.M.; Proulx, R.J.; Karpenko, M.; Gong, Q. Riemann—Stieltjes optimal control problems for uncertain dynamic systems. *J. Guid. Control Dyn.* **2015**, *38*, 1251–1263. [CrossRef]
7. Ozaki, N.; Campagnola, S.; Funase, R.; Yam, C.H. Stochastic differential dynamic programming with unscented transform for low-thrust trajectory design. *J. Guid. Control Dyn.* **2018**, *41*, 377–387. [CrossRef]
8. Longuski, J.M.; Guzmán, J.J.; Prussing, J.E. *Optimal Control with Aerospace Applications*; Springer: Berlin/Heidelberg, Germany, 2014.
9. Ross, I.M.; Karpenko, M.; Proulx, R.J. Path constraints in tychastic and unscented optimal control: Theory, application and experimental results. In Proceedings of the American Control Conference, Boston, MA, USA, 6–8 July 2016. [CrossRef]
10. Ricciardi, L.A.; Maddock, C.A.; Vasile, M. Direct solution of multi-objective optimal control problems applied to spaceplane mission design. *J. Guid. Control Dyn.* **2019**, *42*, 30–46. [CrossRef]
11. Uhlmann, J.K. Dynamic Map Building and Localisation: New Theoretical Foundations. Ph.D. Thesis, University of Oxford, Oxford, UK, 1995.
12. Van der Merwe, R.; Wan, E.A. The square-root unscented Kalman filter for state and parameter-estimation. In Proceedings of the IEEE International Conference on Acoustics, Speech and Signal Processing, Salt Lake City, UT, USA, 7–11 May 2001. [CrossRef]
13. Vasile, M. Finite Elements in Time: A Direct Transcription Method for Optimal Control Problems. In Proceedings of the AIAA/AAS Astrodynamics Specialist Conference, Toronto, ON, Canada, 2–5 August 2010. [CrossRef]
14. Vasile, M.; Finzi, A. Direct Lunar Descent Optimisation by Finite Elements in Time Approach. *Int. J. Mech. Control* **2000**, *1*.
15. Darehmiraki, M.; Farahi, M.H.; Effati, S. A Novel Method to Solve a Class of Distributed Optimal Control Problems Using Bezier Curves. *J. Comput. Nonlinear Dyn.* **2016**, *11*, 061008. [CrossRef]
16. Betts, J.T. *Practical Methods for Optimal Control and Estimation Using Nonlinear Programming*; Advances in Design and Control; SIAM: Philadelphia, PA, USA, 2010.
17. Chankong, V.; Haimes, Y.Y. *Multiobjective Decision Making: Theory and Methodology*; Courier Dover Publications: Mineola, NY, USA, 2008.
18. Hillermeier, C. *Nonlinear Multiobjective Optimization*; International Series of Numerical Mathematics; Birkhäuser Basel: Basel, Switzerland, 2001. [CrossRef]
19. Zuiani, F.; Kawakatsu, Y.; Vasile, M. Multi-objective optimisation of many-revolution, low-thrust orbit raising for Destiny mission. In Proceedings of the AAS/AIAA Space Flight Mechanics Conference, Kauai, HI, USA, 10–14 February 2013.
20. Ricciardi, L.A.; Vasile, M. Improved archiving and search strategies for Multi-agent Collaborative Search. In Proceedings of the International Conference on Evolutionary and Deterministic Methods for Design, Optimization and Control with Applications to Industrial and Societal Problems, Glasgow, UK, 14–16 September 2015.
21. Ricciardi, L.A.; Vasile, M.; Maddock, C. Global solution of multi-objective optimal control problems with Multi Agent Collaborative Search and Direct Finite Elements Transcription. In Proceedings of the IEEE Congress on Evolutionary Computation, Vancouver, BC, Canada, 24–29 July 2016. [CrossRef]
22. Ricciardi, L.A.; Vasile, M.; Toso, F.; Maddock, C.A. Multi-objective optimal control of the ascent trajectories of launch vehicles. In Proceedings of the AIAA/AAS Astrodynamics Specialist Conference, Napa, CA, USA, 15–18 February 2016.
23. Pascoletti, A.; Serafini, P. Scalarizing vector optimization problems. *J. Optim. Theory Appl.* **1984**, *42*, 499–524. [CrossRef]
24. Eichfelder, G. *Adaptive Scalarization Methods in Multiobjective Optimization*; Springer: Berlin/Heidelberg, Germany, 2008. [CrossRef]
25. Chasalow, S.D.; Brand, R.J. Algorithm AS 299: Generation of Simplex Lattice Points. *J. R. Stat. Soc. Ser. C (Appl. Stat.)* **1995**, *44*, 534–545. [CrossRef]

26. Bruinsma, S.; Arnold, D.; Jäggi, A.; Sánchez-Ortiz, N. Semi-empirical thermosphere model evaluation at low altitude with GOCE densities. *J. Space Weather Space Clim.* **2017**, *7*, A4. [CrossRef]
27. Picone, J.M.; Hedin, A.E.; Drob, D.P.; Aikin, A.C. NRLMSISE-00 empirical model of the atmosphere: Statistical comparisons and scientific issues. *J. Geophys. Res. Space Phys.* **2002**, *107*, SIA 15-1–SIA 15-16. [CrossRef]
28. Bowman, B.; Tobiska, W.K.; Marcos, F.; Huang, C.; Lin, C.; Burke, W. A new empirical thermospheric density model JB2008 using new solar and geomagnetic indices. In Proceedings of the AIAA/AAS Astrodynamics Specialist Conference, Cambridge, MD, USA, 29 June–2 July 2008.
29. Bruinsma, S. The DTM-2013 thermosphere model. *J. Space Weather Space Clim.* **2015**, *5*, A1. [CrossRef]
30. Niederreiter, H. *Random Number Generation and Quasi-MONTE Carlo Methods*; SIAM: Philadelphia, PA, USA, 1992.
31. Zipfel, P. *Modeling and Simulation of Aerospace Vehicle Dynamics, Second Edition*; AIAA Education Series; AIAA: Reston, VI, USA, 2007. [CrossRef]
32. Stindt, T.; Merrifield, J.; Fossati, M.; Ricciardi, L.A.; Maddock, C.A.; West, M.; Kontis, K.; Farkin, B.; McIntyre, S. Aerodynamic database development for a future reusable space launch vehicle, the Orbital 500R. In Proceedings of the International Conference on Flight Vehicles, Aerothermodynamics and Re-entry Missions and Engineering, Monopoli, Italy, 30 September–3 October 2019.

Article

Multi-Objective Optimization of Plastics Thermoforming

António Gaspar-Cunha [1,*], Paulo Costa [1], Wagner de Campos Galuppo [1], João Miguel Nóbrega [1], Fernando Duarte [1] and Lino Costa [2]

[1] IPC—Institute of Polymer and Composites, University of Minho, 4800-050 Guimarães, Portugal; byic.mail@gmail.com (P.C.); wagnergaluppo@gmail.com (W.d.C.G.); mnobrega@dep.uminho.pt (J.M.N.); fduarte@dep.uminho.pt (F.D.)

[2] ALGORITMI Center, University of Minho, 4800-050 Guimarães, Portugal; lac@dps.uminho.pt

* Correspondence: agc@dep.uminho.pt

Abstract: The practical application of a multi-objective optimization strategy based on evolutionary algorithms was proposed to optimize the plastics thermoforming process. For that purpose, in this work, differently from the other works proposed in the literature, the shaping step was considered individually with the aim of optimizing the thickness distribution of the final part originated from sheets characterized by different thickness profiles, such as constant thickness, spline thickness variation in one direction and concentric thickness variation in two directions, while maintaining the temperature constant. As far we know, this is the first work where such a type of approach is proposed. A multi-objective optimization strategy based on Evolutionary Algorithms was applied to the determination of the final part thickness distribution with the aim of demonstrating the validity of the methodology proposed. The results obtained considering three different theoretical initial sheet shapes indicate clearly that the methodology proposed is valid, as it provides solutions with physical meaning and with great potential to be applied in real practice. The different thickness profiles obtained for the optimal Pareto solutions show, in all cases, that that the different profiles along the front are related to the objectives considered. Also, there is a clear improvement in the successive generations of the evolutionary algorithm.

Keywords: plastics thermoforming; sheet thickness distribution; evolutionary algorithms; multi-objective optimization

Citation: Gaspar-Cunha, A.; Costa, P.; Galuppo, W.d.C.; Nóbrega, J.M.; Duarte, F.; Costa, L. Multi-Objective Optimization of Plastics Thermoforming. *Mathematics* **2021**, *9*, 1760. https://doi.org/10.3390/math9151760

Academic Editors: Hongyu Liu and Ali Farajpour

Received: 31 May 2021
Accepted: 21 July 2021
Published: 26 July 2021

Publisher's Note: MDPI stays neutral with regard to jurisdictional claims in published maps and institutional affiliations.

Copyright: © 2021 by the authors. Licensee MDPI, Basel, Switzerland. This article is an open access article distributed under the terms and conditions of the Creative Commons Attribution (CC BY) license (https://creativecommons.org/licenses/by/4.0/).

1. Introduction

Thermoforming is a thermoplastic processing technique commonly used in the rigid packaging industry. A variety of thermoplastic materials can be used in this process, including semi-crystalline polymers, such as High Density Polyethylene (HDPE) and Polypropylene (PP) and amorphous polymers, such as Acrylonitrile Butadiene Styrene (ABS), Polystyrene (PS) and High Impact Polystyrene (HIPS). HIPS is a lightweight and inexpensive thermoplastic often used in thermoformed food and pharmaceutical packaging containers.

Thermoforming comprises a sequence of interdependent operations and is characterized by being sensitive to the intrinsic properties of thermoplastics, namely the lower heat conduction and the deformation capability strongly dependent on temperature. In general, the thermoforming comprises: a heating stage, which aims to allow the sheet to acquire the required deformability; a sheet deformation stage in order to reproduce the contours of the piece and finally, a cooling stage, which allows the part to be extracted from the mold without distorting. In this way, the final performance of thermoformed products results from the sum of all actions that occur in these three main stages. Since there are processing variables associated with each of the three stages, including the material properties as a function of temperature, optimizing the thermoforming process is a complex task.

Generally, the optimization of thermoforming, like in the other real word optimization problems, consists of relating the effect of the operating variables of each stage with the performance of the part. Since thermoformed parts are characterized by having a non uniform thickness that can hinder their performance, the thickness distribution of the final part is one of the most used variables to characterize the performance of the part. Furthermore, the effects of processing parameters on the thermoforming of polymeric sheets are highly nonlinear and fully coupled, which increases the difficulty of the process design. The following studies aim at optimizing the sheet deformation stage with the objective of obtaining thermoformed parts with the most uniform thickness distribution.

Yang and Hung [1] proposed an inverse Artificial Neural Network (ANN) with the aim of predicting the optimum processing conditions (decision variables), including sheet temperature, vacuum pressure, plug speed and displacement inside the mold. The network inputs are the thickness distribution at different positions of molded PET parts, and the outputs are the processing conditions obtained by the ANN presented, which show a good agreement between the computed result and experimental data. However, the ANN was trained using experimental data and the authors were not clear about the optimization method used. Also, the fact that different inputs for this inverse ANN can produce identical results at the output side was not discussed.

Chang et al. [2] used a similar inverse ANN to obtain the optimal processing parameters of polypropylene foam thermoforming. The studied variables included the mold temperature, plug speed and displacement, vacuum pressure and time and the heat transfer coefficient of the plug. Experimental data from tests carried out on a lab-scale thermoforming machine were used to train the ANN, used as an inverse model of the process. Product dimensions were used as the inverse model inputs and the corresponding processing parameters as outputs. The feasibility of the proposed method was demonstrated by experimental manufacturing of cups with optimal geometry derived from the computational method. In almost all points, the deviations between predicted and measured points were all below 3.5%. Like in the previous paper, the authors did not take into account that different inputs can produce identical results.

Leite et al. [3,4] developed models to predict and optimize the thermoforming using ANN defining the processing parameters set as the networks' inputs and deviations in part thickness as the outputs. For the ANN data, thermoformed samples were experimentally produced using a 1 mm polystyrene sheet, using a fractional factorial design (2^{k-p}). The studied processing variables included heating time and the electric heating power of the heater panel vacuum time pressure. Preliminary computational studies were carried out with various ANN structures and configurations with the test data, until reaching satisfactory models and, afterward, multi-criteria optimization models were developed. The validation tests were developed with the models' predictions and solutions showed that the estimates for them have prediction errors within the limit of values found in the samples produced. Thus, it was demonstrated that, within certain limits, the ANN models are valid to simulate the vacuum thermoforming process using multiple parameters, decision variables and objectives, by means of reduced data quantity.

Sasimowski [5] used experimental data to determine a utility function by regression analysis in order to calculate the thickness distribution of the thermoformed parts. This utility function was used to optimize the operating conditions of the machine, namely heating time, heater temperature, pre-blow time, vacuum time and cooling time. In all of these works, the ANN was used to predict the behavior evolution of the thermoformed parts thickness during information. Thus, no specific optimization method was described/applied to optimize the results that the trained ANN predicted, being these modeling results were used to optimize the process based on empirical knowledge.

In all of these works, it was not clear what is the optimization method used. In fact, ANN is not an optimization method by itself, but a way of predicting output values from a set of input data, obtained experimentally or computationally, very similar to a regression model. The ANN, in the present context, works as a modeling program with the aim of predicting the process performance. To optimize the process, it is necessary to link a modeling routine with an optimization method through an optimization methodology able to deal with the objectives to be optimized, as will be described in Section 3. Additionally, most of these works are based on experimental data, which makes the process unrealistic given the time required to do the experiments.

In addition to the general studies on the optimization of thermoforming, it is possible to find in the literature studies aiming at controlling the different stages of the process, namely the heating and forming stages.

Regarding the heating stage, the main objective is controlling the sheet temperature and the temperature fields developing during this step. This is very complex since the heating methods are indirect, that is, the sheet temperature is controlled by adjusting the heaters variables, and not directly. Furthermore, determining an optimum processing window in thermoforming process with the aim of achieving high quality parts is critical. In practice, the infrared heating stage is crucial since the final thickness distribution of the thermoformed part is closely related to the temperature of the sheet.

Wang and Nied [6] used a numerical approach based on the finite element method to obtain inverse solutions for the thermoforming processes. This was done by specifying a desired final thickness distribution and iteratively solving the system for the temperature field needed to obtain the desired result. A uniform initial temperature distribution is used as the initial guess for the iterative optimization procedure. Subsequently, updated non-uniform initial temperatures are obtained, based on the complete process of simulation for achieving the final thickness distribution during thermoforming. Suitable inverse solutions are achieved once the desired thickness distribution is obtained within a specified tolerance. The sensitivity of final part thickness to perturbations in temperature distribution is also investigated and shown to be a potential problem for precise thickness control in industrial applications.

Bordival et al. [7] developed an automatic optimization method of the ovens geometric parameters to be used in thermoforming. The first time, a simple analytical model, coupled to a nonlinear constraint optimization method (Sequential Quadratic Programming) allows one to find the best set of parameters, according to a cost function representing, for example, the heat flux uniformity. Then, with these optimized parameters, an accurate raytracing method is used to compute the irradiation resulting from the interaction between lamps and the thermoplastic sheet. Finally, a control volume method is implemented to solve the three-dimensional transient heat transfer equation, where the radiation source is approximated by a diffusion Rosseland model.

Chy and co-authors [8,9] presented a method to control the surface temperature of a plastic sheet using model predictive control (MPC) to solve the inverse heating problem (IHP). The model was implemented on a complex thermoforming oven with a large number of inputs and outputs for precise control of sheet temperatures under hard constraints on heater temperature and their rates. Even though the MPC controller can handle a multivariable process, the large number of computations makes it difficult to apply to large systems such as multi-zone temperature control in a thermoforming machine.

Li and co-authors [10,11] suggested a methodology to compute and optimize the sheet temperature by controlling the heater temperature. The steady-state optimum distribution of heater power is first ascertained by a numerical optimization to obtain a uniform sheet temperature. The time-dependent optimal heater input is then determined to decrease the temperature difference through the direction of the thickness using the response surface method and the D-optimal method. The results show that the time-dependent optimum heater power distribution gives an acceptable uniform sheet temperature in the forming temperature range by the end of the heating process.

More recently, Erchiqui and co-authors [12–15] proposed the application of two different meta-heuristic algorithms, Simulated Annealing (SA) and Evolutionary Algorithms (EA) to detect, from a fixed and random set of temperatures of the radiant zones of the oven, the best temperatures that must be assigned to the heating zones in order to ensure a uniform sheet temperature. For numerical heating analysis, the nonlinear heat conduction problem is solved by a specific 3D volumetric enthalpy-based computational method.

However, these studies only considered the effects of the heating phase on the final thickness distribution of the thermoformed parts, having a starting point a sheet with constant thickness, that is, the aim is to determine how to heat the different zones of the sheet in order to induce uniform thickness on the final part.

Furthermore, an important limitation of the methodologies proposed in the literature lies in the fact that the process is intrinsically multi-objective, and the different stages cannot be considered independent. In reality, the results of one stage of the process depend strongly on the remaining stages, the final part thickness distribution does not depend only on the heating, but also on the initial thickness of the sheet used. Therefore, the main aim of the work is to apply multi-objective strategies to optimize the plastics thermoforming process. The intention is to propose a more global methodology that can take into account the different steps and characteristics of this process, as described in the next section. However, due to its complexity, only the stage of the sheet deformation will be considered here. More specifically, the aim is to study the influence of the initial sheet thickness distribution on the optimization of the process. For that purpose, three different initial thickness distribution sheets will be considered. As far as we know, there is no work in the literature that addresses the problem from this point of view. From an industrial perspective, the fabrication of the sheets to be thermoformed must be changed if economic and/or environmental gains can be obtained.

The paper is organized as follows: in Section 2 the details of thermoforming related to optimization are explained, Section 3 addresses the concepts of multi-objective optimization and the algorithm used, in Section 4 the results and discussion for the optimization of thermoforming are presented and in Section 5 the conclusions are stated.

2. Thermoforming

2.1. The Process

Figure 1 illustrates schematically the thermoforming phases and the optimization sequence. The phase order is indicated by open arrows and consist basically of: (A) producing a plastic sheet, typically by extrusion; (B) heating the sheet until it can be deformed without breaking; (C) shaping the sheet against the contours of a mold by applying a pressure difference on both sides, either by vacuum or pressurized air; (D) cooling the product obtained to make it possible to remove it from the mold and (E) remove the part from the mold. Each one of these phases has particular characteristics that must be considered when optimizing the process, either in terms of what concerns the operating conditions (e.g., heating and cooling times, air pressure and oven temperatures) and/or design parameters (e.g., sheet thicknesses, heater location, heating methods and mold geometry) [16,17].

Figure 2 shows schematically how the shaping process evolves. The heated sheet is forced to deform by the action of a pressure differential between both sides of the sheet. Initially, this deformation is uniformly distributed, but when it touches the cold mold surfaces the plastic material is frozen and only the remaining part of the sheet continues to deform. This implies that the last part of the sheet touching the mold will present the lowest thickness, since that the total volume of the sheet is conserved. This corresponds to the region of the lower corners in Figure 2B.

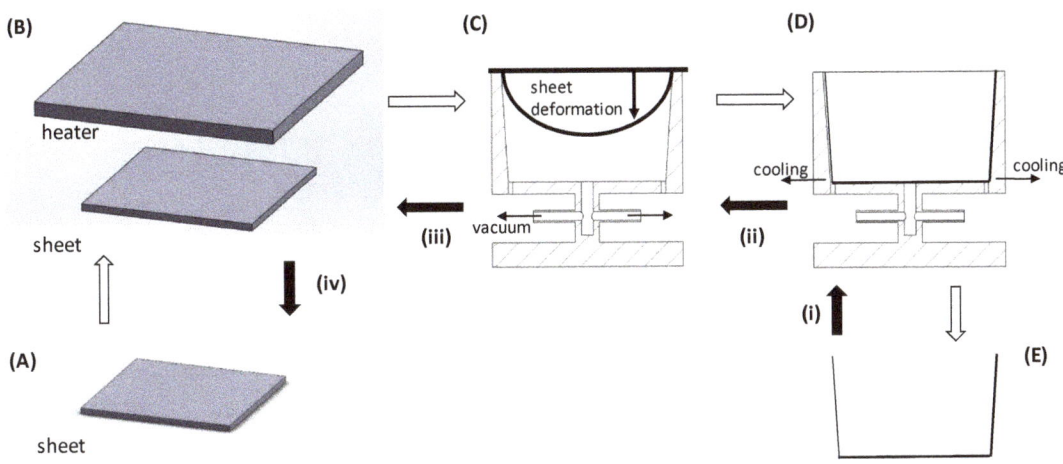

Figure 1. Thermoforming process and optimization. Thermoforming phases: (**A**) sheet; (**B**) heating; (**C**) shaping; (**D**) cooling; (**E**) final part. Optimization steps: (**i**) part properties; (**ii**) cooling effects; (**iii**) final part thickness distribution; (**iv**) sheet thickness distribution.

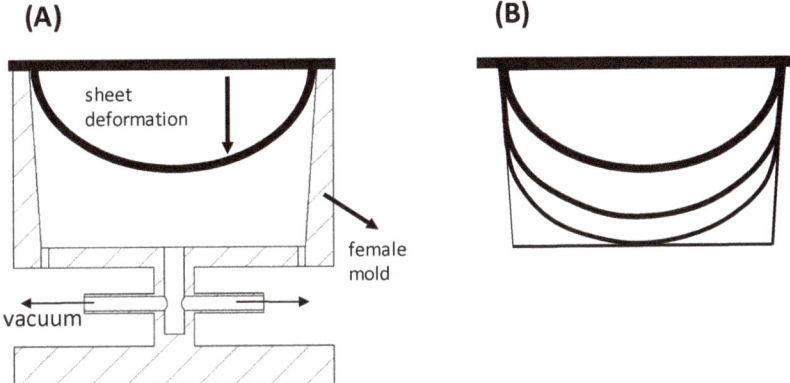

Figure 2. Thermoforming process: (**A**) sheet deformation in the mold; (**B**) evolution of the part thickness during deformation.

Therefore, the most important objective when producing a thermoformed part is to guarantee that its final thickness is as uniform as possible. This requirement is necessary in order to be possible to accomplish two main objectives, mainly in the shaping phase: to minimize the amount of material necessary and to maximize the mechanical behavior and/or other required characteristics.

To produce a part with uniform thickness, it is possible to act in different phases of the process. First, as shown in Figure 3, it is possible to produce sheets with different thicknesses at different regions of the part to be produced. Therefore, in the regions where more deformation occurs during the shaping phase, and, as a consequence, the thickness of the part will be smaller, it is possible to increase the thickness of the original sheet. Three situations are possible, as illustrated in Figure 3: (A) constant sheet thickness, (B) variable sheet thickness in one direction (x) and (C) variable sheet thickness in two directions (x and y). However, the processes used to produce the sheets are different, which must be taken into account in any optimization process since the costs can be very different as well. While (A) and (B) can be produced by extrusion, (C) must be produced by injection molding, given the specific and complicated geometry. While the sheet with constant thickness (A) can be produced in the usual way, using a flat die and a calender, to produce sheet (B) it is necessary to change the die and the calender rolls in order to induce different thicknesses

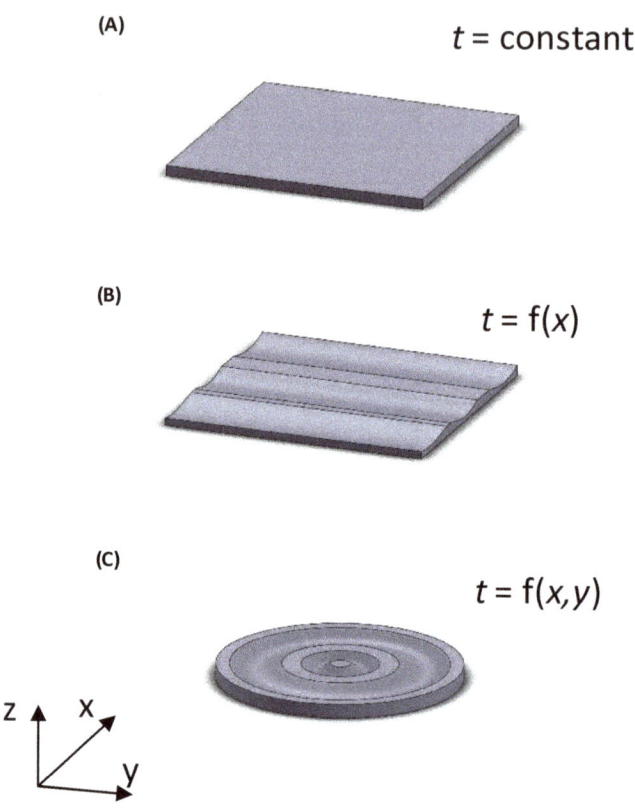

Figure 3. Types of sheets that can be used: (**A**) constant thickness; (**B**) spline thickness variation on x direction and (**C**) concentric thickness variation (direction x,y).

Secondly, the sheet deformation during the shaping phase depends on the mechanical properties of the polymeric material used at the heating temperature. This temperature dependency makes it possible to obtain different sheet deformations in order to control the thickness distribution, that is, in the regions where more deformation is required, the temperature must be lower [11,18]. Figure 4 illustrates schematically the process of heating the plastic sheet. Several parameters can be considered: (i) the number of heaters, for example, for sheets with higher thickness two heaters can be used, one each side; (ii) the distance between the heaters bank and the sheet; (iii) the dimensions of each individual heater and (iv) temperature of each individual heater. This approach is illustrated in Figure 4, where a mesh of heaters (e.g., ceramic heaters) can be used, each one with the possibility of having different temperature distributions. In any case, it is fundamental to control the correct temperature in the entire sheet surface. In practice, and given the lower heat conductivity of plastics, a temperature gradient along the thickness of the sheet will arise. Furthermore, and if radiation heating is used, due to geometric reasons, the temperature at the center of the sheet will be higher than in its edge [19,20]. From the practical point of view and after the heating phase, the temperature of the entire sheet must be in the range of the forming temperatures. This temperature window is intrinsic to each polymer.

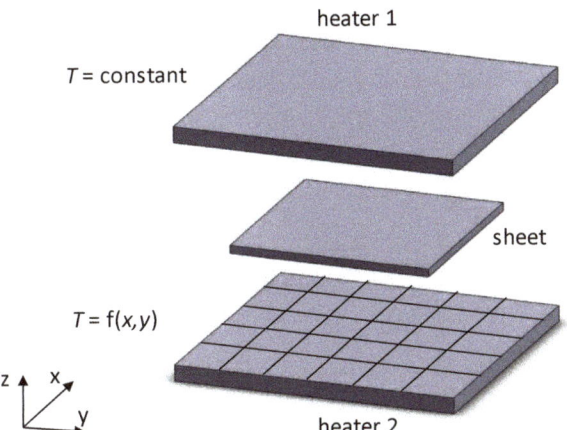

Figure 4. Heating phase: one or two heaters with constant or variable temperature.

Third, a straightforward way of minimizing the difference of thicknesses in the part thermoformed is related to the process of shaping. For that purpose, several thermo-forming variants can be used: female mold (as illustrated in Figure 2), male mold, plug assisted, bubble forming, bubble forming and plug assisted and other combinations of the above [21–25].

2.2. Numerical Modeling

During the inflation process, the polymer sheet deforms due to the pressure difference imposed on both sides. Additionally, due to the deformation process, internal stresses are generated, which will limit the deformation promoted by the pressure difference. If the balance from those two loads is not null the velocity of the sheet changes, that is, there is a non-null acceleration.

For the adopted numerical approach, the polymer sheet is discretized into different computational cells as illustrated in Figure 5A. Moreover, the sheet is assumed to be thin enough to allow resorting to a membrane formulation, thus it is discretized into planar triangular cells each one with thickness tk (Figure 5B).

Considering the conditions above, the conservation of linear momentum is given by:

$$\rho \frac{\partial^2 u}{\partial t^2} = \nabla \cdot \sigma + \rho g, \qquad (1)$$

where t is time, ρ is the density, u is the displacement vector, σ is the Cauchy stress tensor and g is the gravitational acceleration vector. Following the classical Finite Volume Method approach, Equation (1) is integrated into the space in a computational cell with volume Ω and surrounded by the surface Γ, which, after applying the Gauss Divergence theorem results in:

$$\underbrace{\int_\Omega \rho \frac{\partial^2 u}{\partial t^2} d\Omega}_{\text{Inertia}} = \underbrace{\oint_\Gamma n \cdot \sigma d\Gamma}_{\text{Surface Forces}} + \underbrace{\int_\Omega \rho g d\Omega}_{\text{Body Forces}}, \qquad (2)$$

which shows that the acceleration of the computational cell is affected by the surface and body forces. For the body forces, only the weight due to gravity is considered, which can be obtained by:

$$F_{Body} = \rho g V_c \qquad (3)$$

where V_c is the computational cell volume.

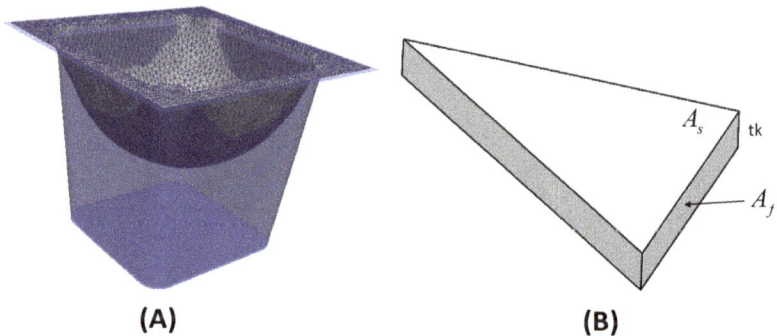

Figure 5. Geometry and illustrative mesh (**A**) and computational cell (**B**) of the polymeric sheet.

The surface forces are of two types, namely external and internal forces. Following the membrane formulation, the external forces are due to the pressure difference between the inner and outer surfaces and the internal forces result from the stress induced by the material deformation. Based on Equation (2), where, in each time step, the acceleration of each computational cell is an explicit function of the applied loads, the surface forces are the sum of the internal and external forces and can be calculated by the following equation:

$$F_{Surface} = \oint_\Gamma \mathbf{n} \cdot \boldsymbol{\sigma} d\Gamma = \int_{A_S} p \mathbf{n}_s dA_S + \int_{A_f} \mathbf{n}_f \cdot \boldsymbol{\tau} dA_f = F_{pressure} + F_{internal} \quad (4)$$

where p is the imposed pressure difference, $\boldsymbol{\tau}$ the internal stress vector, A_S the cell surface area (see Figure 5), with normal vector $\mathbf{n}_s$ and A_f the cross-sectional cell area, with normal vector $\mathbf{n}_f$. The internal forces are calculated at each cell cross section along the thickness. Due to the quick deformation process that takes place in thermoforming, which has been shown to be mostly elastic since the original shape of the sheet if the plastic part is heated [26], for this problem, the relation between the internal stresses was assumed to be based on the elasticity theory:

$$\boldsymbol{\tau} = \mathbb{C} : \boldsymbol{\epsilon} \quad (5)$$

where $\mathbb{C}$ is the stiffness fourth-order constitutive tensor and $\boldsymbol{\epsilon}$ the strain tensor. In this, $\boldsymbol{\tau}$ and $\boldsymbol{\epsilon}$ must be work conjugate pairs of stresses and strains respectively. Considering the membrane formulation, the elastic solid undergoes through large displacements, large rotations and large strains and non-linearity cannot be neglected, the proper work conjugate pair of stresses and strains are the Second Piola–Kirchhoff stress tensor, for $\boldsymbol{\tau}$ and the Green–Lagrange strain tensor, for $\boldsymbol{\epsilon}$, which is given by:

$$\boldsymbol{\epsilon} = \frac{1}{2}\left(\mathbf{F}^T \cdot \mathbf{F} - \mathbf{I}\right) \quad (6)$$

In Equation (6), $\mathbf{F}$ is the deformation gradient tensor, given by the derivative of each component of the deformed position $\mathbf{x}$ vector with respect to each component of the reference position $\mathbf{X}$ vector, and $\mathbf{I}$ is the unit tensor. This formulation is equivalent to the hyperelastic Saint Venant–Kirchhoff model [27].

On the implemented explicit numerical procedure, at each time step all the loads are calculated with Equations (3)–(7) and the acceleration of each cell is given by Equation (2). This allows one to calculate the new cell position and advance to the next time step.

On all the numerical studies performed in this work an elastic modulus of 10 MPa was employed, which is representative of HIPS at the usually employed processing temperature of 140 °C [28]. Moreover, the material was assumed to be incompressible, thus, a Poisson's ratio of 0.5 was considered.

2.3. Experimental Assessment

The modeling computational results were assessed using the circular cup illustrated in Figure 6 [29]. The experimental work was performed on a modified Illig thermoforming machine that allows individual monitoring and control of the main thermoforming variables. The heater bank included 110 Elstein 125 W ceramic radiators (each 120 × 60 mm). Square portions of the sheet were cut and manually clamped (fixed) on all four sides, in such a way that the material was not allowed to slide into the mold cavity. Once the heating stage is completed and a uniform sheet temperature of 150 °C is reached, the frame (holding the sheet) moves toward the forming station. The lower mold section moves vertically, clamping the sheet circularly. The parts were vacuum-formed at 0.7 bar. After forming and cooling, the cups were extracted, cut into four equal portions along a line passing by the center of the base midpoint and their thicknesses along that line and circumferentially were measured with a caliper (precision of 0.01 mm). At least five cups were produced for each set of variables. The results are generally presented along the arc length but correspond to averages in the circumferential direction. The material used in the sheets is a mixture of 50% Polystyrene (PS Laquerene 1540) with 50% Impact Polystyrene (HIPS Laquerene 7240), both produced by Elf-Atochem.

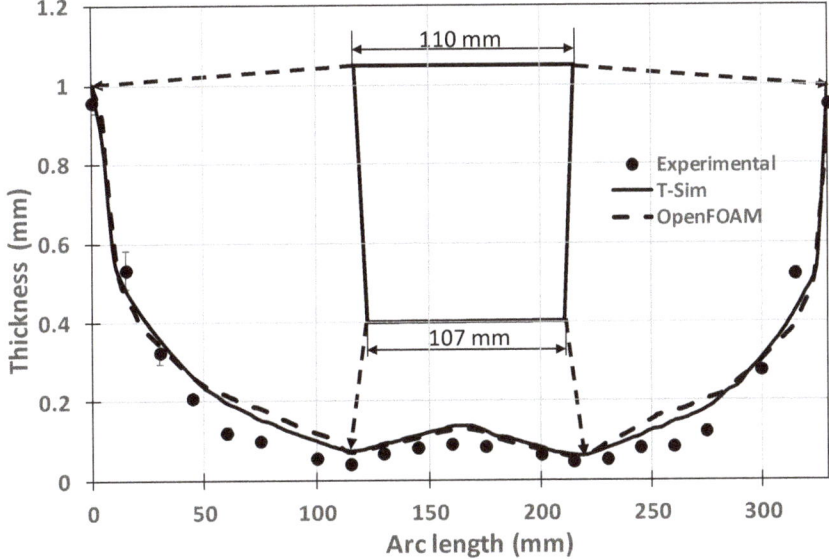

Figure 6. Experimental assessment of the computational results.

The theoretical results were computed using a commercial software T-SIM using a K-BKZ viscoelastic model whose material properties are described in [29] and by using the OpenFOAM considering the hyperelastic Saint Venant–Kirchhoff model for the material. As can be seen, the computational results show a good agreement with the experimental measurements, the behavior of both theoretical models being very similar. In fact, regarding the simulation of the thickness distribution in thermoforming, it is possible to find, throughout the literature, several works that use hyperplastic models (e.g., Ghobadnam [30], Bernard et al. [31], Oueslati et al. [32], Jeet et al. [33]) and viscoelastic models (e.g., Cha et al. [34], Atami et al. [28]). Although researchers are increasingly turning to viscoelastic models for thermoforming applications, according to O'Connor et al. [35], there is little agreement on the best model to use for any given polymer material.

Furthermore, the use of hyperelastic models is based on the experience reported by Schmidt and Carley [26] and Rosenzweig et al. [36]. They described the "plastic memory", where blown bubbles returned completely to their original flat sheet form, either by suddenly releasing the forming pressure or by annealing spheroidal bubbles at a proper temperature. The observation of complete recovery has led the authors to conclude that elastic-like behavior can be ascribed to the processes of blowing soft plastic sheets.

Therefore, and considering that the main objective of this work is to evaluate the optimization process and not to predict more accurately the thickness distribution, and for computation time reasons, a hyperelastic model was chosen to predict the plastic sheet behavior during the forming process.

3. Multi-Objective Optimization

3.1. Multi-Objective Evolutionary Algorithms

In a Multi-Objective Optimization Problem (MOP), the goal is to minimize all objectives simultaneously, that is, to find feasible solutions where every objective function is minimized. A multi-objective optimization problem with m objectives and n decision variables can be formulated as follows:

$$\begin{aligned} \min_{x \in \Omega} \quad & f(x) \equiv (f_1(x), \ldots, f_m(x)) \\ \text{subject to} \quad & g_i(x) \leq 0, i \in \{1, \ldots, p\} \\ & h_j(x) = 0, j \in \{1, \ldots, q\} \\ & l \leq x \leq u \end{aligned} \quad (7)$$

where x is the decision vector, i.e., $x \in \Omega \subseteq R^n$, f is the objective vector of m objective functions, that is, $f \in \Omega \subseteq R^m$, g_i are p inequality constraint functions and h_j are q equality constraint functions and l and u are the vectors of the lower and upper bounds on decision variables, respectively.

Solutions are compared in terms of Pareto dominance, that is, a solution x is said to dominate a solution y ($x \prec y$), if and only if $f_i(x) \leq f_i(y)$, for all $i \in \{1, \ldots, m\}$ and $f_j(x) < f_j(y)$ for at least one $j \in \{1, \ldots, m\}$. If all objectives do not conflict with each other, there exists a unique solution that minimizes all the objectives. In general, there are multiple conflicting objectives giving rise to a set of optimal solutions—the Pareto optimal set, instead of a single optimal solution. A feasible solution, x^* is Pareto optimal if and only if there is no other solution $y \in \Omega$, $y \neq x^*$, that $y \prec x^*$. These solutions are incomparable to each other since none of these solutions can be said to be better than others.

In order to facilitate the decision-making process, an *a posteriori* method in which the search for an approximation (as close and diverse as possible) to the Pareto optimal set is performed before the decision-making process. Thus, the decision-maker can select the most suitable solution from this set according to their preferences. The Pareto optimal set provides valuable information with respect to the trade-offs between alternatives.

There is a large variety of approaches to solving MOPs. The designated classical (or traditional methods) are known as scalarization methods in which the MOP is reformulated and solved as a single objective optimization problem [37]. However, usually, scalarization functions involve several parameters that can be difficult to define in order to obtain different approximations to the Pareto optimal solutions. Other approaches are based on Evolutionary Algorithms (EAs). EAs are meta-heuristics that mimic the process of evolution of a population of individuals over generations. The fittest individuals to the environment will survive over time and therefore will be likely to reproduce. The offspring are generated by genetic operators, such as crossover and mutation and inherit the parent characteristics. Basically, in EAs, a population of individuals representing a candidate solution to the problem evolves using two mechanisms: Selection and Variation. The quality of each individual is measured using a fitness function that, in optimization problems, is related to the objective or objectives functions for single or multi-objective optimization problems, respectively. The selection mechanism guarantees that the best individuals have a higher probability of being selected for generating offspring. The variation mechanism provides the generation of new individuals by application of genetic operators.

Multi-Objective Evolutionary Algorithms (MOEAs) are EAs for solving MOPs. There are advantages to using this type of algorithms. They work with a population of candidate solutions, which makes it possible to approximate the entire Pareto optimal set in a single run. The performance of any MOEA is strongly related to the efficacy of its Selection mechanism that guides the search in the objective space, balancing convergence and diversity and also the variation mechanism that is responsible for the generation of offspring. A common approach to simulating natural selection in MOEAs consists of assigning fitness values to individuals in the population according to its quality for the MOP being solved. In terms of the type of fitness function used, MOEAs can be classified into three different types: dominance-, scalarizing- and indicator-based algorithms.

Dominance-based approaches calculate an individual's fitness on the basis of the Pareto dominance relation [38] or according to different criteria [39]. Scalarizing-based approaches [40] incorporate traditional mathematical techniques based on the aggregation of multiple objectives into a single parameterized function. Indicator-based approaches use performance indicators for fitness assignment; pairs of individuals are compared using some quality measure such as the hypervolume indicator [41,42]. The fitness value reflects the loss in quality if a given solution is removed [42,43].

3.2. SMS-EMOA

The multi-objective evolutionary optimization algorithm used in this work is based on the SMS-EMOA [32] and implemented in MATLAB. The outline of the SMS-EMO is given by Algorithm 1.

Algorithm 1 SMS-EMOA

$P_0 \leftarrow initialize()$ % Initialize population at random with μ individuals
$k \leftarrow 0$
Repeat
    $q_{k+1} \leftarrow generate(P_k)$ % generate offspring by genetic operators
    $P_{k+1} \leftarrow selection(P_k \cup \{q_{k+1}\})$ % select μ best individuals
$k \leftarrow k+1$
Repeat until the stopping criterion is fulfilled

The search starts from an initial population of μ individuals randomly generated satisfying the boundary constraints of the decision variables. In each generation, one parent is selected at random from the population and a single offspring is produced by application of a variation procedure. In this procedure, a Gaussian mutation with covariance matrix adaptation [44] is applied to the parent to produce a single offspring. Next, a deterministic selection procedure selects the μ best individuals to the next generation. The selection involves the non-dominated ranking of the population and the computation of the hypervolume contribution of each individual of the population.

The non-dominated sorting procedure is illustrated in Figure 7. First, the $P_k \cup \{q_{k+1}\}$ individuals are ranked according to a non-dominated sorting procedure defining f fronts of sets of non-dominated individuals [28]. A rank is assigned to each front representing its level of domination. All solutions belonging to each front are incomparable. The first front F_1 contains the non-dominated solutions in $P_k \cup \{q_{k+1}\}$. The second front F_2 contains all non-dominated solutions in $P_k \cup \{q_{k+1}\} \setminus F_1$. This procedure is repeated until all solutions in $P_k \cup \{q_{k+1}\}$ are included in a front. Any solution in front F_{i+1} is dominated by at least one solution of front F_i for $i \geq 1$.

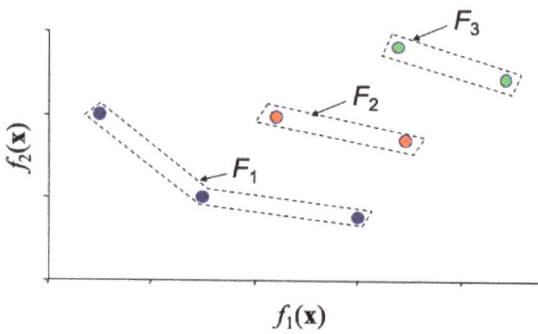

Figure 7. Non-dominated sorting procedure.

Afterward, the hypervolume contribution [39] of each individual in the last front $P_k \cup \{q_{k+1}\}$ is computed. Hypervolume definition guarantees that any non-dominated solution will not be replaced by a dominated solution since non-dominated solutions will have a higher hypervolume contribution than dominated ones. Hypervolume allows one to obtain a well-distributed set of solutions in the objective space as well as to guide the search toward the Pareto optimal front.

The hypervolume contribution computation is straightforward for problems with two objectives [41]. The approximations to the ideal vector (z_i^*) and the nadir vector (z_i^{nad}) computed in the current population are used to normalize objectives functions to the same order of magnitude in the interval [0,1]. The normalized objective function f_i^{norm} for the i-th objective function is computed by:

$$f_i^{norm} = \frac{f_i - z_i^*}{z_i^{nad} - z_i^*}. \tag{8}$$

The solutions of the worst-ranked non-dominated front are sorted in ascending order according to the values of the first normalized objective function. A sequence that is additionally sorted in descending order is obtained since the solutions are mutually non-dominated. Then, the hypervolume contributions of the solutions can be obtained by computing the rectangle area as illustrated in Figure 8. Given a sorted front with r solutions $F = \{s_1, s_2 \ldots, s_r\}$, the hypervolume contribution of solution s_i ($I(s_i)$) is computed by [39]:

$$I(s_i) = (f_1(s_{i+1}) - f_1(s_i)) \times (f_2(s_{i-1}) - f_2(s_i)) \tag{9}$$

where $i = 2, \ldots, r-1$.

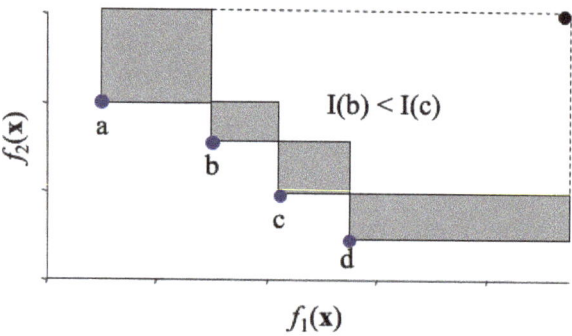

Figure 8. Hypervolume contribution.

In Figure 8, it can be seen that the hypervolume contribution of solution b is inferior to c, as the area covered by b is smaller, which implies that this solution will not be selected in the next generation.

The μ best individuals in terms of the domination ranking are selected to be the progenitors of the next generation. On the last front, the individual with the worst hypervolume contribution is discarded. This process is repeated until the stopping criterion is fulfilled.

4. Case Study
4.1. The Optimization Problem to Solve

In the present study, the cup illustrated in Figure 9 was thermoformed with constant temperature, a female mold and three types of sheets (as shown in Figure 3): constant thickness, linear spline variation and concentric spline variation. The aim was to determine the sheet thickness profile in order to: (i) minimize the initial sheet volume, as it implies less material use (f_1); (ii) minimize the minimum thickness found in the cells of the mesh used in the modeling calculations without hindering its mechanical behavior, as it is related with the capacity of the polymer sheet deformability, representing indirectly a measure of the thickness heterogeneity (f_2); and (iii) minimize the thickness heterogeneity, that is, the difference between the thickness of the part and a reference thickness, as defined by Equation (10) (f_3).

$$f_3 = \frac{1}{M} \sum_{i=1}^{M} \frac{|t_0 - t_i|}{t_0} \qquad (10)$$

where M is the number of points located in a line defining the center of the cup in direction x, t_0 is a reference thickness defined by the user and t_i are the thicknesses of the M points.

This is a bound constrained multi-objective optimization problem with the following decision variables and objectives limitations (dimensions in meters):

$$\begin{aligned} 2.0 \times 10^{-3} \leq x_i \leq 4.0 \times 10^{-3} \\ \text{Minimum thickness} \geq 1.0 \times 10^{-4} \end{aligned} \qquad (11)$$

where x_i is the sheet thickness of the constant thickness along the x direction. The thickness along the x direction is then imposed using a spline variation based on the 10 control points (see Figure 3), or the thickness of the five control points determining the concentric thickness variation, from the center to the border of the circle represented in Figure 3C. First, three bi-objective problems were considered (Cases 1 to 3), one for each case of sheet thickness variation and taking into account objectives f_1 and f_2, with the aim of showing the effectiveness of the methodology in solving this problem. Then, two bi-objective problems using objectives f_1 and f_3 were considered (Cases 4 and 5, respectively with spline and concentric variations), the aim being to approach the solutions to a more realistic industrial situation.

The multi-objective evolutionary algorithm used in this work is based on the SMS-EMOA. Considering the characteristics of the multi-objective optimization problem being solved, different configurations and search mechanisms can be adopted for SMS-EMOA. Thus, as described in Section 3.2, Gaussian mutation with covariance matrix adaptation was selected as the variation procedure, since the problem being solved comprises continuous variables. simultaneously, the hypervolume metric was chosen to provide well-distributed alternative solutions in the objective space. The configuration, including the parameter values of the algorithm, takes into account the considerable computational effort required to compute the objective function values. The population size was set to 20 individuals. The selection was carried out using a uniform distribution and variation was performed by the CMA evolution strategy operator [8], which is designed to work with real number representations. The maximum number of generations was set to 20.

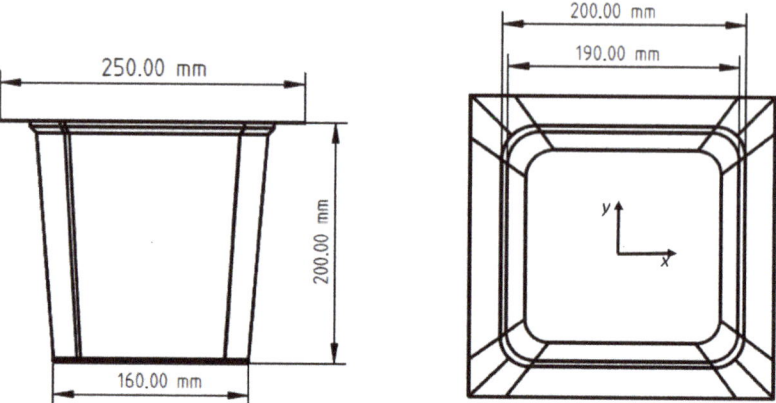

Figure 9. Part dimensions: square cup with rounded vertices.

4.2. Results and Discussion

For Case 1, the sheet with constant thickness, both objectives (volume and thermo-formed part minimum thickness) are in harmony, which means that the solutions of all generations are located in a line and the single optimal solution is the one more near the minimum of these two objectives. The same does not occur for the other two cases.

Figure 10 shows the random initial population and the non-dominated solutions of the last generation for optimization Case 2, which comprises seven solutions. In this case, the sheet thickness is a spline generated from the 10 decision variables represented by black dots in the graphs, provided in Figure 11. It is clear that there is improvement along the 20 generations. The Pareto optimal front of the last generation presents some gaps between the solutions due to the thermoforming problem characteristics, as the location of the 10 points used to generate the sheet spline thickness are fixed and equally spaced along the x axis, which limits the search space.

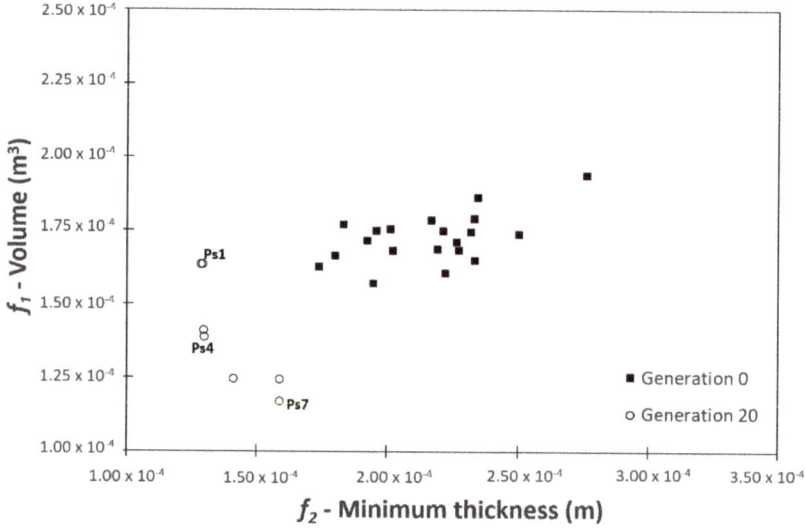

Figure 10. Initial population and non-dominated solutions of final population (20th generation).

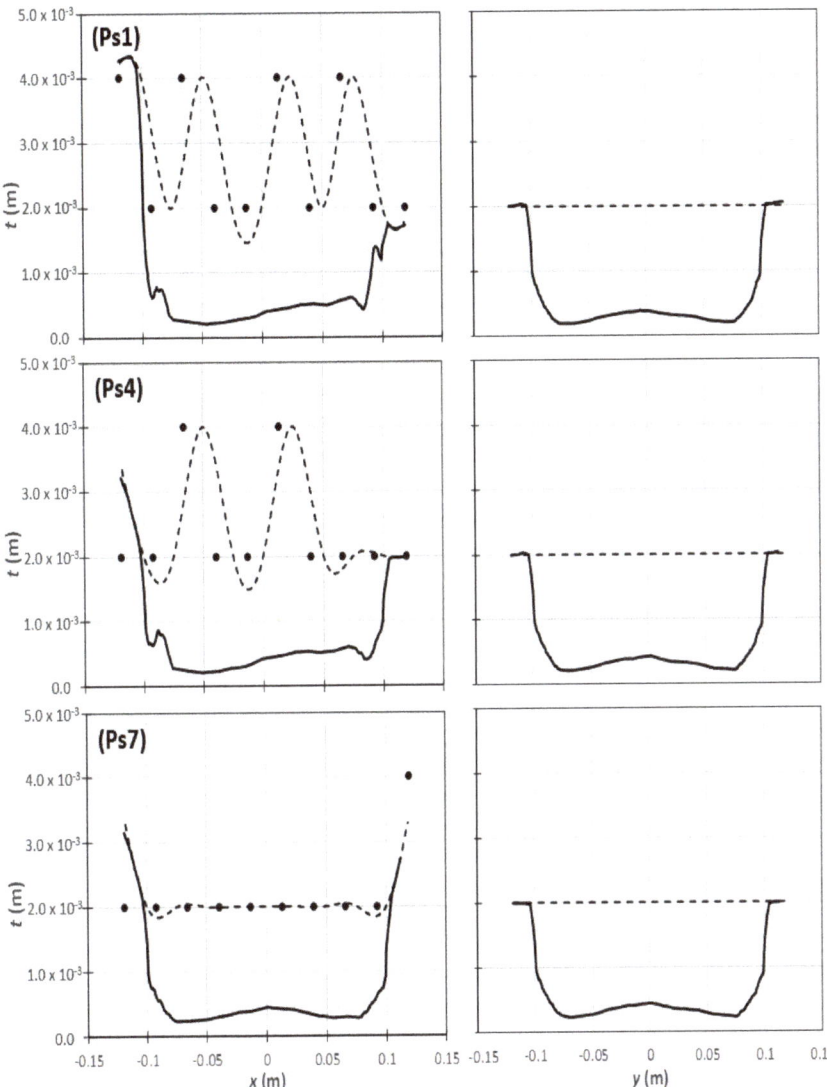

Figure 11. Thickness profile for solutions Ps1, Ps4 and Ps7 (Figure 9): left—black points are the decision variables, dashed line is the spline and the continuous line the part thickness profile a $x = 0$; right—sheet and part thickness profiles for $x = 0$.

The sheet and final part thickness profiles of solutions Ps1, Ps4 and Ps7 are illustrated in Figure 11. As can be seen in the graphs, the thickness profiles perpendicular to the spline (Figure 11-left), when moving from solutions Ps1 to Ps7 the final part profile, are more uniform. From the point of view of the Decision-Maker, this seems to indicate that Ps7 is the best solution for practical purposes. This evidences the great advantage of performing this type of multi-objective optimization. Not only does the algorithm converge to better solutions, but it can also provide the Decision-Maker a set of solutions from which he can choose the best one to be applied in the real thermoforming practice, in this case, solution Ps7. Another characteristic of the solutions shown is that for x equal to zero

(Figure 11-left) the sheet spline thickness is the same in all cases ($t \approx 2.0 \times 10^{-3}$ m), which produces identical part profiles in the transversal direction, as illustrated by the graphs of Figure 11-right.

Figure 12 shows the Pareto optimal solutions for all the cases studied. As can be seen, in Case 3 (concentric spline), the optimization converges to five non-dominated solutions, identified by black dots. The final part thickness profile of four solutions for this case, Pc1, Pc2, Pc4 and Pc5 are represented in Figure 13 (solution Pc3 was not represented due to its similarity with solution Pc2). The black dots identify the location of the points used to generate the symmetrical concentric spline represented by a dashed line, the decision variables. Again, from solution Pc1 to solution Pc5, the profile obtained is more uniform, as in Case 2. Also, it is important to note that, in this case, the part thickness profile is the same in all directions, as the sheet thickness presents an axisymmetric distribution.

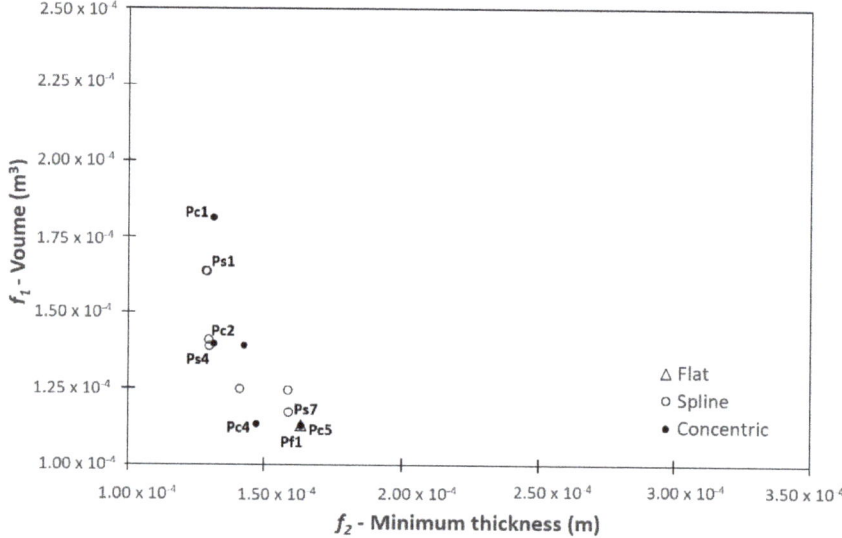

Figure 12. Non-dominated solutions for all cases: flat thickness, spline and concentric spline.

Figure 14 shows the part thickness profile of the unique solution found for Case 1, Pf1. As can be seen, the profile obtained is very similar to that of solutions Ps7 and Pc5 of the previous cases studied. This confirms that the optimization strategy is working and identify solutions with physical meaning, since different starting points, corresponding to diverse initial sheet thickness profiles, led to identical solutions when those solutions are located in the same region of the search space, as can be seen in Figure 12.

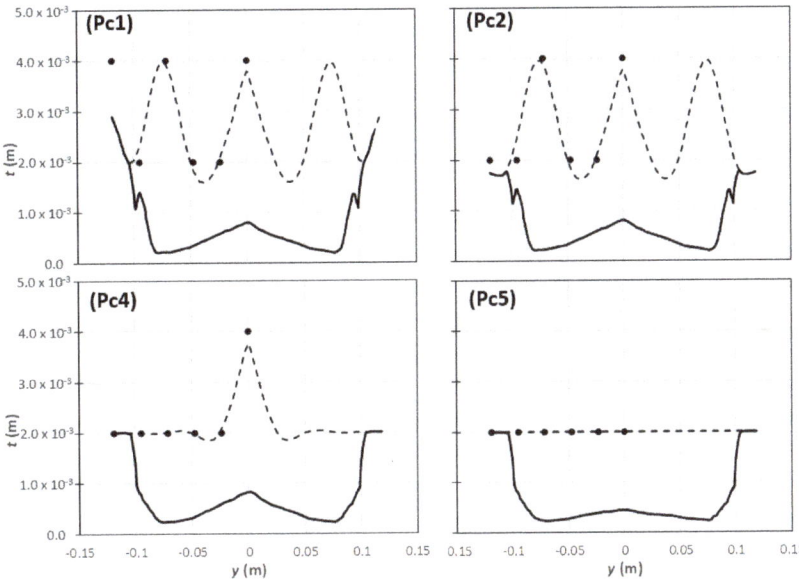

Figure 13. Thickness profile for solutions Pc1, Pc2, Pc4 and Pc5 (Figure 9): black points are the decision variables, dashed line is the concentric spline and the continuous line the part thickness profile a $x = 0$.

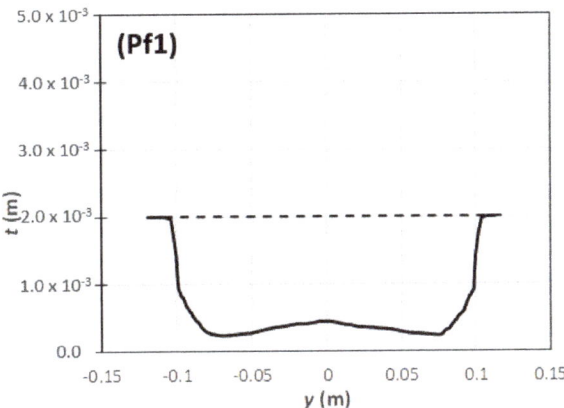

Figure 14. Thickness profile for solution Pf1 (Figure 9): dashed line is the constant sheet thickness and the continuous line the part thickness profile a $x = 0$.

Finally, Figure 15 shows the non-dominated solutions of the 20th generation for Cases 4 and 5, considering spline and concentric sheet thickness variation and t_0 equal to 0.5 mm. As previously, the gaps between the solutions are due to the problem constraints related to a limited search space. It is clear that, as expected, the concentric variation produces much better results concerning the uniformity of the thickness. Solutions P's3 and P'c3 are the same as those obtained previously, that is, solutions Ps7 and Pc5. However, the solutions found depend strongly on the value chosen for t_0. A practical alternative consists of finding the desired thickness profile for the final part through a mechanical analysis, for example, as suggested in the scheme of Figure 1.

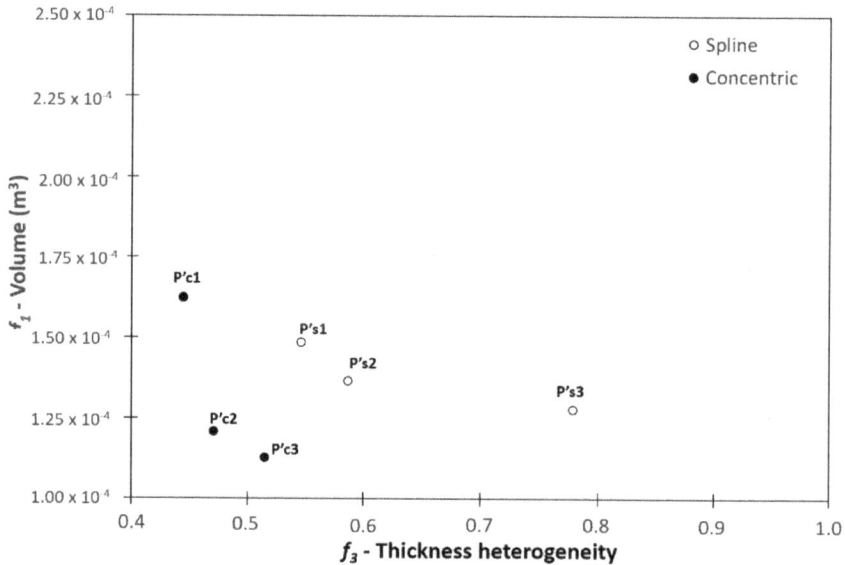

Figure 15. Non-dominated solutions of the 20th generation for Cases 4 (Spline) and 5 (Concentric), t_0 equal to 0.5 mm.

Also, it is important to note that technically and economically it is difficult to produce concentric initial sheet variations, as it implies high costs, mainly when the parts are to be produced in big quantities, as is commonly the case.

5. Conclusions

Faced with the complexity of the real optimization problem in the field of plastics thermoforming described in this paper, the applicability of a multi-objective optimization strategy was proposed to deal with the process forming phase. The aim was to determine the better sheet thickness distribution that allows the production of parts with the least amount of material while assuring the appropriate characteristics of the final part. A reduction of circa 30% in the volume of the material used can be obtained when comparing the solutions of the initial population and the ones in the Pareto front.

The main scientific contribution of this work is to add a new approach to control the thickness distribution of thermoformed parts, in addition to the variations of technique and differential heating previously mentioned. Since the final thickness of the parts depends on the initial thickness of the sheet, several solutions are proposed for adjusting the initial thickness of the sheet in order to achieve the most favorable thickness.

The results obtained showed that the methodology proposed was able to capture the singular features of the process allowing us to conclude that the strategy proposed might be successfully applied in the optimization of the various steps of plastics thermoforming. This work constitutes a further step to support design approaches associated with this important plastics processing technology.

Author Contributions: Conceptualization, methodology, global writing, data curation, supervision and funding acquisition, A.G.-C.; modeling software and calculations, P.C.; optimization software, calculations and writing, L.C.; modeling software, W.d.C.G.; modeling writing and supervision, J.M.N.; state-of-the-art and thermoforming process, F.D. All authors have read and agreed to the published version of the manuscript.

Funding: This research was funded by NAWA-Narodowa Agencja Wymiany Akademickiej, under grant PPN/ULM/2020/1/00125 and European Union's Horizon 2020 research and innovation programme under the Marie Skłodowska-Curie Grant Agreement No 734205–H2020-MSCA-RISE-2016. The authors also acknowledge the funding by FEDER funds through the COMPETE 2020 Programme and National Funds through FCT (Portuguese Foundation for Science and Technology) under the projects UID-B/05256/2020, UID-P/05256/2020, UIDB/00319/2020, MORPHING.TECH— Direct digital Manufacturing of automatic programmable and Continuously adaptable patterned surfaces with a discrete and patronized composition (POCI-01-0247-FEDER-033408).

Acknowledgments: The authors would like to acknowledge the Minho University cluster under the project NORTE-07-0162-FEDER-000086 and the Minho Advanced Computing Center (MACC) for providing HPC resources that contributed to the research results reported within this paper.

Conflicts of Interest: The authors declare no conflict of interest.

References

1. Yang, C.; Hung, S.-W. Modeling and Optimization of a Plastic Thermoforming Process. *J. Reinf. Plast. Compos.* **2004**, *23*, 109–121. [CrossRef]
2. Chang, Y.-Z.; Wen, T.-T.; Liu, S.-J. Derivation of optimal processing parameters of polypropylene foam thermoforming by an artificial neural network. *Polym. Eng. Sci.* **2004**, *45*, 375–384. [CrossRef]
3. Leite, W.D.O.; Rubio, J.C.C.; Cabrera, F.M.; Carrasco, A.; Hanafi, I. Vacuum Thermoforming Process: An Approach to Modeling and Optimization Using Artificial Neural Networks. *Polymers* **2018**, *10*, 143. [CrossRef] [PubMed]
4. Leite, W.; Rubio, J.; Mata, F.; Hanafi, I.; Carrasco, A. Dimensional and Geometrical Errors in Vacuum Thermoforming Prod-ucts: An Approach to Modeling and Optimization by Multiple Response Optimization. *Meas. Sci. Rev.* **2018**, *18*, 113–122. [CrossRef]
5. Sasimowski, E. The use of utility function for optimization of thermoforming. *Polimery* **2018**, *63*, 807–814. [CrossRef]
6. Wang, C.-H.; Nied, H.F. Temperature Optimization for Improved Thickness Control in Thermoforming. *J. Mater. Process. Manuf. Sci.* **1999**, *8*, 113–126. [CrossRef]
7. Bordival, M.; Andrieu, S.; Schmidt, F.; Maoult, Y.L.; Monteix, S. Optimization of Infrared Heating System for the Ther-Moforming Process. In Proceedings of the ESAFORM 2005-8th ESAFORM Conference on Material Forming, Cluj-Napoca, Romania, 27–29 April 2005; hal-01788422. pp. 925–928.
8. Chy, M.M.; Boulet, B. A Conjugate Gradient Method for the Solution of the Inverse Heating Problem in Thermoforming. In Proceedings of the IEEE Industry Applications Society Annual Meeting, Houston, TX, USA, 3–7 October 2010; pp. 1–8.
9. Chy, M.M.; Boulet, B.; Haidar, A. A Model Predictive Controller of Plastic Sheet Temperature for a Thermoforming Process. In Proceedings of the 2011 American Control Conference, San Francisco, CA, USA, 29 June–1 July 2011; pp. 4410–4415.
10. Li, Z.; Heo, K.; Seol, S. Time-dependent Optimal Heater Control in Thermoforming Preheating Using Dual Optimization Steps. *Int. J. Precis. Eng. Manuf.* **2008**, *9*, 51–56.
11. Li, Z.-Z.; Ma, G.; Xuan, D.-J.; Seol, S.-Y.; Shen, Y.-D. A study on control of heater power and heating time for thermoforming. *Int. J. Precis. Eng. Manuf.* **2010**, *11*, 873–878. [CrossRef]
12. Erchiqui, F.; Nahas, N.; Nourelfath, M.; Souli, M. Metaheuristic algorithms for optimisation of infrared heating in thermoforming process. *Int. J. Metaheuristics* **2011**, *1*, 199–221. [CrossRef]
13. Bachir Cherif, K.; Rebaine, D.; Erchiqui, F.; Fofana, I. *Metaheuristics as a Solving Approach for the Infrared Heating in the Thermoforming Process*; GERAD HEC, GERAD-G-2015-139; GERAD: Montreal, QC, Canada, 2015.
14. Cherif, K.B.; Rebaine, D.; Erchiqui, F.; Fofana, I.; Nahas, N. Numerically Optimizing the Distribution of the Infrared Radiative Energy on a Surface of a Thermoplastic Sheet Surface. *J. Heat Transf.* **2018**, *140*, 102101. [CrossRef]
15. Erchiqui, F. Application of genetic and simulated annealing algorithms for optimization of infrared heating stage in ther-moforming process. *Appl. Therm. Eng.* **2018**, *128*, 1263–1272. [CrossRef]
16. Throne, J.L. *Technology of Thermoforming*; Hanser Publishers: Munich, Germany, 1966.
17. DiRaddo, R.W.; Meddad, A. Sensitivity of operating conditions and material properties for thermoforming process. *Plast. Rubber Compos.* **2000**, *29*, 163–167. [CrossRef]
18. Duarte, F.M.; Covas, J. On the Use of the Heating Stage to Control the Thickness Distribution in Thermoformed Parts. *Int. Polym. Process.* **2004**, *19*, 186–198. [CrossRef]
19. Duarte, F.; Covas, J.A. IR sheet heating in roll fed thermoforming: Part 1-Solving direct and inverse heating problems. *Plast. Rubber Compos.* **2002**, *31*, 307–317. [CrossRef]
20. Schmidt, F.M.; Le Maoult, Y.; Monteix, S. Modelling of infrared heating of thermoplastic sheet used in thermoforming process. *J. Mater. Process. Technol.* **2003**, *143*, 225–231. [CrossRef]
21. McCool, R.; Martin, P.J. The role of process parameters in determining wall thickness distribution in plug-assisted thermoforming. *Polym. Eng. Sci.* **2010**, *50*, 1923–1934. [CrossRef]
22. Marathe, D.; Rokade, D.; Busher, A.L.; Jadhav, K.; Mahajan, S.; Ahmad, Z.; Gupta, S.; Kulkarni, S.; Juvekar, V.; Lele, A. Effect of Plug Temperature on the Strain and Thickness Distribution of Components Made by Plug Assist Thermoforming. *Int. Polym. Process.* **2016**, *31*, 166–178. [CrossRef]

23. Martin, P.; Duncan, P. The role of plug design in determining wall thickness distribution in thermoforming. *Polym. Eng. Sci.* **2007**, *47*, 804–813. [CrossRef]
24. Sasimowski, E. A pressure-bubble vacuum forming process for polystyrene sheet. *Adv. Sci. Technol. Res. J.* **2017**, *11*, 180–186. [CrossRef]
25. Ayadi, A.; Lacrampe, M.-F.; Krawczak, P. Bubble assisted vacuum thermoforming: Considerations to extend the use of in-situ stereo-DIC measurements to stretching of sagged thermoplastic sheets. *Int. J. Mater. Form.* **2019**, *13*, 59–76. [CrossRef]
26. Schmidt, R.L.; Carley, J.F. Biaxial stretching of heat-softened plastic sheets using an inflation technique. *Int. J. Eng. Sci.* **1975**, *13*, 563–578. [CrossRef]
27. Tuković, Ž.; Karač, A.; Cardiff, P.; Jasak, H.; Ivanković, A. OpenFOAM Finite Volume Solver for Fluid-Solid Interaction. *Trans. FAMENA* **2018**, *42*, 1–31. [CrossRef]
28. Atmani, O.; Abbès, F.; Li, Y.; Batkam, S.; Abbès, B. Experimental investigation and constitutive modelling of the deformation behaviour of high impact polystyrene for plug-assisted thermoforming. *Mech. Ind.* **2020**, *21*, 607. [CrossRef]
29. Duarte, F.M. Study and Optimization of Plastics Sheet Thermoforming. Ph.D. Thesis, University of Minho, Guimarães, Portugal, 2003.
30. Ghobadnam, M.; Mosaddegh, P.; Rejani, M.R.; Amirabadi, H.; Ghaei, A. Numerical and experimental analysis of HIPS sheets in thermoforming process. *Int. J. Adv. Manuf. Technol.* **2014**, *76*, 1079–1089. [CrossRef]
31. Bernard, C.A.; Correia, J.P.M.; Bahlouli, N.; Ahzi, S. Numerical Simulation of Plug-Assisted Thermoforming Process: Application to Polystyrene. *Key Eng. Mater.* **2013**, *554–557*, 1602–1610. [CrossRef]
32. Oueslati, Z.; Rachik, M.; Lacrampe, M.F. Transversely Isotropic Hyperelastic Constitutive Models for Plastic Thermoforming Simulation. *Key Eng. Mater.* **2013**, *554–557*, 1715–1728. [CrossRef]
33. Patil, J.P.; Nandedkar, V.; Saha, S.; Mishra, S. A numerical approach on achieving uniform thickness distribution in pressure thermoforming. *Manuf. Lett.* **2019**, *21*, 24–27. [CrossRef]
34. Cha, J.; Kim, M.; Park, D.; Go, J.S. Experimental determination of the viscoelastic parameters of K-BKZ model and the influence of temperature field on the thickness distribution of ABS thermoforming. *Int. J. Adv. Manuf. Technol.* **2019**, *103*, 985–995. [CrossRef]
35. O'Connor, C.P.J.; Martin, P.J.; Sweeney, J.; Menary, G.; Caton-Rose, P.; Spencer, P.E. Simulation of the plug-assisted ther-moforming of polypropylene using a large strain thermally coupled constitutive model. *J. Mater. Process. Technol.* **2013**, *213*, 1588–1600. [CrossRef]
36. Rosenzweig, N.; Narkis, M.; Tadmor, Z. Wall thickness distribution in thermoforming. *Polym. Eng. Sci.* **1979**, *19*, 946–951. [CrossRef]
37. Miettinen, K. *Nonlinear Multiobjective Optimization*; Springer: Berlin/Heidelberg, Germany, 1998; Volume 12.
38. Deb, K.; Pratap, A.; Agarwal, S.; Meyarivan, T. A fast and elitist multiobjective genetic algorithm: NSGA-II. In *IEEE Transactions on Evolutionary Computation*; Springer: Berlin/Heidelberg, Germany, 2002; Volume 6, pp. 182–197.
39. Zitzler, E.; Laumanns, M.; Thiele, L. SPEA2: Improving the strength Pareto evolutionary algorithm. *TIK-Report* **2001**, *103*. [CrossRef]
40. Li, H.; Zhang, Q. Multiobjective optimization problems with complicated Pareto sets, MOEA/D and NSGA-II. *IEEE Trans. Evol. Comput.* **2009**, *13*, 284–302. [CrossRef]
41. Zitzler, E.; Künzli, S. Indicator-based selection in multiobjective search. In Proceedings of the Conference on Parallel Problem Solving from Nature, Birmingham, UK, 18–22 September 2004.
42. Beume, N.; Naujoks, B.; Emmerich, M. SMS-EMOA: Multiobjective selection based on dominated hypervolume. *Eur. J. Oper. Res.* **2007**, *181*, 1653–1669. [CrossRef]
43. Emmerich, M.; Beume, N.; Naujoks, B. An EMO Algorithm Using the Hyper-Volume Measure as Selection Criterion. In *Evolutionary Multi-Criterion Optimization*; LNCS; Springer: Berlin, Germany, 2005; Volume 3410, pp. 62–76.
44. Voß, T.; Hansen, N.; Igel, C. Improved step size adaptation for the MO-CMA-ES. In Proceedings of the 6th Annual Conference on Cyber and Information Security Research, Oak Ridge, TN, USA, 21–23 July 2010; Association for Computing Machinery (ACM): New York, NY, USA, 2010; pp. 487–494.

Multi-Objective Optimum Design and Maintenance of Safety Systems: An In-Depth Comparison Study Including Encoding and Scheduling Aspects with NSGA-II

Andrés Cacereño *, David Greiner and Blas J. Galván

Instituto Universitario de Sistemas Inteligentes y Aplicaciones Numéricas en Ingeniería (SIANI), Universidad de Las Palmas de Gran Canaria (ULPGC), Campus Universitario de Tafira Baja, 35017 Las Palmas de Gran Canaria, Spain; david.greiner@ulpgc.es (D.G.); blas.galvan@ulpgc.es (B.J.G.)
* Correspondence: acacereno@iusiani.ulpgc.es; Tel.: +34-928-457-405

Abstract: Maximising profit is an important target for industries in a competitive world and it is possible to achieve this by improving the system availability. Engineers have employed many techniques to improve systems availability, such as adding redundant devices or scheduling maintenance strategies. However, the idea of using such techniques simultaneously has not received enough attention. The authors of the present paper recently studied the simultaneous optimisation of system design and maintenance strategy in order to achieve both maximum availability and minimum cost: the Non-dominated Sorting Genetic Algorithm II (NSGA-II) was coupled with Discrete Event Simulation in a real encoding environment in order to achieve a set of non-dominated solutions. In this work, that study is extended and a thorough exploration using the above-mentioned Multi-objective Evolutionary Algorithm is developed using an industrial case study, paying attention to the possible impact on solutions as a result of different encodings, parameter configurations and chromosome lengths, which affect the accuracy levels when scheduling preventive maintenance. Non-significant differences were observed in the experimental results, which raises interesting conclusions regarding flexibility in the preventive maintenance strategy.

Keywords: multi-objective evolutionary algorithms; availability; design; preventive maintenance scheduling; encoding; accuracy levels

1. Introduction

System Reliability ($R(t)$) can be defined as the probability of failure free operation under specified conditions over an intended period of time [1]. System Maintainability ($M(t)$) can be defined as the probability of being restored to a fully operational condition within a specific period of time [2]. These definitions lead to interest both in time taken for a system to failure (Time To Failure) and in time taken to repair the system (Time To Repair). System Availability ($A(t)$) can be defined as the fraction of the total time in which systems are able to perform their required function [2]. The concepts of Reliability and Maintainability are related to Availability in order to define the way in which the system is able to achieve the function for which it was designed, over a period of time. Therefore, whereas Reliability is a concept related to non-repairable systems, Availability is a concept related to repairable systems because it encompasses the complete failure-recovery cycle over the mission time. From above, it can be seen that repairable systems operate during a certain period of time (Time To Failure) until a failure occurs. After that, a period of time is needed to recover the system operating status (Time To Repair). This creates an interest in Time To Failure and Time To Repair, which can be modelled as random variables that can be represented by continuous probability distributions.

Optimisation through Evolutionary Algorithms is useful when complex problems have to be solved; that is, problems in which the number of potential solutions is usually high and whilst achieving the best solution is a daunting task and achieving the best solution can be extremely difficult, achieving at least a good solution (if not exactly the best) can be a manageable task [3]. Typical reliability optimisation problems have considered the possibility of maximising the system Availability. There are many techniques commonly used to improve it and the present paper focuses on two of them. On the one hand, modifying the structural design through redundancies improves the system Availability. A redundancy is a component added to a subsystem from a series-parallel configuration in order to increase the number of alternative paths [4]. On the other hand, an overall improvement of system availability is possible through preventive maintenance [5]. The unavailability of a system due to its failure can occur at any time, which necessitates a significant effort to recover the operating state. Conversely, a programmed shutdown to perform a preventive maintenance task represents a controlled situation with materials, spares and human teams available.

These techniques were jointly explored by the authors of the present paper preliminarily in [6], coupling Multi-objective Evolutionary Algorithm and Discrete Simulation and dealing with the joint optimisation of systems design (considering whether or not to include redundant devices) and their preventive maintenance strategy (choosing optimum periodical preventive maintenance times in relation to each system device). This allowed necessary preventive maintenance activities to be carried out. The system Availability and operation Cost were the objectives to maximise and minimise, respectively. A simulation approach was used in which each solution proposed by the Multi-objective Evolutionary Algorithm (using real encoding) was evaluated through Discrete Simulation; the technique used to build and modify (depending on the periodical preventive maintenance times) the Functionability Profile. This is a concept presented by Knezevic [7], which explains the history of the system varying among operating and recovery times. It is a powerful modelling technique, which permits analysis of complex systems with a reality more realistic representation of their behaviour. In the previous study, the performance of several configurations of the Multi-objective Evolutionary Algorithm Non-dominated Sorting Genetic Algorithm II (NSGA-II) [8] in a real encoding environment was explored. In the present paper an in-depth encoding comparative study is developed. Firstly, the real encoding case study is extended, and secondly, some binary encoding alternatives are explored, looking for possible advantages and disadvantages to encode this kind of real world problems. Moreover, an accuracy-level experiment for the preventive maintenance strategy is considered. The first part of the study determines the optimum periodical Time To Start a Preventive Maintenance activity measured in hours. Two more levels are explored in this study: days and weeks. There are preventive maintenance activities whose accuracy level in time can be of little importance. It may not be important to determine the exact instant for their development, being enough to define the day or the week. Therefore, the effect of several chromosome lengths is explored looking to improve the evolutionary process. Summarising the contributions of the present study:

- In this work, seven encoding alternatives are thoroughly explored: Real encoding (with Simulated Binary Crossover), Binary encoding (with 1 point Crossover), Binary encoding (with 2 point Crossover), Binary encoding (with Uniform Crossover), Gray encoding (with 1 point Crossover), Gray encoding (with 2 point Crossover) and Gray encoding (with Uniform Crossover). Their performances are compared using the Hypervolume indicator and statistical significance tests.
- Additionally, three accuracy levels on time are explored for the binary encoding; hours, days and weeks, in order to analyse the effect of chromosome length in the evolutionary search and final non-dominated set of solutions. Their performances are compared using the Hypervolume indicator and statistical significance tests.
- The methodology is applied in an industrial test case, obtaining an improved non-dominated front of optimum solutions that could be considered as both a benchmark case and reference solution.

The paper is organised as follows. Section 2 explores the related literature. Section 3 shows an outline of the methodology. Section 4 presents the case study and Section 5 the description of experiments carried out (encodings and accuracy levels). In Section 6, results are shown and discussed, and finally, Section 7 introduces the conclusions.

2. Literature Review

Approaches dealing with reliability systems design (redundancy allocation) are presented first, and approaches dealing with preventive maintenance optimum design are presented secondly in this section. Finally, the third subsection deals with the simultaneous optimisation of reliability systems design and preventive maintenance.

2.1. Redundancy Allocation of Reliability Systems Design Optimisation

Improving the Reliability or Availability for series-parallel systems through redundancy allocation using Multi-objective Evolutionary Algorithms has been considered in several studies. Many authors have developed Genetic Algorithms-based approaches to solve the problem. Bussaca et al. [9] utilized a Multi-objective Genetic Algorithm to optimise the design of a safety system through redundancy allocation, considering components with constant failure rates. They employed profit during the mission time (formed by the profit from plant operation minus costs due to purchases and installations, repairs and penalties during downtime) and the reliability at mission time as objective functions. Marseguerra et al. [10] presented an approach that couples a Multi-objective Evolutionary Algorithm and Monte Carlo simulation for optimal networks design aimed at maximising the network reliability estimate and minimising its uncertainty. Tian and Zuo [11] proposed a multi-objective optimisation model for redundancy allocation for multi-state series-parallel systems. They used physical programming as an approach to optimise the system structure and a Genetic Algorithm to solve it. The objectives consisted of maximising the system performance and minimising its cost and weight. Huang et al. [12] proposed a niched Pareto Genetic Algorithm to solve reliability-based design problems aiming to achieve a high number of feasible solutions. They used reliability and cost as objective functions. Zoulfaghari et al. [13] presented a Mixed Integer Nonlinear Programming model to availability optimisation of a system taking into account both repairable and non-repairable components. They developed a Genetic Algorithm looking for maximum availability and minimum cost. Taboada et al. [14] introduced a Genetic Algorithm to solve multi-state optimisation design problems. The universal moment generating function was used to evaluate the reliability indices of the system. They used reliability, cost and weight as objective functions.

The use of the NSGA and NSGA-II algorithms has been extensive; some applications ranging between the years 2003–2021 are shown as follows. Taboaba et al. [15] presented two methods to reduce the size of the Pareto optimal set for multi-objective system-reliability design problems: on the one hand using a pseudo ranking scheme and on the other hand using data mining clustering techniques. To demonstrate the performance of the methods, they solved the redundancy allocation problem using the Non-dominated Sorting Genetic Algorithm (NSGA) to find the Pareto optimal solutions, and then the methods were successfully applied to reduce the Pareto set. They studied reliability, cost and weight as objective functions. However, the most widely used standard Genetic Algorithm is the second version of the NSGA Multi-objective Evolutionary Algorithm. Greiner et al. [16] introduced new safety systems multi-objective optimum design methodology (based on fault trees evaluated by the weight method) using not only the standard NSGA-II but also the Strength Pareto Evolutionary Genetic Algorithm (SPEA2) and the controlled elitist-NSGA-II, with minimum unavailability and cost criteria. Salazar et al. [17] developed a formulation to solve several optimal system design problems. They used the NSGA-II to achieve maximum reliability and minimum cost. Limbourg and Kochs [18] applied a specification method originating from software engineering named Feature Modelling and a NSGA-II with an external repository. They maximised the life distribution and minimised the system cost. Kumar et al. [19] proposed a multi-objective formulation of multi-level redundancy allocation optimisation problems and a methodology to solve them. They proposed a hierarchical Genetic Algorithm framework by introducing hierarchical genotype encoding for design variables. Not only the NSGA-II but also the SPEA2 Multi-objective Evolutionary Algorithm were used, considering reliability and cost as objective functions. Chambari et al. [20] modelled the redundancy allocation problem taking into account non-repairable series-parallel systems, non-constant failure rates and imperfect switching of redundant cold-standby components. They used NSGA-II as well as Multi-objective Particle Swarm Optimisation (MOPSO) to solve the problem with reliability and cost as objective functions. Safari [21] proposes a variant of the NSGA-II method to solve a novel formulation for redundancy allocation. He considered for the redundancy strategy both active and cold-standby redundancies with reliability and cost as objective functions. Ardakan et al. [22] solved the redundancy allocation by using mixed redundancies combining both active and cold-standby redundancies. Under this approach, the redundancy strategy is not predetermined. They studied reliability and cost as objective functions using the NSGA-II method. Ghorabaee et al. [23] considered reliability and cost as objective functions to solve the redundancies allocation problem using NSGA-II. They introduced modified methods of diversity preservation and constraints handling. Amiri and Khajeh [24] considered repairable components to solve the redundancy allocation problem in a series-parallel system. They used the NSGA-II method with availability and cost as objective functions. Jahromi and Feizabadi [25] presented a formulation for the redundancy allocation of non-homogeneous components considering reliability and cost as objective functions. The NSGA-II method was used to achieve the Pareto optimal front. Kayedpour et al. [26] developed an integrated algorithm to solve reliability design problems taking into account instantaneous availability, repairable components and a selection of configuration strategies (parallel, cold or warm) based on Markov processes and the NSGA-II method. As objective functions, they considered availability and cost. Sharifi et al. [27] presented a new multi-objective redundancy allocation problem for systems where the subsystems were considered weighted-k-out-of-n parallel. They used NSGA-II with reliability and cost as objective functions. Chambari et al. [28] proposed a bi-objective simulation-based optimisation model to redundancy allocation in series-parallel systems with homogeneous components to maximise the system reliability and minimise the cost and using NSGA-II. They considered optimal component types, the redundancy level, and the redundancy strategy (active, cold standby, mixed or K-mixed) with imperfect switching.

Other Multi-objective evolutionary and bio-inspired methods have been used in a lesser measure. Zhao et al. [29] optimised the design of series-parallel systems using the Ant Colony Algorithm in a multi-objective approach, considering reliability, cost and weight as objective functions. Chiang and Chen [30] proposed a Multi-objective Evolutionary Algorithm based on simulated annealing to solve the availability allocation and optimisation problem of a repairable series-parallel system. They applied their algorithm to two study cases presented in references [31] (with objective functions availability and cost) and [9]. Khalili-Damghani et al. [32] proposed a novel dynamic self-adaptive Multi-objective Particle Swarm Optimisation method to solve two-states reliability redundancy allocation problems. They contemplated reliability, cost and weight as objective functions. Jiansheng et al. [33] proposed an Artificial Bee Colony Algorithm to solve the redundancy allocation problem for repairable series-parallel systems where uncertainty in failure rates, repair rates and other relative coefficients involved were considered. They used availability and cost as objective functions. Samanta and Basu [34] proposed an Attraction-based Particle Swarm Optimisation to solve availability allocation problems for systems with repairable components, where they introduced fuzzy theory to manage uncertainties. They used availability and cost as objective functions.

2.2. Preventive Maintenance Strategy Optimisation

Improving the Reliability or Availability for series-parallel systems through the preventive maintenance strategy has been widely studied using Multi-objective Evolutionary Algorithms, too. Many authors used Genetic Algorithms as an optimisation method. Muñoz et al. [35] presented an approach based on Genetic Algorithms and focused on the global and constrained optimisation of surveillance and maintenance of components based on risk and cost criteria. Marseguerra et al. [36] coupled Genetic Algorithms and Monte Carlo simulation in order to optimise profit and availability. The Monte Carlo simulation was used to model system degradation while the Genetic Algorithm was used to determine the optimal degradation level beyond which preventive maintenance has to be performed. Gao et al. [37] studied the flexible job shop scheduling problem with availability constraints affecting maintenance tasks. They used a Genetic Algorithm looking for minimum makespan (time that elapses from the start of work to the end), time expended on a machine and the total time expended on all machines. Sánchez et al. [38] used Genetic Algorithms for the optimisation of testing and maintenance tasks with unavailability and cost as objective functions. They considered the epistemic uncertainty in relation to imperfect repairs. Wang and Pham [39] used a Genetic Algorithm to estimate the preventive maintenance interval allowing for imperfect repairs and the number of preventive maintenance activities before component needs to be replaced. They used availability and cost as objective functions. Ben Ali et al. [40] developed an elitist Genetic Algorithm to deal with the production and maintenance-scheduling problem, minimising makespan and cost. Gao et al. [5] studied preventive maintenance considering the dynamic interval for multi-component systems. They solved the problem using Genetic Algorithms with availability and cost as objective functions. An et al. [41] built a novel integrated optimisation model including the flexible job-shop scheduling problem to reduce energy consumption in the manufacturing sector. They considered degradation effects and imperfect maintenance. They proposed a Hybrid Multi-objective Evolutionary Algorithm taking into account the makespan, total tardiness, total production cost and total energy consumption as objective functions. Bressi et al. [42] proposed a methodology to minimise the present value of the life cycle maintenance costs and maximise the life cycle quality level of the railway track-bed considering different levels of reliability. They used a Genetic Algorithm to achieve optimal solutions.

The use of the NSGA-II algorithm and other non-dominated criterion-based approaches has been extensive with applications ranging across the years 2005-2021. Martorell et al. [43] proposed a methodology to take decisions and determine technical specifications and maintenance looking for increasing reliability, availability and maintainability. They used SPEA2 as an optimisation method. Oyarbide-Zubillaga et al. [44] coupled Discrete Event Simulation and Genetic Algorithms (NSGA-II) to determine the optimal frequency for preventive maintenance of systems under cost and profit criteria. Berrichi et al. [45] proposed a new method to deal with simultaneous production and maintenance scheduling. They used the Weighted Sum Genetic Algorithm (WSGA) and NSGA-II as optimisation methods to compare their performances. They worked with makespan and unavailability due to maintenance tasks as objective functions. Moradi et al. [46] studied simultaneous production and preventive maintenance scheduling to minimise the global time invested in production tasks and unavailability due to preventive maintenance activities. They used four Genetic Algorithms: NSGA-II, NRGA (Non-ranking Genetic Algorithm), CDRNSGA-II (NSGA-II with Composite Dispatching Rule and active scheduling) and CDRNRGA (NRGA with Composite Dispatching Rule and active scheduling). Hnaien and Yalaoui [47] considered a similar problem, minimising the makespan and the delay between the real and the theoretical maintenance frequency for two machines. They used NSGA-II and SPEA2, including two new versions based on the Johnson Algorithm. Wang and Liu [48] considered the optimisation of parallel-machine-scheduling integrated with two kinds of resources (machines and moulds) and preventive maintenance planning. They used makespan and availability as objective functions and NSGA-II as an optimisation method. Piasson et al. [49] proposed a model to solve the problem of optimising the reliability-centred maintenance planning of an electric power distribution system. They used NSGA-II to achieve the Pareto optimal front and, as objective functions, they minimised the cost due to maintenance activities and maximised the index of reliability of the whole system. Sheikhalishahi et al. [50] presented an open shop scheduling model that considers human error and preventive maintenance. They considered three objective functions: human error, maintenance and production factors. They used NSGA-II and SPEA2 as optimisation methods. As well as that, they used another Evolutionary Algorithm, the Multi-Objective Particle Swarm Optimisation (MOPSO) method. Boufellouh and Belkaid [51] proposed a bi-objective model that determines production scheduling, maintenance planning and resource supply rate decisions in order to minimise the makespan and total production costs (including total maintenance, resource consumption and resource inventory costs). They used NSGA-II and BOPSO (Bi-Objective Particle Swarm Optimization) as Evolutionary Optimisation Algorithms. Zang and Yang [52] proposed a multi-objective model of maintenance planning and resource allocation for wind farms using NSGA-II. They considered the implementation of maintenance tasks using the minimal total resources and at the minimal penalty cost.

Other Multi-objective Evolutionary methods have been used. Berrichi et al. [53] solved the joint production and preventive maintenance-scheduling problem using the Ant Colony Algorithm with availability and cost as objective functions. Suresh and Kumarappan [54] presented a model for the maintenance scheduling of generators using hybrid Improved Binary Particle Swarm Optimisation (IBPSO). As objective functions, they used a reduction in the loss of load probability and minimisation of the annual supply reserve ratio deviation for a power system. Li et al. [55] presented a novel Discrete Artificial Bee Colony Algorithm for the flexible job-shop scheduling problem considering maintenance activities. They used as objective functions the makespan, the total workload of machines and the workload of the critical machine.

2.3. Redundancy Allocation and Preventive Strategy Optimisation

Therefore, it is possible to improve the availability of repairable systems by dealing with their design and employing a preventive maintenance strategy. However, only a few works have been developed to look at the simultaneous optimisation of both from a multi-objective point of view. In Galván et al. [56], a methodology for Integrated Safety System Design and Maintenance Optimisation based on a bi-level evolutionary process was presented. While the inner loop is devoted to searching for the optimum maintenance strategy for a given design, the outer loop searches for the optimum system design. They used Genetic Algorithms as an optimisation method and cost and unavailability as objective functions. Okasha and Frangopol [57] considered the simultaneous optimisation of design and maintenance during the life cycle using Genetic Algorithms. They studied system reliability, redundancy and life-cycle cost as objective functions. Adjoul et al. [58] described a new approach to simultaneous optimisation of design and maintenance of multi-component industrial systems improving their performances with reliability and cost as objective functions. They used a two level Genetic Algorithm; the first optimises the design based on reliability and cost, and the second optimises the dynamic maintenance plan.

This work studies the simultaneous optimisation of design and preventive maintenance strategy coupling Multi-objective Evolutionary Algorithms and Discrete Simulation: a technique that has achieved good results in the Reliability field. Coupling Multi-objective Evolutionary Algorithms with Discrete Simulation has been studied, on the one hand, to supply redundancy allocation [10] and, on the other hand, to determine the preventive maintenance strategy [36,44]. Moreover, only a few works have been developed looking at design and corrective maintenance strategy simultaneously [59,60]. Nevertheless, to the knowledge of the authors of this work, coupling Multi-objective Evolutionary Algorithms and Discrete Simulation has not yet been explored for both joint optimisation of the design and preventive maintenance strategy as in the current study where, additionally, a variety of encoding schemes were considered in the analysis. In the studies cited in the present literature review, some used real encoding, while others used binary or integer encoding; however, not one of them studied the possible impact of such encoding schemes on performance. Moreover, an accuracy experiment is developed in the present paper, including the time unit needed to carry out the preventive maintenance activities. As shown in the above literature review, NSGA-II is one of the state-of-the-art Multi-objective Evolutionary Algorithms more commonly used to deal with redundancy allocation and scheduling preventive maintenance problems. Therefore, this method is thorough explored in the present paper.

3. Methodology and Description of the Model

3.1. Extracting Availability and Cost from Functionability Profiles

Availability can be computed by using the unconditional failure $w(t)$ and repair $v(t)$ intensities, as is explained in Ref. [2]. These are described by the following Equation (1), where $f(t)$ is the failure density function of a system, $\int_0^t f(t-u)v(u)du$ is the failure probability of the cited system in interval $[0,t)$ having worked continuously since being repaired in $[u, u+du)$ given that it was working at $t=0$ and $\int_0^t g(t-u)w(u)du$ is the repair probability in interval $[0,t)$, given that it has been in the failed state since the previous failure in $[u, u+du)$ and that it was working at $t=0$.

$$w(t) = f(t) + \int_0^t f(t-u)v(u)du$$
$$v(t) = \int_0^t g(t-u)w(u)du$$
(1)

When the system devices have exponential failure and repair intensities (constant failure and repair rates), it is not relatively straightforward to find its availability using the solutions of Equation (1). However, when the system devices do not have exponential failure and/or repair intensities, finding the system availability using Equation (1) is difficult, so a simulation approach may be more suitable. In this paper, the system availability is characterised in a simulation approach by using the system Functionability Profile, a concept introduced by Knezevic [7] and defined as the inherent capacity of systems to achieve the required function under specific features when they are used as specified. Speaking in general, all systems achieve their function at the beginning of their lives. However, irreversible changes occur over time and variations in system behaviour happen. The deviation of the variations in relation to the satisfactory features reveals the occurrence of system failure, which causes the transition from operating state to failure state. After failing, recovery activities (corrective maintenance) can recover its capacity to fulfil the required function when the system is repairable.

Additional tasks to maintain the operating status could be carried out. These are called preventive maintenance activities. These are generally less complex than a repair and should be done prior to failure. From the Functionability Profile point of view, the states of a repairable system fluctuate between operation and failure over the mission time. The shape of cited changes is called the Functionability Profile because it shows the states over the mission time. Therefore, Functionability Profiles depend on operation times (either Time To Failure or Time To Start a scheduled Preventive Maintenance activity) $(t_{f1}, t_{f2}, ..., t_{fn})$ and recovery times (either Time To Repair after failure or Time To Perform a scheduled Preventive Maintenance activity) $(t_{r1}, t_{r2}, ..., t_{rn})$. It is obvious that, after a period of operation, a period of recovery is needed.

In the present paper, the system Functionability Profile is built by using Discrete Event Simulation in a simulation approach as will be explained later. Once known, the system Functionability Profile will be able to compute the system Availability through the relation between the system operation times and the system recovery times. The system will be able to fulfil its purpose during t_f times, so it is possible to evaluate its Availability at mission time by using Equation (2).

$$A = \frac{\sum_{i=1}^{n} t_{fi}}{\sum_{i=1}^{n} t_{fi} + \sum_{j=1}^{m} t_{rj}} \qquad (2)$$

where:
- n is the total number of operation times,
- t_{fi} is the i-th operation time in hours (Time To Failure or Time To Start a Preventive Maintenance activity),
- m is the total number of recovery times,
- t_{rj} is the j-th recovery time in hours (Time To Repair or Time To Perform a Preventive Maintenance activity).

Operation and recovery times are random variables so they may be treated statistically. They can be defined as probability density functions through their respective parameters. There are several databases such as OREDA [61], which supply the characteristic parameters for the referred functions, so operation and recovery times can be characterised for system devices. When systems are operating, earnings are generated due to the availability of the system. Conversely, when systems have to be recovered, economic cost is invested in order to regain the operation status. In this paper, the economic cost is a variable directly related to recovery times, which are related to corrective and preventive maintenance activities; quantities computed by Equation (3).

$$C = \sum_{i=1}^{q} cc_i + \sum_{j=1}^{p} cp_j \qquad (3)$$

where:
- C is the system operation cost quantified in economic units,
- q is the total number of corrective maintenance activities,
- cc_i is the cost due to the i-th corrective maintenance activity,
- p is the total number of preventive maintenance activities,
- cp_j is the cost due to the j-th preventive maintenance activity.

In this work, costs derived from maintenance activities depend on fixed quantities per hour (corrective and preventive) so the global cost is directly related to the recovery times. Preventive maintenance activities are scheduled shutdowns so recovery times will be shorter and more economical than recovery times due to corrective maintenance activities (for reasons explained before, such as access to human personnel who are willing and/or trained, and/or the availability of spare parts). It will be necessary to carry out preventive maintenance activities to avoid long recovery times. These should ideally be done before the failure but as close as possible to it. Therefore, the basic Functionability Profile of the system (i.e., the system Functionability Profile, which represents the continuous cycle of failure-repair within the mission time, without considering preventive maintenance activities) should be modified by including preventive maintenance activities for system devices. This approach makes it possible to maximise the system Availability and minimise the costs due to recovery times.

3.2. Building Functionability Profiles to Evaluate the Objective Functions

It is necessary to characterise both the system Availability and the Cost from the system Functionability Profile in order to optimise the system design and preventive maintenance strategy. The Functionability Profiles of all system devices are built by using Discrete Event Simulation. Finally, the system Functionability Profile is built from these Functionability Profiles. With this purpose, it is necessary to have information about how to characterise operation Times To Failure *(TF)* and Times To Repair after failure *(TR)*, which are related to the parameters of their probability density functions. The Functionability

Profiles for system devices are built by generating random times, which are obtained from the respective probability density functions, both for Times To Failure *(TF)* and Times To Repair *(TR)*. In order to modify the Functionability Profiles relating to the preventive maintenance activities, Times To Start a Preventive Maintenance *(TP)* have to be used. This information is supplied via each solution provided by the behaviour Algorithm (each individual of the population) which is used to build the Functionability Profile of each device through Discrete Simulation. Moreover, recovery times in relation to the preventive maintenance activities or Times To Perform a Preventive Maintenance activity *(TRP)* have to be introduced by generating random times within the limits previously fixed. The process is explained below:

1. System mission time must be defined and then, the process continues for all devices.
2. The device Functionability Profile *(PF)* must be initialised.
3. The Time To Start a Preventive Maintenance activity *(TP)* proposed by the Multi-objective Evolutionary Algorithm is extracted from the individual (candidate solution) that is being evaluated and a Time To Perform a Preventive Maintenance activity *(TRP)* is randomly generated, within the limits previously fixed.
4. With reference to the failure probability density function related to the device, a Time To Failure *(TF)* is randomly generated, within the limits previously fixed.
5. If $TP < TF$, the preventive maintenance activity is performed before the failure. In this case, as many operating times units as TP units followed by as many recovery times units as TRP units are added to the device Functionability Profile. Each time unit represented in this way (both operating times and recovery times) is equivalent to one hour, day or week of real time, depending on the accuracy level chosen.
6. If $TP > TF$, the failure occurs before carrying out the preventive maintenance activity. In this case, taking into consideration the repair probability density function related to the device, the Time To Repair after the failure *(TR)* is randomly generated, within the limits previously fixed. Then, as many operating times units as TF units followed by as many recovery times units as TR units are added to the device Functionability Profile. Each time unit represented in this way (both operating times and recovery times) is equivalent to one hour, day or week of real time, depending on the accuracy level chosen.
7. Steps 4 to 6 have to be repeated until the end of the device mission time.
8. Steps 2 to 7 have to be repeated until the construction of the Functionability Profiles of all devices.
9. After building the Functionability Profiles of the devices, the system Functionability Profile is built by referring to the series (AND) or parallel (OR) distribution of the system devices.

Once the system Functionability Profile is built, the values of the objective functions can be computed by using both Equation (2) (evaluating the system Availability in relation to the time in which the system is operating and being recovered) and Equation (3) (evaluating the system operation Cost depending on the cost of the time units in relation to the development of corrective or preventive maintenance).

3.3. Multi-Objective Optimisation Approach

The optimisation method used, the Non-dominated Sorting Genetic Algorithm II (NSGA-II) belongs to the Evolutionary Algorithms *(EA)* paradigm. These kind of methods use a population of individuals with a specific size. Each individual is a multidimensional vector, called a chromosome, representing a possible candidate solution to the problem, while the vector components are called genes or decision variables. NSGA-II uses the concept of Pareto dominance as the basis of its selection criterion. Extended information on Evolutionary Optimisation Algorithms can be found in Ref. [3] and related to Multi-objective Evolutionary Algorithms in Ref. [62]. In this work, each individual in the population consists of a string (of real numbers or integers, depending on the encoding used, with values between 0 and 1) in which the system design alternatives and the periodic Times

To Start a Preventive Maintenance activity related to each device included in the system design is codified. Optimisation problems can be minimised or maximised for one or more of the objectives. In most cases, real world problems present various objectives in conflict with each other. These problems are called "Multi-objective" and their solutions arise from a solutions set which represents the best compromise among the objectives (Pareto optimal set) [62,63]. These kind of problems are described by Equation (4) (considering a minimisation problem) [3].

$$\min_x f(x) = \min_x [f_1(x), f_2(x), ..., f_k(x)] \quad (4)$$

In problems defined in this way, the k functions have to be minimised simultaneously. In the present paper, the objective functions are, on the one hand, the system Availability (first objective function, which is maximised and which is mathematically expressed by Equation (2); maximising Availability is similar to minimise Unavailability) and, on the other hand, the operation Cost (second objective function, which is minimised and which is mathematically expressed by Equation (3)).

4. The Case Study

The case study is based on the case presented in Ref. [6]. It consists of simultaneously optimising the design and the preventive maintenance strategy for an industrial fluid injection system, which is composed of cut valves (V_i) and impulsion pumps (P_i), taking Availability and operation Cost as objective functions. The desired outcome is maximum Availability and minimum operation Cost. The higher the investment in preventive maintenance, the better the system Availability. Conversely, this policy implies the growth of unwanted Cost. The system scheme is shown in Figure 1.

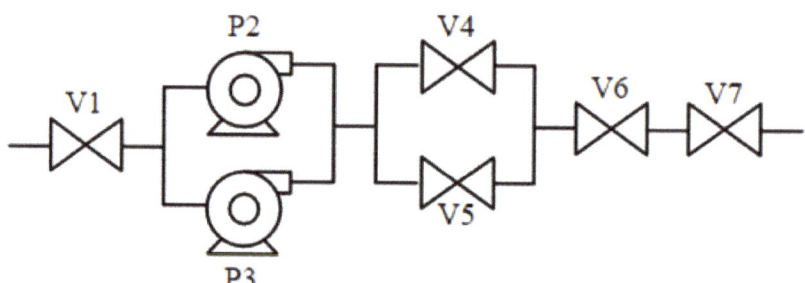

Figure 1. Case study: fluid injection system.

Some considerations are taken as follows:

- The number of redundant devices is limited as shown in Figure 1,
- two states are considered for each device: operation or failed state,
- the devices are independent of each other,
- a repair starts just after the failure of the device,
- a repair returns the device to the as-good-as-new state.

The data used in this work are shown in Table 1. Extended information regarding the parameters is supplied in Appendix A.

Table 1. Data set for system devices.

Parameter	Value	Source
Life Cycle	700,800 h	-
Corrective Maintenance Cost	0.5 units	Machinery & Reliability Institute
Preventive Maintenance Cost	0.125 units	Machinery & Reliability Institute
Pump TF_{min}	1 h	Machinery & Reliability Institute
Pump TF_{max}	70,080 h	Machinery & Reliability Institute
Pump $TF\ \lambda$	159.57×10^{-6} h	OREDA 2009
Pump TR_{min}	1 h	Machinery & Reliability Institute
Pump TR_{max}	24.33 h	$\mu + 4\sigma$
Pump $TR\ \mu$	11 h	OREDA 2009
Pump $TR\ \sigma$	3.33 h	$(\mu - TRmin)/3$
Pump TP_{min}	2920 h	Machinery & Reliability Institute
Pump TP_{max}	8760 h	Machinery & Reliability Institute
Pump TRP_{min}	4 h	Machinery & Reliability Institute
Pump TRP_{max}	8 h	Machinery & Reliability Institute
Valve TF_{min}	1 h	Machinery & Reliability Institute
Valve TF_{max}	70,080 h	Machinery & Reliability Institute
Valve $TF\ \lambda$	44.61×10^{-6} h	OREDA 2009
Valve TR_{min}	1 h	Machinery & Reliability Institute
Valve TR_{max}	20.83 h	$\mu + 4\sigma$
Valve $TR\ \mu$	9.5 h	OREDA 2009
Valve $TR\ \sigma$	2.83 h	$(\mu - TRmin)/3$
Valve TP_{min}	8760 h	Machinery & Reliability Institute
Valve TP_{max}	35,040 h	Machinery & Reliability Institute
Valve TRP_{min}	1 h	Machinery & Reliability Institute
Valve TRP_{max}	3 h	Machinery & Reliability Institute

The data were obtained from specific literature [61], expert judgement (based on professional experience from the Machinery & Reliability Institute (MRI), Alabama, USA) and mathematical relations. In this sense, the TR σ for valves and pumps were set in relation to the μ of their respective normal distribution functions and their TR_{min} previously established. In relation to the TR_{max}, it is known that 99.7% of the values of a normally distributed variable are included in the interval $\mu \pm 3\sigma$. The interval was extended to $\mu \pm 4\sigma$, taking into account anecdotal further values. As shown above, optimisation objectives consist of maximising the system Availability and minimising the operation Cost due unproductive system phases (both because the system is being repaired and because the system is being maintained). To do that:

- It is necessary to establish the optimum period to perform a preventive maintenance activity for the system devices, and
- It is necessary to decide whether to include redundant devices such as P2 and/or V4 by evaluating design alternatives. Including redundant devices will improve the system Availability but it will also increase the system operation Cost.

5. Description of the Experiments Carried Out

Two sets of experiment comparisons have been developed: first, comparing several encodings (real, binary and gray), and second, comparing several accuracies in the binary encoding. Finally, a description of the NSGA-II configuration is shown in the last subsection.

5.1. Comparing Encodings

From the optimisation point of view, it was explained before that the Evolutionary Algorithm *(EA)* uses a population of individuals called chromosomes, which represent possible solutions to the problem through their decision variables. The encoding of the system parameters is a crucial aspect of the algorithm. This has a significant influence on whether or not the algorithm works [3]. In the present paper, we intend to check whether there is a significant difference between the performances of the different encodings. Depending on the encoding type, each individual is codified as follows:

- **Real encoding:** This is formed by strings of real numbers (with 0 as a minimum value and 1 as a maximum value) following the shape $[B_1\ B_2\ T_1\ T_2\ T_3\ T_4\ T_5\ T_6\ T_7]$. The presence of redundant devices, P2 and V4, is defined by B_1 and B_2, respectively, and the optimum Time To Start a Preventive Maintenance activity in relation to each system device is represented by T_1 to T_7. The values of the decision variables must be scaled and rounded, i.e., B_1 and B_2 are rounded to the nearest integer (0 implies that the respective device is not selected whereas 1 implies the opposite). T_1 to T_7 are scaled among the respective TP_{min} and TP_{max} (depending on the type of device) and rounded to the nearest integer using Equation (5), where TP is the true value of the Time To Start a Preventive Maintenance activity, measured in hours (e.g., the decision variable T_1 represents the Time To Start a Preventive Maintenance for the valve V1 whose TP_{min} and TP_{max} has a value of 8760 h and 35,040 h, respectively. If the value of the decision variable T_1 is 0.532, the value of the Time To Start a Preventive Maintenance activity will be $8760 + 0.532 \cdot (35{,}040 - 8760) \approx 22{,}741$ h.

$$TP = round(TP_{min} + X \cdot (TP_{max} - TP_{min})) \quad (5)$$

- **Binary encoding:** This is formed by strings of binary numbers that vary between 0 and 1, where the total number of bits is 103 and they are:
    1. B_1: This denotes the presence of the pump P2 in the system design (0 implies that the respective device is not selected whereas 1 implies the opposite).
    2. B_2: This denotes the presence of the valve V4 in the system design (0 implies that the respective device is not selected whereas 1 implies the opposite).
    3. T_3 to T_{17}: These denote the Time To Start a Preventive Maintenance activity to the valve V1. A binary scale that allows representation of the numbers from TP_{min} to TP_{max} is needed. TP_{min} has a value of 8760 h and TP_{max} has a value of 35,040 h so $35{,}040 - 8760 = 26{,}280$ steps needed, where the step 0 represents a time of 8760 h and the step 26,279 represents a time of 35,040 h. A binary scale with at least 26,280 steps involves using 15 bits (as 26,280 steps lies between $2^{14} = 16{,}384$ and $2^{15} = 32{,}768$). Since 26,280 steps are needed and 32,768 are possible on the scale, an equivalent relation must be used. Each step in the scale of 32,768 steps represents $26{,}768 \div 32{,}768 = 0.8020019531$ steps in the scale of 26,768 steps. Therefore, it is possible to calculate the true Time To Start a Preventive Maintenance activity (in hours) using the scale change shown by Equation (6), where B represents the decimal value of the binary string T_3 to T_{17} (e.g., if the values of the decision variables in binary encoding are 1 0 1 1 0 1 1 0 0 0 1 1 1 0 1, the decimal value in the scale of 32,768 steps will be 23,325. If 26,768 steps are scaled, the number achieved is $23{,}325 \times 0.8020019531 \approx 18{,}707$ steps. Therefore, the true Time To Start a Preventive Maintenance activity amounts to $18{,}707 + 8760 = 24{,}467$ h).

$$TP = round(TP_{min} + B \cdot (0.8020019531)) \quad (6)$$

    4. T_{18} to T_{30}: These denote the Time To Start a Preventive Maintenance activity to the pump P2. A binary scale that allows representation of the numbers from TP_{min} to TP_{max} in needed. TP_{min} has a value of 2920 h and TP_{max} has a value of 8760 h so $8760 - 2920 = 5840$ steps needed, where the step 0 represents the time of 2920 h and the step 5839 represents the time of 8760 h. A binary scale with at least 5840 steps involves using 13 bits (as 5840 steps lies between $2^{12} = 4096$ and $2^{13} = 8192$). Since 5840 steps are needed and 8142 are possible on the scale, an equivalent relation must be used. Each step in the scale of 8142 represents $5840 \div 8192 = 0.712890625$ steps on a scale of

5840 steps. Therefore, it is possible to calculate the true Time To Start a Preventive Maintenance activity (hours) using the scale change shown by Equation (7), where B represents the decimal value of the binary string T_{18} to T_{30} (e.g., if the values of the decision variables in binary encoding are 1 0 1 1 0 1 1 0 0 0 1 1 1, the value on a scale of 8192 steps will be 5831. If 5840 steps are scaled, the number achieved is 5831 × 0.712890625 ≈ 4157 steps. Therefore, the true Time To Start a Preventive Maintenance activity amounts to 4157 + 2920 = 7077 h).

$$TP = round(TP_{min} + B \cdot (0.712890625)) \quad (7)$$

5. T_{31} to T_{43}: These denote the Time To Start a Preventive Maintenance activity to the pump P3. The behaviour of its encoding is similar to the behaviour explained for the pump P2.
6. T_{44} to T_{58}: These denote the Time To Start a Preventive Maintenance activity to the valve V4. The behaviour of its encoding is similar to the behaviour explained for the valve V1.
7. T_{59} to T_{73}: These denote the Time To Start a Preventive Maintenance activity to the valve V5. The behaviour of its encoding is similar to the behaviour explained for the valve V1.
8. T_{74} to T_{88}: These denote the Time To Start a Preventive Maintenance activity to the valve V6. The behaviour of its encoding is similar to the behaviour explained for the valve V1.
9. T_{89} to T_{103}: These denote the Time To Start a Preventive Maintenance activity to the valve V7. The behaviour of its encoding is similar to the behaviour explained for the valve V1.

- **Gray encoding:** When binary encoding is used, close numbers could bring big scheme modifications (e.g., the code for 15 is 0 1 1 1 while the code for 16 is 1 0 0 0, which represents changes in four (all) bits). Conversely, very similar numbers can represent numbers that are very apart (e.g., the code for 0 is 0 0 0 0 while the code for 16 is 1 0 0 0). In addition to standard binary encoding, Gray encoding is used. Gray code is a binary code where the difference among neighbouring numbers always differs by one bit [64–66].

Therefore, the performance of real, standard binary and Gray encodings can be compared.

5.2. Comparing Accuracies

Once the encoding experiment had been developed, a second experiment was executed. This consisted of studying the possible impact of the size of the chromosome. In the encoding experiment, the hour was utilised by the chromosomes as a measure of time. In this case, based on the idea that the exact hour to develop a preventive maintenance task is not necessary, the day and the week were used as measures of time. Therefore, in these cases, the solution regarding preventive maintenance strategy consisted of supplying the Time To Start a Preventive Maintenance activity with the day or the week as a time unit, respectively. The consequence was a reduction in the size of the chromosome, which was applied to the binary encoding as follows:

- **Binary encoding—Days:** This is formed by strings of binary numbers that vary between 0 and 1, where the total number of bits is 73 and they are:
    1. B_1: This denotes the presence of the pump P2 in the system design (0 implies that the device is not selected whereas 1 implies the opposite).
    2. B_2: This denotes the presence of the valve V4 in the system design (0 implies that the device is not selected whereas 1 implies the opposite).
    3. T_3 to T_{13}: These denote the Time To Start a Preventive Maintenance activity to the valve V1. A binary scale that allows representation of the numbers from TP_{min} to TP_{max} expressed in days as a time unit is needed. TP_{min} has a value of 8760 h (equivalent to 365 days) and TP_{max} has a value of 35,040 h (equivalent to

1460 days) so 1460 − 365 = 1095 steps needed, where the step 0 represents the time of 365 days and the step 1094 represents the time of 1460 days. A binary scale with at least 1095 steps involves using 11 bits (due to the fact that 1095 steps lies between $2^{10} = 1024$ and $2^{11} = 2048$). Since 1095 steps are needed and 2048 are possible on the scale, an equivalent relation must be used. Each step on the scale of 2048 steps represents $1095 \div 2048 = 0.5346679688$ steps on a scale of 1095 steps. Therefore, it is possible to achieve the true Time To Start a Preventive Maintenance activity (days) using the scale change shown by Equation (8), where B represents the decimal value of the binary string T_3 to T_{13} (e.g., if the values of the decision variables in binary encoding are 1 0 1 1 0 1 1 0 0 0 1, the decimal value on the scale of 2048 steps will be 1457. Scaling to a scale of 1095 steps, the number achieved is $1457 \times 0.5346679688 \approx 779$ steps. Therefore, the true Time To Start a Preventive Maintenance activity amounts to 779 + 365 = 1144 days).

$$TP = round(TP_{min} + B \cdot (0.5346679688)) \qquad (8)$$

4. T_{14} to T_{21}: These denote the Time To Start a Preventive Maintenance activity to the pump P2. A binary scale that allows representation of the numbers from TP_{min} to TP_{max} expressed in days as a time unit is needed. TP_{min} has a value of 2920 h (equivalent to 122 days) and TP_{max} has a value of 8760 (equivalent to 365 days) so 365 − 122 = 243 steps needed, where the step 0 represents the time of 122 days and the step 242 represents the time of 365 days. A binary scale with at least 243 steps involves using 8 bits (as 243 steps lies between $2^7 = 128$ and $2^8 = 256$). Since 243 steps are needed and 256 are possible on the scale, an equivalent relationship must be used. Each step in the scale of 256 represents $243 \div 256 = 0.94921875$ steps in the scale of 243 steps. Therefore, it is possible to achieve the true Time To Start a Preventive Maintenance activity (days) using the scale change shown by Equation (9), where B represents the decimal value of the binary string T_{14} to T_{21} (e.g., if the values of the decision variables in binary encoding are 1 0 1 1 0 1 1 0, the value on the scale of 256 steps will be 182). Scaling to a scale of 243 steps, the number achieved is $182 \times 0.94921875 \approx 173$ steps. Therefore, the true Time To Start a Preventive Maintenance activity amounts to 173 + 122 = 295 days).

$$TP = round(TP_{min} + B \cdot (0.712890625)) \qquad (9)$$

5. T_{22} to T_{29}: These denote the Time To Start a Preventive Maintenance activity to the pump P3. The behaviour of its encoding is similar to the behaviour explained for the pump P2.
6. T_{30} to T_{40}: These denote the Time To Start a Preventive Maintenance activity to the valve V4. The behaviour of its encoding is similar to the behaviour explained for the valve V1.
7. T_{41} to T_{51}: These denote the Time To Start a Preventive Maintenance activity to the valve V5. The behaviour of its encoding is similar to the behaviour explained for the valve V1.
8. T_{52} to T_{62}: These denote the Time To Start a Preventive Maintenance activity to the valve V6. The behaviour of its encoding is similar to the behaviour explained for the valve V1.
9. T_{63} to T_{73}: These denote the Time To Start a Preventive Maintenance activity to the valve V7. The behaviour of its encoding is similar to the behaviour explained for the valve V1.

- **Binary encoding—Weeks:** It is formed by strings of binary numbers that vary between 0 and 1, where the total number of bits is 54 and they are:
    1. B_1: This denotes the presence of the pump P2 in the system design (0 implies that the device is not selected whereas 1 implies the opposite).

2. B_2: This denotes the presence of the valve V4 in the system design (0 implies that the device is not selected whereas 1 implies the opposite).
3. T_3 to T_{10}: These denote the Time To Start a Preventive Maintenance activity to the valve V1. A binary scale that allows representation of the numbers from TP_{min} to TP_{max} expressed in weeks as a time unit is needed. TP_{min} has a value of 8760 h (equivalent to 52 weeks) and TP_{max} has a value of 35,040 h (equivalent to 209 weeks) so $209 - 52 = 157$ steps needed, where step 0 represents a time of 52 weeks and step 156 represents a time of 209 weeks. A binary scale with at least 157 steps involves using 8 bits (as 157 steps lies between $2^7 = 128$ and $2^8 = 256$). Since 157 steps are needed and 256 are possible on the scale, an equivalent relationship must be used. Each step on a 256-steps scale represents $157 \div 256 = 0.61328125$ steps on the 157-steps scale. Therefore, it is possible to achieve the true Time To Start a Preventive Maintenance activity (weeks) using the scale change shown by Equation (10), where B represents the decimal value of the binary string T_3 to T_{10} (e.g., if the values of the decision variables in binary encoding are 1 0 1 1 0 1 1 0, the decimal value in the scale of 256 steps will be 182. Working with a scale of 157 steps, the number achieved is $182 \times 0.61328125 \approx 112$ steps. Therefore, the true Time To Start a Preventive Maintenance activity amounts to $112 + 52 = 164$ weeks).

$$TP = round(TP_{min} + B \cdot (0.61328125)) \qquad (10)$$

4. T_{11} to T_{16}: These denote the Time To Start a Preventive Maintenance activity to the pump P2. A binary scale that allows representation of the numbers from TP_{min} to TP_{max} expressed in weeks as a time unit is needed. TP_{min} has a value of 2920 h (equivalent to 17 weeks) and TP_{max} has a value of 8760 (equivalent to 52 weeks) so $52 - 17 = 35$ steps needed, where step 0 represents the time of 17 weeks and step 34 represents the time of 52 weeks. A binary scale with at least 35 steps involves using 6 bits (as 35 steps lies between $2^5 = 32$ and $2^6 = 64$). Since 35 steps are needed and 64 are possible on the scale, an equivalent relationship must be used. Each step on the scale of 64-steps scale represents $35 \div 64 = 0.546875$ steps in the 35-steps scale. Therefore, it is possible to achieve the true Time To Start a Preventive Maintenance activity (weeks) using the scale change shown by Equation (11), where B represents the decimal value of the binary string T_{11} to T_{16} (e.g., if the values of the decision variables in binary encoding are 1 0 1 1 0 1, the value in the scale of 64 steps will be 45. Scaling on the 35-steps scale, the number achieved is $45 \times 0.546875 \approx 25$ steps. Therefore, the true Time To Start a Preventive Maintenance activity amounts to $45 + 17 = 62$ weeks).

$$TP = round(TP_{min} + B \cdot (0.546875)) \qquad (11)$$

5. T_{17} to T_{22}: These denote the Time To Start a Preventive Maintenance activity to the pump P3. The behaviour of its encoding is similar to the behaviour explained for the pump P2.
6. T_{23} to T_{30}: These denote the Time To Start a Preventive Maintenance activity to the valve V4. The behaviour of its encoding is similar to the behaviour explained for the valve V1.
7. T_{31} to T_{38}: These denote the Time To Start a Preventive Maintenance activity to the valve V5. The behaviour of its encoding is similar to the behaviour explained for the valve V1.
8. T_{39} to T_{46}: These denote the Time To Start a Preventive Maintenance activity to the valve V6. The behaviour of its encoding is similar to the behaviour explained for the valve V1.

9. T_{47} to T_{54}: These denote the Time To Start a Preventive Maintenance activity to the valve V7. The behaviour of its encoding is similar to the behaviour explained for the valve V1.

5.3. NSGA-II Configuration

The parameters used to configure the NSGA-II evolutionary process are shown in Table 2.

Table 2. Parameters to configure the evolutionary process.

Method	Encoding	Crossover	Time Unit	Population	PrM	disM	PrC	disC
NSGA-II	Real	SBX	Hour	50-100-150	0.5-1-1.5	20	1	20
	Standard Binary	1 point (1PX)	Hour	50-100-150	0.5-1-1.5	-	1	-
	Standard Binary	2 Point (2PX)	Hour	50-100-150	0.5-1-1.5	-	1	-
	Standard Binary	Uniform (UX)	Hour	50-100-150	0.5-1-1.5	-	1	-
	Gray	1 Point (1PX)	Hour	50-100-150	0.5-1-1.5	-	1	-
	Gray	2 Point (2PX)	Hour	50-100-150	0.5-1-1.5	-	1	-
	Gray	Uniform (UX)	Hour	50-100-150	0.5-1-1.5	-	1	-
	Standard Binary	1 Point (1PX)	Day	50-100-150	0.5-1-1.5	-	1	-
	Standard Binary	2 Point (2PX)	Day	50-100-150	0.5-1-1.5	-	1	-
	Standard Binary	Uniform (UX)	Day	50-100-150	0.5-1-1.5	-	1	-
	Standard Binary	1 Point (1PX)	Week	50-100-150	0.5-1-1.5	-	1	-
	Standard Binary	2 Point (2PX)	Week	50-100-150	0.5-1-1.5	-	1	-
	Standard Binary	Uniform (UX)	Week	50-100-150	0.5-1-1.5	-	1	-

Depending on the encoding applied, specific parameters have to be set, which are described below:

- Crossover: Type of crossover during the evolutionary process. The Simulated Binary Crossover (SBX) is used for real encoding while one point (1PX), two point (2PX) and uniform crossover (UX) are used for binary and Gray encodings.
- Population size (N): The population sizes used are 50, 100 and 150 individuals.
- Mutation Probability (PrM): This is the expectation of the number of genes mutating. The central value is equivalent to 1/decision variables (for the case study, the number of decision variables is 9 using real encoding, 103 using standard and Gray encoding with the hour as a time unit, 73 using standard binary encoding with the day as a time unit and 54 using standard binary encoding with the week as a time unit). Two more probabilities, one above and the other below the central value (1.5/decision variables and 0.5/decision variables, respectively) are set.
- Mutation Distribution (disM): This is the mutation distribution index when the Simulated Binary Crossover is used. It is set to the common value of 20.
- Crossover Probability (PrC): The probability of applying a crossover operator is set to 1 in all cases.
- Crossover Distribution (disC): The crossover distribution index when the Simulated Binary Crossover is used. It is set to the common value of 20.

Each configuration was executed 10 times (for statistical purposes) with 10,000,000 evaluations used as the stopping criterion. Scale factors in relation to the value of the objective functions were used with the purpose of achieving a dispersed non-dominated front with the unit as maximum value. The values were obtained through a practical approach in which the values of the scale factors are extracted from the values of the objective functions when the optimisation process starts. This approach is based on the assumption that the values of the objective functions will improve over the evolutionary process. The scale factors were used as follows:

- The scale factor used to compute the Cost was 1700 economic units.
- The scale factor used to compute the system Unavailability was 0.003.

Finally, a two dimensional reference point is needed to compute the Hypervolume indicator. The cited point has to cover the values limited by the scale factors, which restricts the values of the objective functions to a maximum of one. The reference point is set to (2,2). The Software Platform PlatEMO [67] (programmed in MATLAB) is used to optimise the application case. The Design and Maintenance Strategy analysis software (as explained in Section 3) has been developed and implemented into the platform to solve the case study shown in Section 4.

6. Results

Due to the complexity of the problem, a general-purpose calculation cluster was used during the computational process. The cluster is composed of 28 calculation nodes and one access node. Each calculation node consists of two Intel Xeon E5645 Westmere-EP processors with 6 cores each and 48 GB of RAM memory, allowing 336 executions to be run simultaneously.

Once the results were obtained, valuable information emerged. For each analysed case, the following information is provided: Firstly, information related to the computational process is given with the purpose of showing the complexity of the problem and computational cost. It consists of the time taken for 10 executions of the nine configurations (three population sizes and three mutation rates) related to each analysed case. Secondly, the values of the main measures obtained for the final evaluation are shown. These measures are the Average, Median, Minimum Value, Maximum Value and Standard Deviation of the Hypervolume indicator [68] (HV). Thirdly, in order to establish the existence of significant differences among the performance of the analysed case, a rigorous statistical analysis is carried out. The Friedman's test allows significant differences among results obtained to be detected, and the null hypothesis (H_0) to be rejected in that case. The p-value is a useful datum, which represents the smallest significant value that can result in the rejection of H_0. The p-value provides information about whether a statistical hypothesis test is significant (or not), and also indicates how significant the result is: The smaller the p-value (<0.05), the stronger the evidence against the null hypothesis. Finally, the Hypervolume is computed for the accumulated best non-dominated solutions obtained (the non-dominated front). These represent the best equilibrium solutions among the objectives and the computational procedure is described in Ref. [69].

Once the configurations had been ordered according to the Friedman's test values, one configuration of each analysed case was used for the final comparison taking two experiments into consideration: one looking at encodings and the other looking at to accuracy. In each case, additional information is given. The Hypervolume indicator average value evolution (in ten executions) is shown for each configuration. Moreover, box plots are given for the Hypervolume values distribution after the stopping criterion is met. In addition, the Friedman's test is used to detect significant differences among the performance of the configurations for each experiment. Finally, the accumulated best non-dominated solutions obtained (non-dominated front) are shown.

In Section 6.1, the results (of optimising the system that illustrate the case study) obtained after using the NSGA-II with different encodings (Real, Standard Binary and Gray) are presented; next, in Section 6.2 different accuracy levels of time (hours, days and weeks) in the conditions explained above are analysed and finally, in Section 6.3, the accumulated non-dominated designs are presented.

6.1. Encoding Experiment

The results of the Real encoding are presented first. Second, results of the binary encoding are shown, followed by the results of the Gray encoding. Finally, a comparison of the best performance cases of each codification is presented in Section 6.1.4.

6.1.1. Real Encoding

The results of using real encoding with simulated binary crossover are shown below. The computational time consumed is shown in Table 3. The average time represents the computational time regarding each one of ten executions and nine different configurations (real time consumed). The sequential time represents the computational time that would have been needed in the case of not using the cluster. The computational time shows the importance of using the cluster, which enables parallel processes.

Table 3. Computational time consumed.

Encoding	Time Unit	Average Time	Sequential Time
Real SBX	Hour	2 days, 18 h and 12 min.	8 months, 4 days, 23 h and 22 min.
Binary 1PX	Hour	2 days, 18 h and 57 min.	8 months, 7 days, 18 h and 48 min.
Binary 2PX	Hour	3 days, 2 h and 39 min.	9 months, 6 days, 5 h and 52 min.
Binary UX	Hour	3 days and 39 min.	8 months, 29 days, 3 h and 8 min.
Gray 1PX	Hour	2 days, 19 h and 35 min.	8 months, 10 days, 3 h and 23 min.
Gray 2PX	Hour	2 days, 19 h and 56 min.	8 months, 11 days, 11 h and 29 min.
Gray UX	Hour	2 days, 19 h and 40 min.	8 months, 10 days, 10 h and 34 min.
Total		2 days, 21 h and 6 min.	4 years, 11 months, 19 days, 8 h and 36 min.

The relationship between method configurations (where N represents the population size and PrM the mutation probability) and identifiers is shown in Table 4. Moreover, statistical information in relation to the Hypervolume value at the end of the evolutionary process is shown (average, median, maximum, minimum and standard deviation, out of 10 independent executions). It is possible to conclude that the configuration with the identifier ID9 (with a population of 150 individuals and mutation probability of 1.5 gene per chromosome) presents the highest Hypervolume average value, the highest Hypervolume median value and the highest Hypervolume maximum value. The configuration with identifier ID3 (population of 150 individuals and mutation probability of 0.5 gene per chromosome) presents the highest Hypervolume minimum value and the configuration with identifier ID4 (population of 50 individuals and mutation probability of one gene per chromosome) presents the lowest Hypervolume standard deviation value.

Table 4. Hypervolume statistics (Real encoding).

Identifier	Configuration	Average	Median	Max.	Min.	St. D.	Friedman's Test Av. Rank
ID1	N = 50 ; PrM = 0.5	2.2821	2.2857	2.3079	2.2531	0.0179	5.900
ID2	N = 100 ; PrM = 0.5	2.2886	2.2831	2.3258	2.2682	0.0170	4.800
ID3	N = 150 ; PrM = 0.5	2.2938	2.2883	2.3187	**2.2730**	0.0144	4.300
ID4	N = 50 ; PrM = 1	2.2927	2.2932	2.3113	2.2710	**0.0117**	**3.999**
ID5	N = 100 ; PrM = 1	2.2863	2.2916	2.3064	2.2584	0.0148	5.200
ID6	N = 150 ; PrM = 1	2.2890	2.2885	2.3157	2.2619	0.0170	5.100
ID7	N = 50 ; PrM = 1.5	2.2857	2.2788	2.3228	2.2701	0.0171	6.100
ID8	N = 100 ; PrM = 1.5	2.2821	2.2826	2.3051	2.2561	0.0162	5.600
ID9	N = 150 ; PrM = 1.5	**2.2968**	**2.2962**	**2.3307**	2.2641	0.0211	4.000
p-value							0.5788

In order to establish the best behaviour amongst the configurations, a statistical significance hypothesis test was conducted. The average ranks computed through the Friedman's test and the *p*-value obtained (a value bigger than 0.05 implies that the null hypothesis cannot be rejected, suggesting that all configurations perform in a similar way) are shown in Table 4. It can be seen that the configuration with identifier ID4 (population of 50 individuals and mutation probability of one gene per chromosome) presents the best average rank (in order to maximise the Hypervolume, the average rank has to be as low as possible) so it was selected for the final comparison study among encoding configurations.

Finally, the best accumulated non-dominated solutions obtained through the final generation of the evolutionary process for all executions and all configurations were used to compute the accumulated Hypervolume (as described in Ref. [69]) whose value was 2.3943. As expected, the value is higher than 2.3307, the maximum value shown in Table 4.

6.1.2. Standard Binary Encoding

The results of using standard binary encoding with one point, two point and uniform crossover are shown below. The computational time consumed by each one is shown in Table 3. The relationship between method configurations and identifiers is shown in Table 5. Moreover, statistical information relating to the Hypervolume value at the final of the evolutionary process is shown.

For the binary encoding with one point crossover (B1PX), it is possible to conclude that the configuration with identifier ID8 (population of 100 individuals and mutation probability of 1.5 gene per chromosome) presents both the highest Hypervolume average value and the highest Hypervolume median value, the configuration with identifier ID9 (population of 150 individuals and mutation probability of 1.5 gene per chromosome) presents the highest Hypervolume maximum value, the configuration with identifier ID6 (population of 150 individuals and mutation probability of one gene per chromosome) presents the highest Hypervolume minimum value, and the configuration with identifier ID2 (population of 100 individuals and mutation probability of 0.5 gene per chromosome) presents the lowest Hypervolume standard deviation.

For the binary encoding with two point crossover (B2PX), it is possible to conclude that the configuration with identifier ID2 (population of 100 individuals and mutation probability of 0.5 gene per chromosome) presents both the highest Hypervolume average value and the highest Hypervolume median value, the configuration with identifier ID3 (population of 150 individuals and mutation probability of 0.5 gene per chromosome) presents the highest Hypervolume maximum value and the configuration with identifier ID5 (population of 100 individuals and mutation probability of one gene per chromosome) presents both the highest Hypervolume minimum value and the lowest Hypervolume standard deviation value.

For the binary encoding with uniform crossover (BUX), it is possible to conclude that the configuration with identifier ID4 (population of 50 individuals and mutation probability of one gene per chromosome) presents the highest Hypervolume average value, the highest Hypervolume maximum value and the highest Hypervolume minimum value. The configuration with identifier ID2 (population of 100 individuals and mutation probability of 0.5 gene per chromosome) presents the highest Hypervolume median value and the configuration with identifier ID6 (population of 150 individuals and mutation probability of one gene per chromosome) presents the lowest Hypervolume standard deviation value.

Table 5. Hypervolume statistics (Binary encoding).

Encoding	Identifier	Configuration	Average	Median	Max.	Min.	St. D.	Friedman's Test Av. Rank
B1PX	ID1	$N = 50$; $PrM = 0.5$	2.2936	2.2919	2.3297	2.2738	0.0169	4.600
	ID2	$N = 100$; $PrM = 0.5$	2.2919	2.2953	2.3057	2.2703	**0.0107**	4.500
	ID3	$N = 150$; $PrM = 0.5$	2.2863	2.2915	2.2984	2.2554	0.0132	5.400
	ID4	$N = 50$; $PrM = 1$	2.2929	2.2918	2.3151	2.2566	0.0171	4.400
	ID5	$N = 100$; $PrM = 1$	2.2865	2.2822	2.3200	2.2674	0.0151	5.799
	ID6	$N = 150$; $PrM = 1$	2.2932	2.2926	2.3216	**2.2761**	0.0150	4.700
	ID7	$N = 50$; $PrM = 1.5$	2.2803	2.2775	2.2994	2.2642	0.0136	7.000
	ID8	$N = 100$; $PrM = 1.5$	**2.2970**	**2.3008**	2.3126	2.2708	0.0134	**3.100**
	ID9	$N = 150$; $PrM = 1.5$	2.2884	2.2871	**2.3394**	2.2616	0.0209	5.500
p-value								0.1228
B2PX	ID1	$N = 50$; $PrM = 0.5$	2.2808	2.2800	2.3173	2.2401	0.0237	6.500
	ID2	$N = 100$; $PrM = 0.5$	**2.3013**	**2.3051**	2.3260	2.2714	0.0172	**3.300**
	ID3	$N = 150$; $PrM = 0.5$	2.2916	2.2875	**2.3627**	2.2630	0.0298	5.400
	ID4	$N = 50$; $PrM = 1$	2.2941	2.2926	2.3156	2.2724	0.0152	4.600
	ID5	$N = 100$; $PrM = 1$	2.2908	2.2929	2.3056	**2.2755**	**0.0089**	4.800
	ID6	$N = 150$; $PrM = 1$	2.2921	2.2905	2.3215	2.2644	0.0160	4.999
	ID7	$N = 50$; $PrM = 1.5$	2.2924	2.2873	2.3483	2.2694	0.0226	5.100
	ID8	$N = 100$; $PrM = 1.5$	2.2931	2.2930	2.3326	2.2606	0.0205	4.900
	ID9	$N = 150$; $PrM = 1.5$	2.2883	2.2849	2.3107	2.2738	0.0145	5.399
p-value								0.4762
BUX	ID1	$N = 50$; $PrM = 0.5$	2.2841	2.2828	2.3023	2.2611	0.0122	5.500
	ID2	$N = 100$; $PrM = 0.5$	2.2954	**2.3000**	2.3141	2.2647	0.0163	**3.300**
	ID3	$N = 150$; $PrM = 0.5$	2.2883	2.2880	2.3265	2.2576	0.0199	5.200
	ID4	$N = 50$; $PrM = 1$	**2.2959**	2.2939	**2.3497**	2.2702	0.0223	3.400
	ID5	$N = 100$; $PrM = 1$	2.2848	2.2866	2.3046	2.2660	0.0140	5.700
	ID6	$N = 150$; $PrM = 1$	2.2830	2.2850	2.2955	2.2687	**0.0085**	5.799
	ID7	$N = 50$; $PrM = 1.5$	2.2893	2.2866	2.3112	2.2622	0.0163	5.100
	ID8	$N = 100$; $PrM = 1.5$	2.2808	2.2800	2.2994	2.2579	0.0138	6.100
	ID9	$N = 150$; $PrM = 1.5$	2.2857	2.2855	2.3036	2.2657	0.0121	4.899
p-value								0.2132

In order to establish the best behaviour amongst the configurations, a statistical significance hypothesis test was conducted. The average ranks computed through the Friedman's test and the *p*-value obtained (a value bigger than 0.05 implies that the null hypothesis cannot be rejected, suggesting that all configurations perform in a similar way) are shown in Table 5. It can be seen that the configuration with identifier ID8 (population of 100 individuals and mutation probability of 1.5 gene per chromosome) presents the best average rank for the binary encoding with one point crossover. It can be seen that the configuration with identifier ID2 (population of 100 individuals and mutation probability of 0.5 gene per chromosome) presents the best average rank both for the binary encoding with two point crossover and for the binary encoding with uniform crossover. These configurations were selected for the final comparison study among encoding configurations which is shown later. The best accumulated non-dominated solutions obtained through the final generation of the evolutionary process for all executions and all configurations were used to compute the accumulated Hypervolume whose values were 2.4142, 2.4298 and 2.3984 for the binary encoding with one point, two point and uniform crossover, respectively. As expected, the values are higher than 2.3394, 2.3627 and 2.3497, the maximum values shown in Table 5, respectively.

6.1.3. Gray Encoding

The results of using Gray encoding with one point, two point and uniform crossover are shown below. The computational time consumed by each one is shown in Table 3. The relationship between method configurations and identifiers is shown in Table 6. Moreover, statistical information in relation to the Hypervolume value at the end of the evolutionary process is shown. For the Gray encoding with one point crossover (G1PX), it is possible to conclude that the configuration with identifier ID2 (population of 100 individuals and mutation probability of 0.5 gene per chromosome) presents the highest Hypervolume average value, the highest Hypervolume median value and the highest Hypervolume maximum value. The configuration with identifier ID4 (population of 50 individuals and mutation probability of one gene per chromosome) presents both the highest Hypervolume minimum value and the lowest Hypervolume standard deviation value.

For the Gray encoding with two point crossover (G2PX), it is possible to conclude that the configuration with identifier ID1 (population of 50 individuals and mutation probability of 0.5 gene per chromosome) presents the highest Hypervolume average value, the highest Hypervolume median value and the highest Hypervolume minimum value. The configuration with identifier ID9 (population of 150 individuals and mutation probability of 1.5 gene per chromosome) presents the highest Hypervolume maximum value and the configuration with identifier ID5 (population of 100 individuals and mutation probability of one gene per chromosome) presents the lowest Hypervolume standard deviation value.

For the Gray encoding with uniform crossover (GUX), it is possible to conclude that the configuration with identifier ID4 (population of 50 individuals and mutation probability of one gene per chromosome) presents both the highest Hypervolume average value and the highest Hypervolume median value. The configuration with identifier ID5 (population of 100 individuals and mutation probability of 1 gene per chromosome) presents the highest Hypervolume maximum value, the configuration with identifier ID6 (population of 150 individuals and mutation probability of one gene per chromosome) presents the highest Hypervolume minimum value and the configuration with identifier ID2 (population of 100 individuals and mutation probability of 0.5 gene per chromosome) presents the lowest Hypervolume standard deviation value.

In order to establish the best behaviour amongst the configurations, a statistical significance hypothesis test was conducted. The average ranks computed through the Friedman's test and the p-value obtained (a value bigger than 0.05 implies that the null hypothesis cannot be rejected, suggesting that all configurations perform in a similar way) are shown in Table 6. It can be seen that the configuration with identifier ID5 (population of 100 individuals and mutation probability of 1 gene per chromosome) presents the best average rank for the Gray encoding with one point crossover. It can be seen that the configuration with identifier ID1 (population of 50 individuals and mutation probability of 0.5 gene per chromosome) presents the best average rank for the Gray encoding with two point crossover. It can be seen that the configuration with identifier ID4 (population of 50 individuals and mutation probability of one gene per chromosome) presents the best average rank for the Gray encoding with uniform crossover. These configurations were selected for the final comparison study among encoding configurations, which is shown later. The best accumulated non-dominated solutions obtained through the final generation of the evolutionary process for all executions and all configurations were used to compute the accumulated Hypervolume whose values were 2.4011, 2.3982 and 2.3829 for the Gray encoding with one point, two point and uniform crossover, respectively. As expected, the values are higher than 2.3556, 2.3364 and 2.3165, the maximum values shown in Table 6, respectively.

Table 6. Hypervolume Statistics (Gray Encoding).

Encoding	Identifier	Configuration	Average	Median	Max.	Min.	St. D.	Friedman's Test Av. Rank
G1PX	ID1	N = 50 ; PrM = 0.5	2.2833	2.2838	2.2929	2.2710	0.0064	5.500
	ID2	N = 100 ; PrM = 0.5	**2.2990**	**2.3010**	**2.3556**	2.2640	0.0252	3.600
	ID3	N = 150 ; PrM = 0.5	2.2815	2.2850	2.2951	2.2626	0.0099	6.200
	ID4	N = 50 ; PrM = 1	2.2865	2.2834	2.3017	**2.2762**	**0.0083**	5.200
	ID5	N = 100 ; PrM = 1	2.2989	2.2986	2.3347	2.2652	0.0186	**3.100**
	ID6	N = 150 ; PrM = 1	2.2812	2.2791	2.3043	2.2611	0.0117	6.000
	ID7	N = 50 ; PrM = 1.5	2.2882	2.2916	2.3171	2.2526	0.0187	4.200
	ID8	N = 100 ; PrM = 1.5	2.2786	2.2804	2.2992	2.2608	0.0137	6.100
	ID9	N = 150 ; PrM = 1.5	2.2874	2.2820	2.3180	2.2706	0.0155	5.100
p-value								0.0943
G2PX	ID1	N = 50 ; PrM = 0.5	**2.2947**	2.2971	2.3192	**2.2757**	0.0128	**3.700**
	ID2	N = 100 ; PrM = 0.5	2.2802	2.2814	2.2953	2.2592	0.0121	6.500
	ID3	N = 150 ; PrM = 0.5	2.2856	2.2895	2.2978	2.2519	0.0136	4.299
	ID4	N = 50 ; PrM = 1	2.2912	2.2868	2.3186	2.2659	0.0192	4.900
	ID5	N = 100 ; PrM = 1	2.2832	2.2835	2.2951	2.2690	**0.0070**	5.600
	ID6	N = 150 ; PrM = 1	2.2899	2.2913	2.3256	2.2640	0.0193	4.100
	ID7	N = 50 ; PrM = 1.5	2.2866	2.2880	2.3132	2.2617	0.0169	4.800
	ID8	N = 100 ; PrM = 1.5	2.2809	2.2813	2.3140	2.2585	0.0148	6.500
	ID9	N = 150 ; PrM = 1.5	2.2920	2.2912	**2.3364**	2.2739	0.0169	4.600
p-value								0.2164
GUX	ID1	N = 50 ; PrM = 0.5	2.2862	2.2892	2.3088	2.2540	0.0180	4.300
	ID2	N = 100 ; PrM = 0.5	2.2828	2.2813	2.2967	2.2696	**0.0079**	5.100
	ID3	N = 150 ; PrM = 0.5	2.2885	2.2922	2.3077	2.2666	0.0136	4.100
	ID4	N = 50 ; PrM = 1	**2.2907**	**2.2939**	2.3134	2.2683	0.0156	**4.000**
	ID5	N = 100 ; PrM = 1	2.2879	2.2858	**2.3165**	2.2611	0.0187	4.699
	ID6	N = 150 ; PrM = 1	2.2835	2.2816	2.3048	**2.2701**	0.0120	5.400
	ID7	N = 50 ; PrM = 1.5	2.2862	2.2852	2.3056	2.2608	0.0143	5.600
	ID8	N = 100 ; PrM = 1.5	2.2852	2.2851	2.3068	2.2524	0.0153	5.100
	ID9	N = 150 ; PrM = 1.5	2.2755	2.2700	2.3064	2.2638	0.0142	6.700
p-value								0.4572

6.1.4. Comparing Encoding Configurations

The total computational time consumed is shown in Table 3. The computational cost shows the importance of using the cluster. Previously, configurations with the best average rank according to the Friedman's test were selected to be compared globally. These configurations are shown in Table 7.

Table 7. Hypervolume statistics (Encoding experiment).

Identifier	Description	Configuration	Average	Median	Max.	Min.	St. D.	Friedman's Test Av. Rank
ID1	Real	N = 50 ; PrM = 1	2.2927	2.2932	2.3113	2.2710	**0.0117**	4.600
ID2	Binary 1P	N = 100 ; PrM = 1.5	2.2970	2.3008	2.3126	2.2708	0.0134	4.000
ID3	Binary 2P	N = 100 ; PrM = 0.5	**2.3013**	**2.3051**	2.3260	2.2714	0.0172	**3.000**
ID4	Binary U	N = 100 ; PrM = 0.5	2.2954	2.3000	2.3141	2.2647	0.0163	4.199
ID5	Gray 1P	N = 100 ; PrM = 1	2.2989	2.2986	**2.3347**	2.2652	0.0186	3.500
ID6	Gray 2P	N = 50 ; PrM = 0.5	2.2947	2.2971	2.3192	**2.2757**	0.0128	4.000
ID7	Gray U	N = 50 ; PrM = 1	2.2907	2.2939	2.3134	2.2683	0.0156	4.699
p-value								0.5979

The Hypervolume average values evolution versus the evaluations number is shown in Figure 2a. The detail for the final evaluations (last million fitness function evaluations, from 9 to 10 million) is shown in Figure 2b. It can be seen that the configuration with identifier ID3 (with binary encoding and two point crossover, population of 100 individuals and mutation probability of 0.5 gene per chromosome) reaches the highest Hypervolume average value.

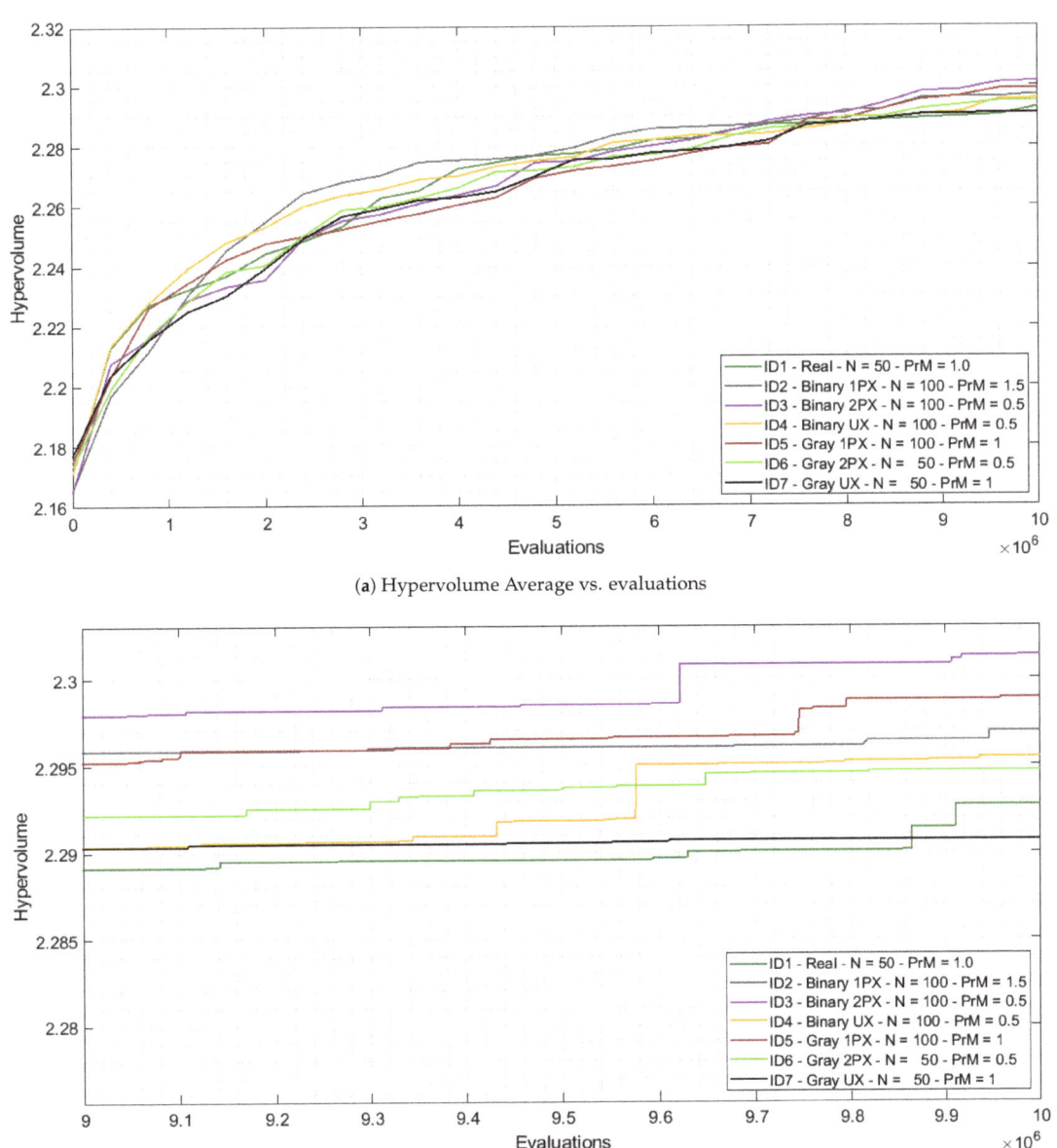

(a) Hypervolume Average vs. evaluations

(b) Hypervolume Average vs. evaluations, detail (last million evaluations displayed)

Figure 2. Hypervolume Average vs. evaluations (encoding experiment).

Box plots of the Hypervolume values distribution at the end of the process are shown in Figure 3. Statistical information in relation to the Hypervolume value at the final of the evolutionary process is shown in Table 7. It can be seen that the configuration with identifier ID3 (with Binary encoding and two point crossover, population of 100 individuals and mutation probability of 0.5 gene per chromosome) presents both the highest Hypervolume average value and the highest Hypervolume median value, the configuration with identifier ID5 (with Gray encoding and one point crossover, population of 100 individuals and mutation probability of one gene per chromosome) presents the highest Hypervolume maximum value, the configuration with identifier ID6 (with Gray encoding and two point crossover, population of 50 individuals and mutation probability of 0.5 gene per chromosome) presents the highest Hypervolume minimum value and the configuration with identifier ID1 (with real encoding, population of 50 individuals and mutation probability of one gene per chromosome) presents the lowest Hypervolume standard deviation value.

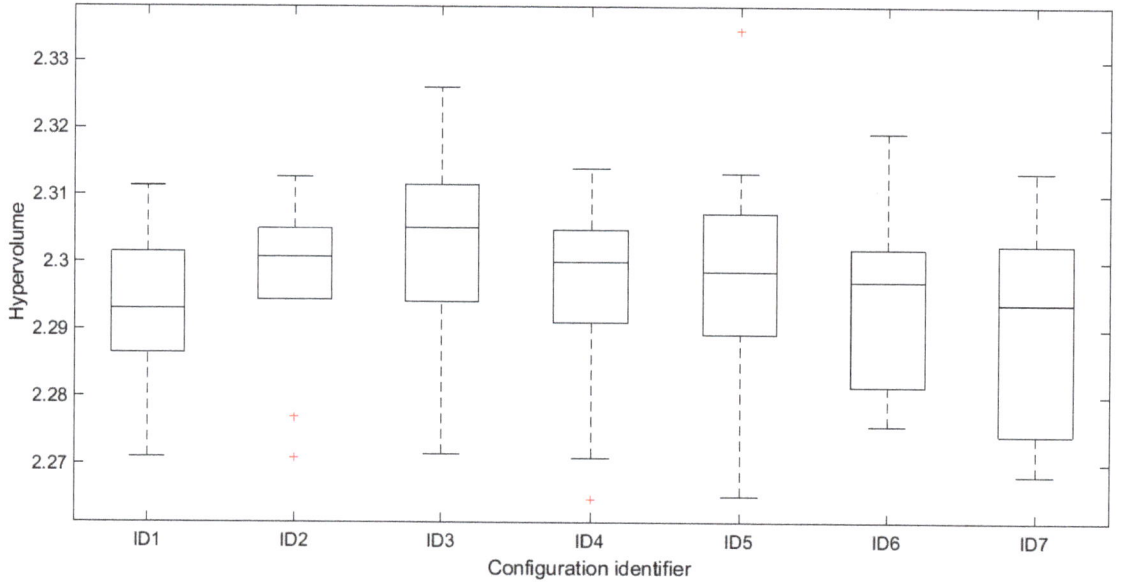

Figure 3. Box plots of the final Hypervolume (encoding experiment).

In order to establish whether one of the configurations performs better than any other, a statistical significance hypothesis test was conducted. The average ranks computed through the Friedman's test are shown in Table 7. It can be seen that the configuration with identifier ID3 (with binary encoding and two point crossover, population of 100 individuals and mutation probability of 0.5 gene per chromosome) presents the best average rank. However, the p-value computed (0.5979) implies that the null hypothesis (H_0) cannot be rejected (p-value > 0.05), so it is possible to conclude that, in the studied conditions, there is no configuration that performs better than another.

The best accumulated non-dominated solutions obtained for all encodings and configurations were used to compute the accumulated Hypervolume, whose value was 2.4553. As expected, the value is higher than 2.4298, the maximum accumulated value obtained after the evolutionary process for the standard binary encoding with two point crossover. This is shown in Table 8.

Table 8. Maximum accumulated Hypervolume value (encoding experiment).

Encoding	Hypervolume Accumulated Value
Real	2.3943
Binary 1 Point Crossover	2.4142
Binary 2 Point Crossover	2.4298
Binary Uniform Crossover	2.3984
Gray 1 Point Crossover	2.4011
Gray 2 Point Crossover	2.3982
Gray Uniform Crossover	2.3829
Global	2.4553

6.2. Accuracy Experiment

In the first experiment, a thorough comparison of the performances of encodings was developed using the hour as a time unit. Although non-significant differences among performances were found, the best average rank using the Friedman's test was presented by the standard binary encoding. For this reason, the results achieved for the standard binary encoding are used in the present experiment to compare the performance using the day and the week as time units.

6.2.1. Standard Binary Encoding (Days)

The results of using standard binary encoding with one point, two point and uniform crossover with the day as a time unit are shown below. The computational time consumed by each one is shown in Table 9.

Table 9. Computational time consumed.

Encoding	Time Unit	Average Time	Sequential Time
Binary 1PX	Hour	2 days, 18 h and 57 min	8 months, 7 days, 18 h and 48 min.
Binary 2PX	Hour	3 days, 2 h and 39 min	9 months, 6 days, 5 h and 52 min.
Binary UX	Hour	3 days and 39 min	8 months, 29 days, 3 h and 8 min.
Binary 1PX	Day	2 days, 22 h and 3 min	8 months, 19 days, 9 h and 14 min.
Binary 2PX	Day	2 days, 19 h and 13 min	8 months, 8 days, 18 h and 18 min.
Binary UX	Day	2 days, 19 h and 13 min	8 months, 8 days, 18 h and 55 min.
Binary 1PX	Week	2 days, 18 h and 3 min	8 months, 4 days, 9 h and 39 min.
Binary 2PX	Week	2 days, 19 h and 21 min	8 months, 9 days, 6 h and 55 min.
Binary UX	Week	2 days, 19 h and 20 min	8 months, 9 days, 4 h and 40 min.
Total		2 days, 20 h and 53 min	6 years, 4 months, 12 days, 22 h and 41 min.

The relationship between method configurations and identifiers is shown in Table 10. Moreover, statistical information in relation to the Hypervolume value at the final of the evolutionary process is shown. For the binary encoding with one point crossover and the day as a time unit (B1PX-D), it is possible to conclude that the configuration with identifier ID6 (population of 150 individuals and mutation probability of one gene per chromosome) presents the highest Hypervolume average value, the highest Hypervolume minimum value and the lowest Hypervolume standard deviation. The configuration with identifier ID9 (population of 150 individuals and mutation probability of 1.5 gene per chromosome) presents the highest Hypervolume median value and the configuration with identifier ID1 (population of 50 individuals and mutation probability of 0.5 gene per chromosome) presents the highest Hypervolume maximum value.

For the binary encoding with two point crossover and the day as a time unit (B2PX-D), it is possible to conclude that the configuration with identifier ID4 (population of 50 individuals and mutation probability of one gene per chromosome) presents the highest Hypervolume average value, the configuration with identifier ID6 (population of 150 individuals and mutation probability of one gene per chromosome) presents the highest Hypervolume median value, the configuration with identifier ID9 (population of 150 individuals and mutation probability of 1.5 gene per chromosome) presents the highest Hypervolume maximum value and the configuration with identifier ID3 (population of 150 individuals and mutation probability of 0.5 gene per chromosome) presents both the highest Hypervolume minimum value and the lowest Hypervolume standard deviation value.

For the binary encoding with uniform crossover and the day as a time unit (BUX-D), it is possible to conclude that the configurations with identifiers ID6 (population of 150 individuals and mutation probability of one gene per chromosome) and ID8 (population of 100 individuals and mutation probability of 1.5 gene per chromosome) present the highest Hypervolume average value, the configuration with identifier ID1 (population of 50 individuals and mutation probability of 0.5 gene per chromosome) presents the highest Hypervolume median value, the configuration with identifier ID9 (population of 150 individuals and mutation probability of 1.5 gene per chromosome) presents the highest Hypervolume maximum value, the configuration with identifier ID5 (population of 100 individuals and mutation probability of one gene per chromosome) presents the highest Hypervolume minimum value, and the configuration with identifier ID7 (population of 50 individuals and mutation probability of 1.5 gene per chromosome) presents the lowest Hypervolume standard deviation value.

Table 10. Hypervolume statistics (Binary encoding-Days).

Encoding	Identifier	Configuration	Average	Median	Max.	Min.	St. D.	Friedman's Test Av. Rank
B1PX-D	ID1	N = 50 ; PrM = 0.5	2.2834	2.2781	**2.3430**	2.2655	0.0220	5.800
	ID2	N = 100 ; PrM = 0.5	2.2942	2.2911	2.3198	2.2788	0.0133	4.100
	ID3	N = 150 ; PrM = 0.5	2.2785	2.2834	2.2934	2.2563	0.0126	6.600
	ID4	N = 50 ; PrM = 1	2.2819	2.2795	2.3127	2.2622	0.0152	5.800
	ID5	N = 100 ; PrM = 1	2.2896	2.2870	2.3188	2.2621	0.0184	4.500
	ID6	N = 150 ; PrM = 1	**2.2964**	2.2941	2.3193	**2.2804**	**0.0103**	**3.200**
	ID7	N = 50 ; PrM = 1.5	2.2922	2.2932	2.3239	2.2707	0.0177	4.400
	ID8	N = 100 ; PrM = 1.5	2.2776	2.2793	2.3057	2.2530	0.0171	6.800
	ID9	N = 150 ; PrM = 1.5	2.2939	**2.2947**	2.3107	2.2758	0.0135	3.800
p-value								0.0246
B2PX-D	ID1	N = 50 ; PrM = 0.5	2.2838	2.2795	2.3043	2.2681	0.0123	5.400
	ID2	N = 100 ; PrM = 0.5	2.2783	2.2785	2.2947	2.2609	0.0106	6.600
	ID3	N = 150 ; PrM = 0.5	2.2867	2.2880	2.3043	**2.2757**	**0.0084**	4.300
	ID4	N = 50 ; PrM = 1	**2.2930**	2.2865	2.3173	2.2746	0.0176	4.200
	ID5	N = 100 ; PrM = 1	2.2874	2.2831	2.3188	2.2652	0.0192	5.399
	ID6	N = 150 ; PrM = 1	2.2916	**2.2937**	2.3150	2.2712	0.0152	**4.100**
	ID7	N = 50 ; PrM = 1.5	2.2877	2.2849	2.3249	2.2693	0.0161	4.600
	ID8	N = 100 ; PrM = 1.5	2.2841	2.2817	2.3136	2.2658	0.0133	5.300
	ID9	N = 150 ; PrM = 1.5	2.2887	2.2826	**2.3372**	2.2651	0.0216	5.100
p-value								0.5612
BUX-D	ID1	N = 50 ; PrM = 0.5	2.2918	**2.2948**	2.3152	2.2643	0.0168	**4.200**
	ID2	N = 100 ; PrM = 0.5	2.2897	2.2865	2.3356	2.2602	0.0198	4.699
	ID3	N = 150 ; PrM = 0.5	2.2844	2.2866	2.3104	2.2522	0.0179	5.600
	ID4	N = 50 ; PrM = 1	2.2798	2.2732	2.3078	2.2601	0.0160	6.400
	ID5	N = 100 ; PrM = 1	2.2897	2.2893	2.3127	**2.2729**	0.0151	4.800
	ID6	N = 150 ; PrM = 1	**2.2923**	2.2907	2.3333	2.2666	0.0219	4.700
	ID7	N = 50 ; PrM = 1.5	2.2894	2.2880	2.3201	2.2691	**0.0138**	5.100
	ID8	N = 100 ; PrM = 1.5	**2.2923**	2.2901	2.3279	2.2637	0.0183	5.000
	ID9	N = 150 ; PrM = 1.5	2.2908	2.2863	**2.3461**	2.2606	0.0249	4.500
p-value								0.8007

In order to establish the best behaviour amongst the range of configurations, a statistical significance hypothesis test was conducted. The average ranks computed through the Friedman's test and the *p*-value obtained (a value bigger than 0.05 implies that the null hypothesis cannot be rejected, suggesting that all configurations perform in a similar way) are shown in Table 10. It can be seen that the configuration with identifier ID6 (population of 150 individuals and mutation probability of one gene per chromosome) presents the best average rank for the binary encoding with one point crossover and the day as a time unit. Moreover, the *p*-value obtained of 0.0246 explains that the null Hypothesis can be rejected so, in this case, the configuration ID6 performs better than any other configuration. In order to find the concrete pairwise comparisons that produce differences, a statistical significance analysis using the Wilcoxon signed-rank test was run as explained by Benavoli et al. [70]. The results of using such a test, in which the configuration ID6 is compared with the rest of configurations, is shown in Table 11. It can be seen that configuration ID6 performs better than the configurations ID3, ID4 and ID8.

Table 11. *p*-values from Wilcoxon signed-rank test.

Comparison	*p*-Value	Conclusion
ID3 versus ID6	0.0059	Significant difference found
ID4 versus ID6	0.0371	Significant difference found
ID6 versus ID8	0.0371	Significant difference found
ID1 versus ID6	0.0840	The null hypothesis cannot be rejected
ID5 versus ID6	0.3223	The null hypothesis cannot be rejected
ID2 versus ID6	0.3750	The null hypothesis cannot be rejected
ID6 versus ID7	0.4922	The null hypothesis cannot be rejected
ID6 versus ID9	0.6250	The null hypothesis cannot be rejected

Regarding the binary encoding with two point crossover and the day as a time unit, the same configuration (ID6) presents the best average rank. Finally, it can be seen that the configuration with identifier ID1 (population of 50 individuals and mutation probability of 0.5 gene per chromosome) presents the best average rank for the binary encoding with uniform crossover and the day as a time unit. In each case, the best configurations were selected for the final comparison study between the accuracy level encodings.

The best accumulated non-dominated solutions obtained through the final generation of the evolutionary process for all executions and all configurations were used to compute the accumulated Hypervolume whose values were 2.4007, 2.3942 and 2.3984 for the binary encoding with one point, two point and uniform crossover with the day as a time unit, respectively. As expected, the values are higher than 2.3430, 2.3372 and 2.3461, the maximum values shown in Table 10, respectively.

6.2.2. Standard Binary Encoding (Weeks)

The results of using standard binary encoding with one point, two point and uniform crossover and the week as a time unit are shown below. The computational time consumed by each one is shown in Table 9. The relationship between method configurations and identifiers is shown in Table 12. Statistical information in relation to the Hypervolume value at the final of the evolutionary process is also shown. For the binary encoding with one point crossover and the week as a time unit (B1PX-W), it is possible to conclude that the configuration with identifier ID1 (population of 50 individuals and mutation probability of 0.5 gene per chromosome) presents the highest Hypervolume average value, the highest Hypervolume median value and the highest Hypervolume maximum value. The configuration with identifier ID5 (population of 100 individuals and mutation probability of one gene per chromosome) presents both the highest Hypervolume minimum value and the lowest Hypervolume standard deviation value.

For the binary encoding with two point crossover and the week as a time unit (B2PX-W), it is possible to conclude that the configuration with identifier ID7 (population of 50 individuals and mutation probability of 1.5 gene per chromosome) presents both the highest Hypervolume average value and the highest Hypervolume maximum value. The configuration with identifier ID1 (population of 50 individuals and mutation probability of 0.5 gene per chromosome) presents both the highest Hypervolume median value and the highest Hypervolume minimum value. Finally, the configuration with identifier ID5 (population of 100 individuals and mutation probability of one gene per chromosome) presents the lowest Hypervolume standard deviation value.

For the binary encoding with uniform crossover and the week as a time unit (BUX-W), it is possible to conclude that the configuration with identifier ID8 (population of 100 individuals and mutation probability of 1.5 gene per chromosome) presents the highest Hypervolume average value, the highest Hypervolume median value and the highest Hypervolume minimum value. The configuration with identifier ID1 (population of 50 individuals and mutation probability of 0.5 gene per chromosome) presents the highest Hypervolume minimum value and the configuration with identifier ID2 (population of 100 individuals and mutation probability of 0.5 gene per chromosome) presents the lowest Hypervolume standard deviation value.

Table 12. Hypervolume statistics (Binary encoding-Weeks).

Encoding	Identifier	Configuration	Average	Median	Max.	Min.	St. D.	Friedman's Test Av. Rank
B1PX-W	ID1	$N = 50$; $PrM = 0.5$	**2.2957**	**2.2935**	2.3430	2.2561	0.0250	4.300
	ID2	$N = 100$; $PrM = 0.5$	2.2872	2.2873	2.3127	2.2665	0.0142	5.100
	ID3	$N = 150$; $PrM = 0.5$	2.2897	2.2808	2.3252	2.2623	0.0212	5.100
	ID4	$N = 50$; $PrM = 1$	2.2812	2.2805	2.2977	2.2520	0.0157	5.799
	ID5	$N = 100$; $PrM = 1$	2.2907	2.2893	2.3007	**2.2812**	**0.0060**	**3.999**
	ID6	$N = 150$; $PrM = 1$	2.2825	2.2825	2.3004	2.2679	0.0101	6.100
	ID7	$N = 50$; $PrM = 1.5$	2.2890	2.2899	2.3110	2.2597	0.0164	4.900
	ID8	$N = 100$; $PrM = 1.5$	2.2874	2.2848	2.3066	2.2690	0.0123	5.000
	ID9	$N = 150$; $PrM = 1.5$	2.2888	2.2892	2.3176	2.2634	0.0155	4.700
p-value								0.7979
B2PX-W	ID1	$N = 50$; $PrM = 0.5$	2.2917	**2.2911**	2.3101	**2.2764**	0.0121	**4.000**
	ID2	$N = 100$; $PrM = 0.5$	2.2858	2.2858	2.3013	2.2671	0.0122	5.300
	ID3	$N = 150$; $PrM = 0.5$	2.2892	2.2869	2.3237	2.2567	0.0250	4.900
	ID4	$N = 50$; $PrM = 1$	2.2797	2.2743	2.3014	2.2703	0.0108	6.299
	ID5	$N = 100$; $PrM = 1$	2.2840	2.2838	2.2965	2.2665	**0.0085**	5.200
	ID6	$N = 150$; $PrM = 1$	2.2793	2.2780	2.3066	2.2502	0.0147	6.200
	ID7	$N = 50$; $PrM = 1.5$	**2.2921**	2.2817	**2.3465**	2.2709	0.0246	4.899
	ID8	$N = 100$; $PrM = 1.5$	2.2895	2.2880	2.3198	2.2546	0.0168	4.001
	ID9	$N = 150$; $PrM = 1.5$	2.2918	2.2854	2.3130	2.2710	0.0145	4.200
p-value								0.4439
BUX-W	ID1	$N = 50$; $PrM = 0.5$	2.2894	2.2847	**2.3336**	2.2659	0.0197	5.500
	ID2	$N = 100$; $PrM = 0.5$	2.2925	2.2922	2.3144	2.2756	**0.0113**	4.399
	ID3	$N = 150$; $PrM = 0.5$	2.2911	2.2932	2.3216	2.2664	0.0156	4.399
	ID4	$N = 50$; $PrM = 1$	2.2858	2.2836	2.3057	2.2660	0.0128	5.499
	ID5	$N = 100$; $PrM = 1$	2.2826	2.2834	2.3144	2.2631	0.0140	6.700
	ID6	$N = 150$; $PrM = 1$	2.2854	2.2857	2.3122	2.2645	0.0164	5.499
	ID7	$N = 50$; $PrM = 1.5$	2.2893	2.2887	2.3094	2.2681	0.0125	4.899
	ID8	$N = 100$; $PrM = 1.5$	**2.3009**	**2.2942**	2.3316	**2.2870**	0.0145	**3.400**
	ID9	$N = 150$; $PrM = 1.5$	2.2921	2.2862	2.3163	2.2774	0.0143	4.700
p-value								0.3128

In order to establish the best behaviour amongst configurations, a statistical significance hypothesis test was conducted. The average ranks computed through the Friedman's test and the p-value obtained (a value bigger than 0.05 implies that the null hypothesis cannot be rejected, suggesting that all configurations perform in a similar way) are shown in Table 12. It can be seen that the configuration with identifier ID5 (population of 100 individuals and mutation probability of one gene per chromosome) presents the best average rank for the binary encoding with one point crossover and the week as a time unit. It can be seen that the configuration with identifier ID1 (population of 50 individuals and mutation probability of 0.5 gene per chromosome) presents the best average rank for the binary encoding with two point crossover and the week as a time unit. It can be seen that the configuration with identifier ID8 (population of 100 individuals and mutation probability of 1.5 gene per chromosome) presents the best average rank for the binary encoding with uniform crossover and the week as a time unit. These configurations were selected for the final comparison study of the accuracy-level configurations.

The best accumulated non-dominated solutions obtained through the final generation of the evolutionary process for all executions and all configurations were used to compute the accumulated Hypervolume whose values were 2.4047, 2.4285 and 2.4037 for the binary encoding with one point, two point and uniform crossover with the week as a time unit, respectively. As expected, the values are higher than 2.3430, 2.3465 and 2.3336, the maximum values shown in Table 12, respectively.

6.2.3. Comparing Accuracy-Level Configurations

The global computational time consumed is shown in Table 9. The computational cost shows the importance of using the cluster. Previously, configurations with the best average rank according to the Friedman's test were selected to be globally compared. These configurations are shown in Table 13.

Table 13. Hypervolume statistics (accuracy experiment).

Id.	Description	Configuration	Average	Median	Max.	Min.	St. D.	Friedman's Test Av. Rank
ID1	Binary 1P (hour)	N = 100 ; PrM = 1.5	2.2970	2.3008	2.3126	2.2708	0.0134	4.700
ID2	Binary 2P (hour)	N = 100 ; PrM = 0.5	**2.3013**	**2.3051**	2.3260	2.2714	0.0172	**3.699**
ID3	Binary U (hour)	N = 100 ; PrM = 0.5	2.2954	2.3000	2.3141	2.2647	0.0163	4.500
ID4	Binary 1P (day)	N = 150 ; PrM = 1.0	2.2964	2.2941	2.3193	2.2804	0.0103	5.000
ID5	Binary 2P (day)	N = 150 ; PrM = 1.0	2.2916	2.2937	2.3150	2.2712	0.0152	5.800
ID6	Binary U (day)	N = 50 ; PrM = 0.5	2.2918	2.2948	2.3152	2.2643	0.0168	5.300
ID7	Binary 1P (week)	N = 100 ; PrM = 1.0	2.2907	2.2893	2.3007	2.2812	**0.0060**	6.200
ID8	Binary 2P (week)	N = 50 ; PrM = 0.5	2.2917	2.2911	2.3101	2.2764	0.0121	5.900
ID9	Binary U (week)	N = 100 ; PrM = 1.5	2.3009	2.2942	**2.3316**	**2.2870**	0.0145	3.900
p-value								0.4053

The Hypervolume average values evolution versus the evaluations number is shown in Figure 4a. The detail for the final evaluations (last million fitness function evaluations, from 9 to 10 million) is shown in Figure 4b. It can be seen that the configuration with identifier ID2 (with Binary encoding, two point crossover, the hour as a time unit, population of 100 individuals and mutation probability of 0.5 gene per chromosome) reaches the highest Hypervolume average value.

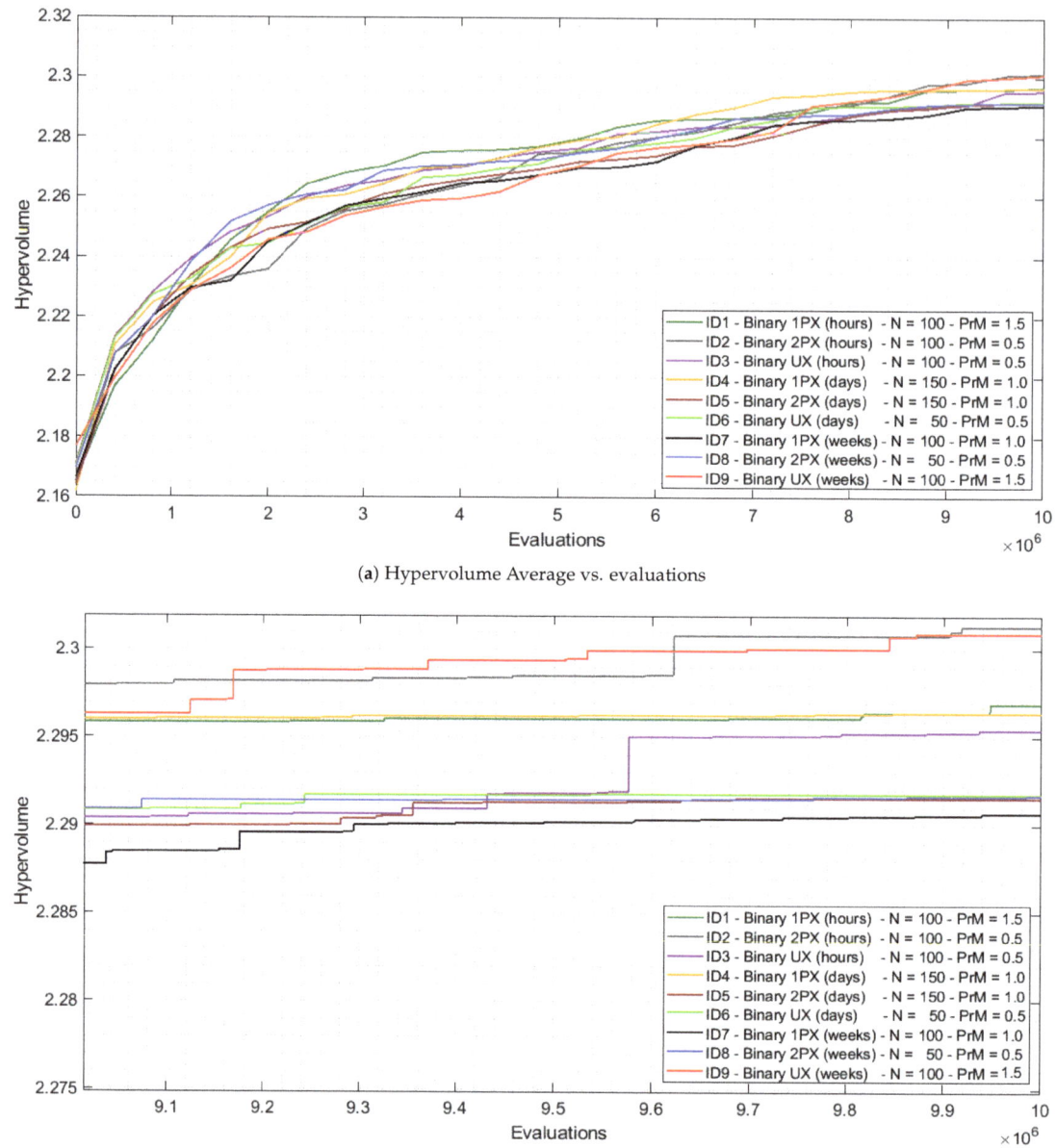

(a) Hypervolume Average vs. evaluations

(b) Hypervolume Average vs. evaluations, detail (last million evaluations displayed)

Figure 4. Hypervolume Average vs. evaluations (accuracy experiment).

Box plots of the Hypervolume values distribution at the end of the process are shown in Figure 5. They are ordered from left to right in relation to time units of hours (H), days (D) and weeks (W) and crossover types of one point (1PX), two point (2PX) and uniform crossover (UX). It can be seen that the medians are ordered from biggest to smallest for each group of crossover. The greater the accuracy the bigger the Hypervolume median value (it is greater for hours than for days and greater for days than for weeks). Statistical information in relation to the Hypervolume value at the end of the evolutionary process is shown in

Table 13. It can be seen that the configuration with identifier ID2 (with binary encoding, two point crossover and the hour as a time unit, population of 100 individuals and mutation probability of 0.5 gene per chromosome) presents both the highest Hypervolume average value and the highest Hypervolume median value, the configuration with identifier ID9 (with binary encoding, uniform crossover and the week as a time unit, population of 100 individuals and mutation probability of 1.5 gene per chromosome) presents both the highest Hypervolume maximum value and the highest Hypervolume minimum value. The configuration with identifier ID7 (with binary encoding, one point crossover and the week as a time unit, population of 100 individuals and mutation probability of one gene per chromosome) presents the lowest Hypervolume standard deviation.

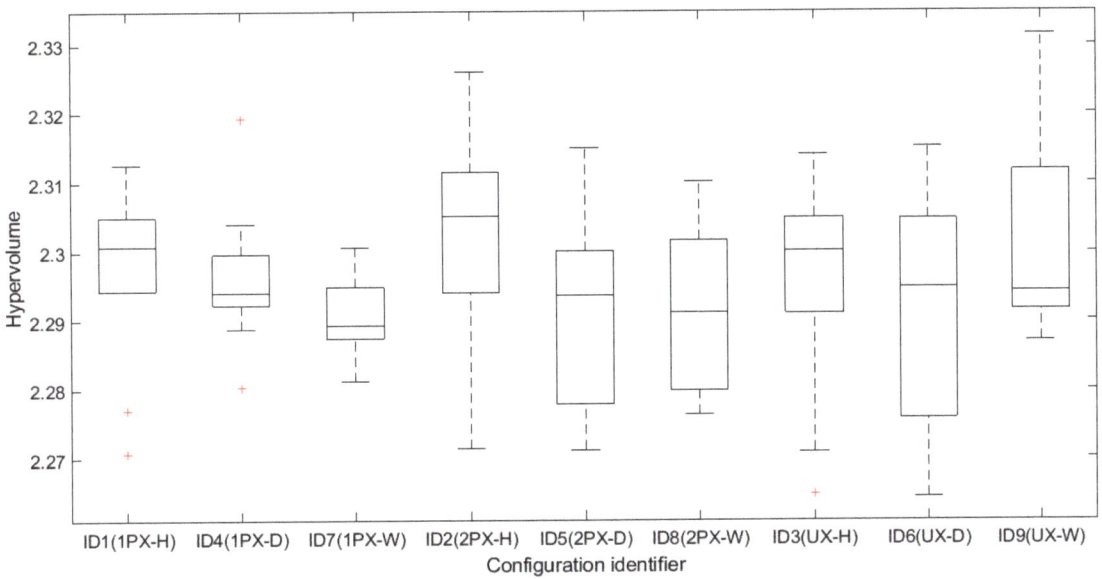

Figure 5. Box plots of the final Hypervolume (accuracy experiment).

In order to establish whether one of the configurations performs better than any other, a statistical significance hypothesis test was conducted. The average ranks computed through the Friedman's test are shown in Table 13. It can be seen that the configuration with identifier ID2 (with Binary encoding, two point crossover and the hour as a time unit, population of 100 individuals and mutation probability of 0.5 gene per chromosome) presents the best average rank. However, the p-value computed (0.4053) implies that the null hypothesis (H_0) cannot be rejected (p-value > 0.05), so it is possible to conclude that, in the studied conditions, there is no configuration that performs better than any other.

The best accumulated non-dominated solutions obtained were used to compute the accumulated Hypervolume, whose value was 2.4046. As expected, the value is higher than 2.4298, the maximum accumulated value obtained after the evolutionary process for the standard binary encoding with two point crossover and the hour as a time unit. This is shown in Table 14.

Table 14. Maximum accumulated Hypervolume value (accuracy experiment).

Encoding	Time Unit	Hypervolume Accumulated Value
Binary 1 Point Crossover	Hour	2.4142
Binary 2 Point Crossover	Hour	2.4298
Binary Uniform Crossover	Hour	2.3984
Binary 1 Point Crossover	Day	2.4007
Binary 2 Point Crossover	Day	2.3942
Binary Uniform Crossover	Day	2.3984
Binary 1 Point Crossover	Week	2.4047
Binary 2 Point Crossover	Week	2.4285
Binary Uniform Crossover	Week	2.4037
Global		2.4646

6.3. Accumulated Non-Dominated Set of Designs

The non-dominated solutions to the problem provided at the end of the evolutionary process for all executions, all configurations, all encodings and time units are shown in Figure 6. All optimum solutions belonging to the achieved non-dominated front are shown in Table 15. Unavailability (Q) is shown as a fraction, Cost is shown in economic units and the rest of the variables represent, for the respective devices, the optimum Time To Start a Preventive Maintenance activity with the hour, day or week as a time unit.

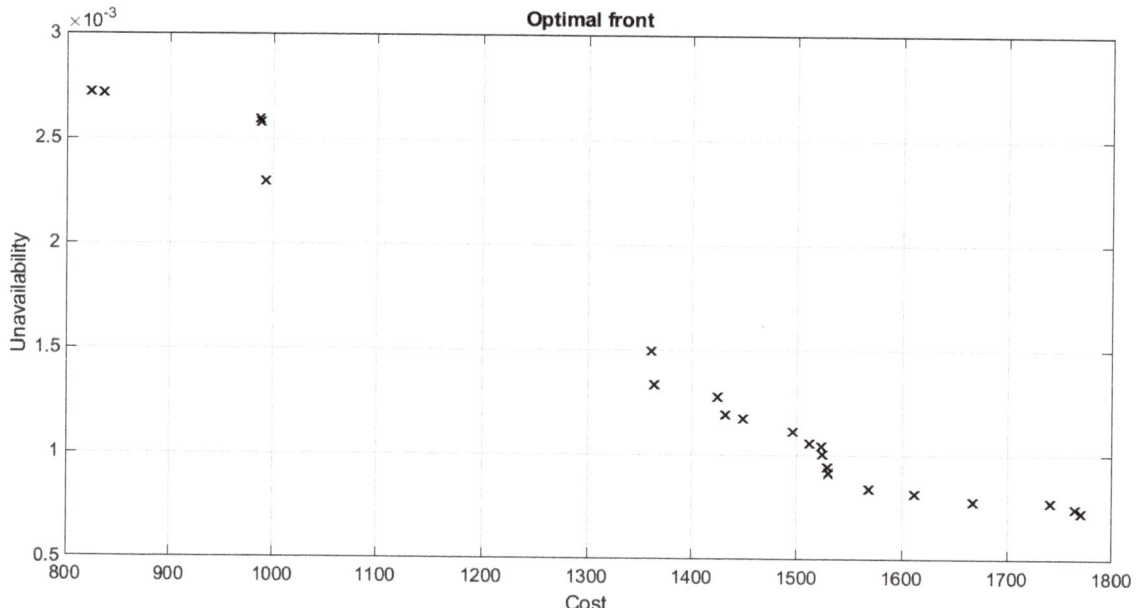

Figure 6. Non-dominated front.

Table 15. Optimum solutions (encoding experiment).

Id	Q	Cost [eu]	Time Unit	V1 [h]	P2 [h]	P3 [h]	V4 [h]	V5 [h]	V6 [h]	V7 [h]
1	0.002720	823.38	Hours	25,408	0	8633	0	34,179	34,903	31,386
2	0.002713	835.75	Hours	29,225	0	8633	0	27,070	33,454	33,690
3	0.002591	986.88	Days	1285	0	353	902	1441	967	1386
4	0.002576	988.00	Weeks	127	0	47	165	206	172	203
5	0.002295	992.75	Days	1435	0	350	830	1088	1459	1454
6	0.001495	1360.50	Hours	34,746	8239	8408	0	32,676	31,769	31,484
7	0.001334	1363.75	Days	1394	360	315	0	1125	1301	1026
8	0.001276	1424.25	Weeks	204	50	37	0	174	173	188
9	0.001189	1431.75	Hours	31,040	8617	8103	0	34,787	31,445	29,929
10	0.001174	1449.00	Days	178	51	50	0	182	171	195
11	0.001112	1496.12	Weeks	112	48	48	0	176	196	144
12	0.001056	1512.50	Days	880	358	354	1186	1020	1346	1098
13	0.001039	1524.00	Weeks	162	45	42	195	147	176	189
14	0.001002	1524.62	Hours	28,042	7228	6549	30,777	20,951	31,779	34,982
15	0.000939	1528.88	Hours	29,939	7028	7904	21,690	25,711	34,814	34,791
16	0.000913	1530.38	Weeks	194	47	49	103	137	145	179
17	0.000835	1567.75	Days	1028	363	340	1280	1430	986	1326
18	0.000813	1611.38	Hours	22,516	5955	6298	17,237	29,568	27,075	22,757
19	0.000776	1667.25	Hours	32,355	7185	8384	27,489	26,949	32,700	32,770
20	0.000771	1741.00	Hours	27,443	7556	8520	30,003	30,056	30,579	21,639
21	0.000742	1764.75	Hours	22,776	8696	8506	12,256	34,763	33,006	29,706
22	0.000725	1770.12	Hours	30,813	7371	8453	29,958	16,345	30,776	25,358

The solution with the lowest Cost (ID1) (823.38 economic units) represents the biggest Unavailability (0.002720). These values are followed by periodic optimum times (using the hour as a time unit in this case) measured from the moment in which the system mission time starts (Time To Perform a Preventive Maintenance activity (TRP) is not included). For solution ID1, it can be seen that periodic optimum Times To Start a Preventive Maintenance activity (TP) for devices P2 and V4 are not supplied. This is because the design alternative does not include such devices. The opposite case shows the biggest Cost (ID22) (1770.12 economic units) and the lowest Unavailability (0.000725). For solution ID22, periodic optimum Times To Start a Preventive Maintenance activity (TP) are supplied for all devices. This is because the design alternative includes devices P2 and V4. Other optimum solutions were found in these two solutions and can be seen in Table 15. The decision makers will need to decide which is the preferable design for them, taking into account their individual requirements.

Moreover, solutions were clustered in Figure 7 according to their final design. Solutions are shown from left to right and in ascending order in relation to the Cost from ID1 to ID22. The solutions contained in Cluster 1 (the solutions 1 to 2, see also Table 15) are the solutions in which non-redundant devices were included in the design. In this case, the system contains exclusively devices placed in series. These solutions present the lowest Cost and the biggest Unavailability. The solutions contained in Cluster 2 (the solutions 3 to 5, see also Table 15) are the solutions in which a redundant valve was included in the design as a parallel device. These solutions present a bigger Cost and a lower Unavailability than the solutions contained in Cluster 1. The solutions contained in Cluster 3 (the solutions 6 to 11, see also Table 15) are the solutions in which a redundant pump was included in the design as a parallel device. These solutions present a higher Cost and a lower Unavailability than the solutions contained in Clusters 1 and 2. Finally, the solutions contained in Cluster 4 (the solutions 12 to 22, see also Table 15) are the solutions in which both a redundant valve and a redundant pump were included in the design as parallel devices. These solutions present the biggest Cost and the lowest Unavailability.

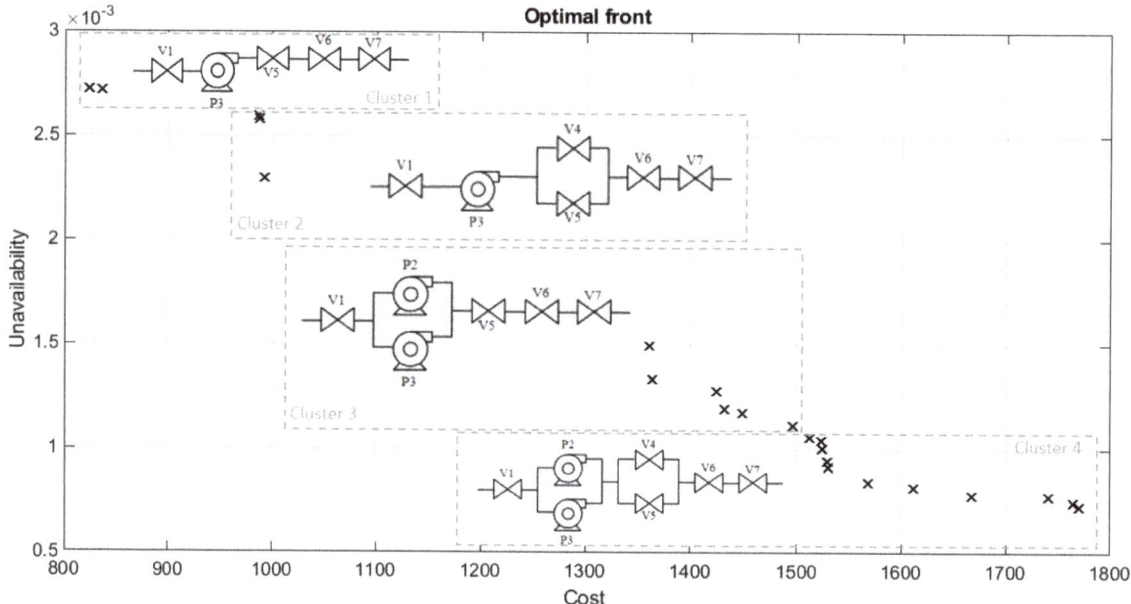

Figure 7. Design alternatives (encoding experiment).

The best accumulated non-dominated solutions obtained were used to compute the accumulated Hypervolume, whose value was 2.4651. As expected, the value is higher than the rest of the maximum accumulated values obtained after the evolutionary process for the encoding experiment and the accuracy experiment. This is shown in Table 16. This value is also higher than the value obtained in [6] and could be considered as an actual benchmark value of the case study.

Table 16. Maximum accumulated Hypervolume value.

Experiment	Hypervolume Accumulated Value
Encoding	2.4553
Accuracy	2.4646
Total	2.4651

7. Conclusions

In the present paper, a methodology presented previously by the authors [6] was used. This consists of coupling a Multi-Objective Evolutionary Algorithm and Discrete Simulation in order to optimise simultaneously both the system design (based on redundancy allocation) and its preventive maintenance strategy (based on determining periodic preventive maintenance activities with regard to each device included in the design), whilst addressing the conflict between Availability and operational Cost. The Multi-Objective Evolutionary Algorithm gives rise to a population of individuals, each encoding one design alternative and its preventive maintenance strategy. Each individual represents a possible solution to the problem, which is then used to modify and evaluate the system Functionability Profile through Discrete Simulation. The individuals evolve generation after generation until the stopping criterion is reached. This process was applied to a technical system in a case study in which two experiments were developed: Firstly, an encoding experiment which consisted of comparing the performance of seven encoding types (real, standard binary with one point, two point and uniform crossover, and Gray with one point, two point and uniform crossover), and secondly, an accuracy level encoding which consisted of comparing the performance of using standard binary encoding with accuracy levels across a range of time unit (hours, days and weeks) with impact in the form of size of the chromosome (the smaller the time unit, the bigger the chromosome). The Multi-objective Evolutionary Algorithm used was NSGA-II and a set of optimum non-dominated solutions were obtained for all cases.

In conclusion, the use of the Multi-Objective Evolutionary Algorithm NSGA-II and Discrete Simulation to address the joint optimisation of the system design and its preventive maintenance strategy provides Availability-Cost-balanced solutions to the real world problem studied, where data based on specific bibliography, mathematical relations and field experience were used. The computational cost reveals the complexity of the process and the necessity of using a computing cluster, which allowed parallel executions.

With regard to the encoding experiment, the best ordered method by the Friedman test case based on the final Hypervolume indicator distributions was the two point crossover standard binary encoding, although no statistically significant differences were observed. With regard to the accuracy experiment, the best ordered method by the Friedman test case based on the final Hypervolume indicator distributions was the two point crossover standard binary encoding with hours as a time unit, although no statistically significant differences were observed. From the authors' point of view, an important conclusion arises from this experiment, which relates to flexibility regarding the time unit to schedule the preventive maintenance activities. Using the hour, the day or the week as a time unit does not affect significantly the performance of the configurations so, in the studied conditions, the preventive maintenance activities can be planned using weeks as a time unit. This allows a better range of time for planning than if days or hours are used as a time unit.

In addition, a higher benchmark value of the case study (in terms of Hypervolume indicator) is attained in this work as a reference.

As future work, the authors consider that these conclusions should be explored in greater depth extending the analysis to other real world problems in the reliability field, as well as comparing this analysis with other state of the art Multi-objective Evolutionary Algorithms.

Author Contributions: All of the authors contributed to publishing this article. Conceptualization A.C., D.G., B.J.G.; methodology, A.C., D.G., B.J.G.; software, A.C., D.G., B.J.G.; validation A.C., D.G., B.J.G.; formal analysis A.C., D.G., B.J.G.; investigation A.C., D.G., B.J.G.; resources D.G., B.J.G.; data curation, A.C., D.G., B.J.G.; writing—original draft preparation A.C., D.G.; writing—review and editing A.C., D.G., B.J.G.; visualization A.C., D.G., B.J.G.; supervision D.G., B.J.G.; funding acquisition D.G., B.J.G. All authors have read and agreed to the published version of the manuscript.

Funding: This research was funded by Pre-doctoral Research Staff of Universidad de Las Palmas de Gran Canaria (ULPGC) grant of Andrés Cacereño. The APC was funded by Instituto Universitario de Sistemas Inteligentes y Aplicaciones Numéricas en Ingeniería (SIANI) of ULPGC.

Institutional Review Board Statement: Not applicable.

Informed Consent Statement: Not applicable.

Data Availability Statement: Not applicable.

Acknowledgments: A. Cacereño is the recipient of a contract from the Program of Training for Pre-doctoral Research Staff of the University of Las Palmas de Gran Canaria. The authors are grateful for this support. We also convey our gratitude to MSc. Ernesto Primera, from the Machinery & Reliability Institute, Alabama, USA. David Greiner also acknowledges partial Project support funding from the Ministerio de Ciencia, Innovacion y Universidades, Gobierno de España, grant contract: PID2019-110185RB-C22, and from the Agencia Canaria de Investigacion, Innovacion y Sociedad de la Informacion, Consejeria de Economia, Conocimiento y Empleo del Gobierno de Canarias, Grant Contract Number: PROID2020010022, and European Regional Development Funds (ERDF/FEDER).

Conflicts of Interest: The authors declare no conflict of interest.

Appendix A. Detailed Information of the Case Study Parameters

- **Life Cycle**. System mission time, expressed in hours.
- **Corrective Maintenance Cost**. The cost involved in developing a repair activity to recover the system following a failure, expressed in economic units per hour.
- **Preventive Maintenance Cost**. The cost involved in developing a Preventive Maintenance activity, expressed in relation to the Corrective Maintenance Cost.
- **Pump TF_{min}**. Minimum operation Time To Failure for a pump without Preventive Maintenance, expressed in hours.
- **Pump TF_{max}**. Maximum operation Time To Failure for a pump without Preventive Maintenance, expressed in hours.
- **Pump $TF\ \lambda$**. Failure rate for a pump, which follows an exponential failure distribution, expressed in hours raised to the power of minus six.
- **Pump TR_{min}**. Minimum Time To Repair or duration of a Corrective Maintenance activity for a pump, expressed in hours.
- **Pump TR_{max}**. Maximum Time To Repair or duration of a Corrective Maintenance activity for a pump, expressed in hours.
- **Pump $TR\ \mu$**. Mean for the normal distribution followed for the Time To Repair assumed for a pump, expressed in hours.
- **Pump $TR\ \sigma$**. Standard deviation for the normal distribution followed for the Time To Repair assumed for a pump, expressed in hours.
- **Pump TP_{min}**. Minimum operation Time To Start a scheduled Preventive Maintenance activity for a pump, expressed in hours.
- **Pump TP_{max}**. Maximum operation Time To Start a scheduled Preventive Maintenance activity for a pump, expressed in hours.
- **Pump TRP_{min}**. Minimum Time To Perform a Preventive Maintenance activity for a pump, expressed in hours.
- **Pump TRP_{max}**. Maximum Time To Perform a Preventive Maintenance activity for a pump, expressed in hours.
- **Valve TF_{min}**. Minimum operation Time To Failure for a valve without Preventive Maintenance, expressed in hours.
- **Valve TF_{max}**. Maximum operation Time To Failure for a valve without Preventive Maintenance, expressed in hours.
- **Valve $TF\ \lambda$**. Failure rate for a valve, which follows an exponential failure distribution, expressed in hours raised to the power of minus six.
- **Valve TR_{min}**. Minimum Time To Repair or duration of a Corrective Maintenance activity for a valve, expressed in hours.
- **Valve TR_{max}**. Maximum Time To Repair or duration of a Corrective Maintenance activity for a valve, expressed in hours.
- **Valve $TR\ \mu$**. Mean for the normal distribution followed for the Time To Repair assumed for a valve, expressed in hours.
- **Valve $TR\ \sigma$**. Standard deviation for the normal distribution followed for the Time To Repair assumed for a valve, expressed in hours.
- **Valve TP_{min}**. Minimum operation Time To Start a scheduled Preventive Maintenance activity for a valve, expressed in hours.
- **Valve TP_{max}**. Maximum operation Time To Start a scheduled Preventive Maintenance activity for a valve, expressed in hours.
- **Valve TRP_{min}**. Minimum Time To Perform a Preventive Maintenance activity for a valve, expressed in hours.
- **Valve TRP_{max}**. Maximum Time To Perform a Preventive Maintenance activity for a valve, expressed in hours.

References

1. Misra, K.B. Reliability Engineering: A Perspective. In *Handbook of Performability Engineering*; Misra, K.B., Ed.; Springer Limited: London, UK, 2008; Chapter 19; pp. 253–289; ISBN 978-1-84800-130-5.
2. Andrews, J.D.; Moss, T.R. *Reliability and Risk Assessment*, 2nd ed.; ASME Press: New York, NY, USA, 2002; p. 540. ISBN 0-7918-0183-7.
3. Simon, D. *Evolutionary Optimization Algorithms. Biologically Inspired and Population-Based Approaches to Computer Intelligence*; John Wiley & Sons, Inc.: Hoboken, NJ, USA, 2013; p. 742. ISBN 978-0-470-93741-9.
4. Kuo, W.; Prasad, R.; Tillman, F.A.; Mwang, C.L. *Optimal Reliability Design: Fundamentals and Applications*, 1st ed.; Cambridge University Press: Cambridge, UK, 2001; p. 389. ISBN 978-0-521-78127-5.
5. Gao, Y.; Feng, Y.; Zhang, Z.; Tan, J. An optimal dynamic interval preventive maintenance scheduling for series systems. *Reliab. Eng. Syst. Saf.* **2015**, *142*, 19–30. [CrossRef]
6. Cacereño, A.; Greiner, D.; Galván, B. Solving Multi-objective Optimal Design and Maintenance for Systems Based on Calendar Times Using NSGA-II. In *Advances in Evolutionary and Deterministic Methods for Design, Optimization and Control in Engineering and Sciences*; Gaspar-Cunha, A., Periaux, J., Giannakoglou, K.C., Gauger, N.R., Quagliarella, D., Greiner, D., Eds.; Springer Nature: Cham, Switzerland, 2021; pp. 245–259.
7. Knezevic, J. *Mantenibilidad*, 1st ed.; Isdefe: Madrid, Spain, 1996; p. 209. ISBN 84-89338-08-6.
8. Deb, K.; Pratap, A.; Agarwal, S.; Meyarivan, T. A Fast and Elitist Multiobjective Genetic Algorithm: NSGA-II. *IEEE Trans. Evol. Comput.* **2002**, *6*, 182–197. [CrossRef]
9. Busacca, P.G.; Marseguerra, M.; Zio, E. Multiobjective optimization by genetic algorithms: Application to safety systems. *Reliab. Eng. Syst. Saf.* **2001**, *72*, 59–74. ISSN 0951-8320. [CrossRef]
10. Marseguerra, M.; Zio, E.; Podofillini, L.; Coit, D.W. Optimal Design of Reliable Network Systems in Presence of Uncertainty. *IEEE Trans. Reliab.* **2005**, *54*, 243–253. [CrossRef]
11. Tian, Z.; Zuo, M.J. Redundancy allocation for multi-state systems using physical programming and genetic algorithms. *Reliab. Eng. Syst. Saf.* **2006**, *91*, 1049–1056. ISSN 0951-8320. [CrossRef]
12. Huang, H.Z.; Qu, J.; Zuo, M.J. Genetic-algorithm-based optimal apportionment of reliability and redundancy under multiple objectives. *IIE Trans.* **2009**, *41*, 287–298. [CrossRef]
13. Zoulfaghari, H.; Hamadani, A.Z.; Ardakan, M.A. Bi-objective redundancy allocation problem for a system with mixed repairable and non-repairable components. *ISA Trans.* **2014**, *53*, 17–24. ISSN 0019-0578. [CrossRef]
14. Taboada, H.A.; Espiritu, J.F.; Coit, D.W. MOMS-GA: A Multi-Objective Multi-State Genetic Algorithm for System Reliability Optimization Design Problems. *IEEE Trans. Reliab.* **2008**, *57*, 182–191. [CrossRef]
15. Taboada, H.A.; Baheranwala, F.; Coit, D.W.; Wattanapongsakorn, N. Practical solutions for multi-objective optimization: An application to system reliability design problems. *Reliab. Eng. Syst. Saf.* **2007**, *92*, 314–322. ISSN 0951-8320. [CrossRef]
16. Greiner, D.; Galván, B.; Winter, G. Safety Systems Optimum Design by Multicriteria Evolutionary Algorithms. In *Evolutionary Multi-Criterion Optimization. Lecture Notes in Computer Science, Proceedings of the Second International Conference, EMO 2003, Faro, Portugal, 8–11 April 2003*; Fonseca, C.M., Fleming, P.J., Zitzler, E., Deb, K., Thiele, L., Eds.; Springer: Berlin, Germany, 2003; pp. 722–736.
17. Salazar, D.; Rocco, C.M.; Galván, B.J. Optimization of constrained multiple-objective reliability problems using evolutionary algorithms. *Reliab. Eng. Syst. Saf.* **2006**, *91*, 1057–1070. ISSN 0951-8320. [CrossRef]
18. Limbourg, P.; Kochs, H.D. Multi-objective optimization of generalized reliability design problems using feature models—A concept for early design stages. *Reliab. Eng. Syst. Saf.* **2008**, *93*, 815–828. ISSN 0951-8320. [CrossRef]
19. Kumar, R.; Izui, K.; Yoshimura, M.; Nishiwaki, S. Multi-objective hierarchical genetic algorithms for multilevel redundancy allocation optimization. *Reliab. Eng. Syst. Saf.* **2009**, *94*, 891–904. ISSN 0951-8320. [CrossRef]
20. Chambari, A.; Rahmati, S.H.A.; Najafi, A.A.; Karimi, A. A bi-objective model to optimize reliability and cost of system with a choice of redundancy strategies. *Comput. Ind. Eng.* **2012**, *63*, 109–119. ISSN 0360-8352. [CrossRef]
21. Safari, J. Multi-objective reliability optimization of series-parallel systems with a choice of redundancy strategies. *Reliab. Eng. Syst. Saf.* **2012**, *108*, 10–20. ISSN 0951-8320. [CrossRef]
22. Ardakan, M.A.; Hamadani, A.Z.; Alinaghian, M. Optimizing bi-objective redundancy allocation problem with a mixed redundancy strategy. *ISA Trans.* **2015**, *55*, 116–128. ISSN 0019-0578. [CrossRef]
23. Ghorabaee, M.K.; Amiri, M.; Azimi, P. Genetic algorithm for solving bi-objective redundancy allocation problem with k-out-of-n subsystems. *Appl. Math. Model.* **2015**, *39*, 6396–6409. ISSN 0307-904X. [CrossRef]
24. Amiri, M.; Khajeh, M. Developing a bi-objective optimization model for solving the availability allocation problem in repairable series–parallel systems by NSGA-II. *J. Ind. Eng. Int.* **2016**, *12*, 61–69. [CrossRef]
25. Jahromi, A.E.; Feizabadi, M. Optimization of multi-objective redundancy allocation problem with non-homogeneous components. *Comput. Ind. Eng.* **2017**, *108*, 111–123. ISSN 0360-8352. [CrossRef]
26. Kayedpour, F.; Amiri, M.; Rafizadeh, M.; Nia, A.S. Multi-objective redundancy allocation problem for a system with repairable components considering instantaneous availability and strategy selection. *Reliab. Eng. Syst. Saf.* **2017**, *160*, ISSN 0951-8320. [CrossRef]
27. Sharifi, M.; Moghaddam, T.A.; Shahriari, M. Multi-objective Redundancy Allocation Problem with weighted-k-out-of-n subsystems. *Heliyon* **2019**, *5*, e02346, ISSN 2405-8440. [CrossRef]

28. Chambari, A.; Azimi, P.; Najafi, A.A. A bi-objective simulation-based optimization algorithm for redundancy allocation problem in series-parallel systems. *Expert Syst. Appl.* **2021**, *173*, 114745, ISSN 0957-4174. [CrossRef]
29. Zhao, J.H.; Liu, Z.; Dao, M.T. Reliability optimization using multiobjective ant colony system approaches. *Reliab. Eng. Syst. Saf.* **2007**, *92*, 109–120. ISSN 0951-8320. [CrossRef]
30. Chiang, C.H.; Chen, L.H. Availability allocation and multi-objective optimization for parallel–series systems. *Eur. J. Oper. Res.* **2007**, *180*, 1231–1244. ISSN 0377-2217. [CrossRef]
31. Elegbede, C.; Adjallah, K. Availability allocation to repairable systems with genetic algorithms: A multi-objective formulation. *Reliab. Eng. Syst. Saf.* **2003**, *82*, 319–330. ISSN 0951-8320. [CrossRef]
32. Khalili-Damghani, K.; Abtahi, A.R.; Tavana, M. A new multi-objective particle swarm optimization method for solving reliability redundancy allocation problems. *Reliab. Eng. Syst. Saf.* **2013**, *111*, 58–75. ISSN 0951-8320. [CrossRef]
33. Jiansheng, G.; Zutong, W.; Mingfa, Z.; Ying, W. Uncertain multiobjective redundancy allocation problem of repairable systems based on artificial bee colony algorithm. *Chin. J. Aeronaut.* **2014**, *27*, 1477–1487.
34. Samanta, A.; Basu, K. An attraction based particle swarm optimization for solving multi-objective availability allocation problem under uncertain environment. *J. Intell. Fuzzy Syst.* **2018**, *35*, 1169–1178. [CrossRef]
35. Muñoz, A.; Martorell, S.; Serradell, V. Genetic algorithms in optimizing surveillance and maintenance of components. *Reliab. Eng. Syst. Saf.* **1997**, *57*, 107–120. ISSN 0951-8320. [CrossRef]
36. Marseguerra, M.; Zio, E.; Podofillini, L. Condition-based maintenance optimization by means of genetic algorithms and Monte Carlo simulation. *Reliab. Eng. Syst. Saf.* **2002**, *77*, 151–166. ISSN 0951-8320. [CrossRef]
37. Gao, J.; Gen, M.; Sun, L. Scheduling jobs and maintenances in flexible job shop with a hybrid genetic algorithm. *J. Intell. Manuf.* **2006**, *17*, 493–507. [CrossRef]
38. Sánchez, A.; Carlos, S.; Martorell, S.; Villanueva, J.F. Addressing imperfect maintenance modelling uncertainty in unavailability and cost based optimization. *Reliab. Eng. Syst. Saf.* **2009**, *94*, 22–32. ISSN 0951-8320. [CrossRef]
39. Wang, Y.; Pham, H. A multi-objective optimization of imperfect preventive maintenance policy for dependent competing risk systems with hidden failure. *IEEE Trans. Reliab.* **2011**, *60*, 770–781. [CrossRef]
40. Ben Ali, M.; Sassi, M.; Gossab, M.; Harrath, Y. Simultaneous scheduling of production and maintenance tasks in the job shop. *J. Intell. Manuf.* **2011**, *49*. [CrossRef]
41. An, Y.; Chen, X.; Zhang, J.; Li, Y. A hybrid multi-objective evolutionary algorithm to integrate optimization of the production scheduling and imperfect cutting tool maintenance considering total energy consumption. *J. Clean Prod.* **2020**, *268*, 121540, ISSN 0959-6526. [CrossRef]
42. Bressi, S.; Santos, J.; Losa, M. Optimization of maintenance strategies for railway track-bed considering probabilistic degradation models and different reliability levels. *Reliab. Eng. Syst. Saf.* **2021**, *207*, 107359, ISSN 0951-8320. [CrossRef]
43. Martorell, S.; Villanueva, J.F.; Carlos, S.; Nebot, Y.; Sánchez, A.; Pitarch, J.L.; Serradell, V. RAMS+C informed decision-making with application to multi-objective optimization of technical specifications and maintenance using genetic algorithms. *Reliab. Eng. Syst. Saf.* **2005**, *87*, 65–75. ISSN 0951-8320. [CrossRef]
44. Oyarbide-Zubillaga, A.; Goti, A.; Sanchez, A. Preventive maintenance optimisation of multi-equipment manufacturing systems by combining discrete event simulation and multi-objective evolutionary algorithms. *Prod. Plan. Control.* **2008**, *19*, 342–355. [CrossRef]
45. Berrichi, A.; Amodeo, L.; Yalaoui, F.; Châtelet, E.; Mezghiche, M. Bi-objective optimization algorithms for joint production and maintenance scheduling: Application to the parallel machine problem. *J. Intell. Manuf.* **2009**, *20*, 389–400. [CrossRef]
46. Moradi, E.; Fatemi Ghomi, S.M.T.; Zandieh, M. Bi-objective optimization research on integrated fixed time interval preventive maintenance and production for scheduling flexible job-shop problem. *Expert Syst. Appl.* **2011**, *38*, 7169–7178. ISSN 0957-4174. [CrossRef]
47. Hnaien, F.; Yalaoui, F. A bi-criteria flow-shop scheduling with preventive maintenance. *IFAC Proc. Vol.* **2013**, *46*, 1387–1392. ISSN 1474-6670, ISBN 9783902823359. [CrossRef]
48. Wang, S.; Liu, M. Multi-objective optimization of parallel machine scheduling integrated with multi-resources preventive maintenance planning. *J. Manuf. Syst.* **2015**, *37*, 182–192. ISSN 0278-6125. [CrossRef]
49. Piasson, D.; Bíscaro, A.A.P.; Leão, F.B.; Sanches Mantovani, J.R. A new approach for reliability-centered maintenance programs in electric power distribution systems based on a multiobjective genetic algorithm. *Electr. Power Syst. Res.* **2016**, *137*, 41–50. ISSN 0378-7796. [CrossRef]
50. Sheikhalishahi, M.; Eskandari, N.; Mashayekhi, A.; Azadeh, A. Multi-objective open shop scheduling by considering human error and preventive maintenance. *Appl. Math. Model.* **2019**, *67*, 573–587. ISSN 0307-904X, doi:10.1016/j.apm.2018.11.015. [CrossRef]
51. Boufellouh, R.; Belkaid, F. Bi-objective optimization algorithms for joint production and maintenance scheduling under a global resource constraint: Application to the permutation flow shop problem. *Comput. Oper. Res.* **2020**, *122*, 104943, ISSN 0305-0548. [CrossRef]
52. Zhang, C.; Yang, T. Optimal maintenance planning and resource allocation for wind farms based on non-dominated sorting genetic algorithm-II. *Renew. Energy* **2021**, *164*, 1540–1549. ISSN 0960-1481. [CrossRef]
53. Berrichi, A.; Yalaoui, F.; Amodeo, L.; Mezghiche, M. Bi-Objective Ant Colony Optimization approach to optimize production and maintenance scheduling. *Comput. Oper. Res.* **2010**, *37*, 1584–1596. ISSN 0305-0548. [CrossRef]

44. Suresh, K.; Kumarappan, N. Hybrid improved binary particle swarm optimization approach for generation maintenance scheduling problem. *Swarm Evol. Comput.* **2013**, *9*, 69–89. ISSN 2210-6502. [CrossRef]
45. Li, J.Q.; Pan, Q.K.; Tasgetiren, M.F. A discrete artificial bee colony algorithm for the multi-objective flexible job-shop scheduling problem with maintenance activities. *Swarm Evol. Comput.* **2014**, *38*, 1111–1132. ISSN 0307-904X. [CrossRef]
46. Galván, B.; Winter, G.; Greiner, D.; Salazar, D.; Méndez, M. New Evolutionary Methodologies for Integrated Safety System Design and Maintenance Optimization. In *Computational Intelligence in Reliability Engineering: Evolutionary Techniques in Reliability Analysis and Optimization*; Levitin, G., Ed.; Springer: Berlin/Heidelberg, Germany, 2007; pp. 151–190.
47. Okasha, N.M.; Frangopol, D.M. Lifetime-oriented multi-objective optimization of structural maintenance considering system reliability, redundancy and life-cycle cost using GA. *Struct. Saf.* **2009**, *31*, 460–474. ISSN 0167-4730. [CrossRef]
48. Adjoul, O.; Benfriha, K.; El Zant, C.; Aoussat, A. Algorithmic Strategy for Simultaneous Optimization of Design and Maintenance of Multi-Component Industrial Systems. *Reliab. Eng. Syst. Saf.* **2021**, *208*, ISSN 0951-8320. [CrossRef]
49. Lins, I.D.; Droguett, E.A.L. Multiobjective optimization of availability and cost in repairable systems design via genetic algorithms and discrete event simulation. *Pesquisa Oper.* **2009**, *29*, 43–66. [CrossRef]
50. Lins, I.D.; López, E. Redundancy allocation problems considering systems with imperfect repairs using multi-objective genetic algorithms and discrete event simulation. *Simul. Model. Pract. Theory* **2011**, *19*, 362–381. ISSN 1569-190X. [CrossRef]
51. SINTEF. *OREDA—Offshore Reliability Data Handbook*, 5th ed.; Det Norske Veritas: Hovik, Norway, 2009; p. 796. ISBN 978-82-14-04830-8.
52. Emmerich, M.; Deutz, A. A tutorial on multiobjective optimization: Fundamentals and evolutionary methods. *Nat. Comput.* **2018**, *17*, 585–609. [CrossRef]
53. Coello, C.A. Multi-objective Evolutionary Algorithms in Real-World Applications: Some Recent Results and Current Challenges. In *Advances in Evolutionary and Deterministic Methods for Design, Optimization and Control in Engineering and Sciences*; Computational Methods in Applied Sciences; Greiner, D., Galván, B., Périaux, J., Gauger, N., Giannakoglou, K., Winter, G., Eds.; Springer International Publishing AG: Cham, Switzerland, 2015; pp. 3–18.
54. Whitley, D.; Rana, S.; Heckendorn, R. Representation Issues in Neighborhood Search and Evolutionary Algorithms. In *Genetic Algorithms and Evolution Strategies in Engineering and Computer Science*; Quagliarella, D., Périaux, J., Poloni, C., Winter, G., Eds.; John Wiley & Sons: Chichester, UK, 1997; pp. 39–57.
55. Savage, C. A Survey of Combinatorial Gray Codes. *SIAM Rev.* **1997**, *39*, 605–629. [CrossRef]
56. Greiner, D.; Winter, G.; Emperador, J.M.; Galván, B. Gray Coding in Evolutionary Multicriteria Optimization: Application in Frame Structural Optimum Design. In *Evolutionary Multi-Criterion Optimization. EMO 2005. Lecture Notes in Computer Science*; Coello, C.A., Hernández Aguirre, A., Zitzler, E., Eds.; Springer: Berlin/Heidelberg, Germany, 2005; Volume 3410._40. [CrossRef]
57. Tian, Y.; Cheng, R.; Zhang, X.; Jin, Y. PlatEMO: A MATLAB Platform for Evolutionary Multi-Objective Optimization [Educational Forum]. *IEEE Comput. Intell. Mag.* **2017**, *12*, 73–87. [CrossRef]
58. Zitzler, E.; Thiele, L.; Laumanns, M.; Fonseca, C.M.; da Fonseca, V.G. Performance Assessment of Multiobjective Optimizers: An Analysis and Review. *IEEE Trans. Evol. Comput.* **2003**, *7*, 117–132. [CrossRef]
59. Fonseca, C.M.; Paquete, L.; López-Ibáñez, M. An Improved Dimension-Sweep Algorithm for the Hypervolume Indicator. In Proceedings of the IEEE International Conference on Evolutionary Computation, Vancouver, BC, Canada, 16–21 July 2006; pp. 1157–1163. [CrossRef]
70. Benavoli, A.; Corani, G.; Mangili, F. Should We Really Use Post-Hoc Tests Based on Mean-Ranks? *J. Mach. Learn. Res.* **2016**, *17*, 1–10.

Article

Population Diversity Control of Genetic Algorithm Using a Novel Injection Method for Bankruptcy Prediction Problem

Nabeel Al-Milli [1,*,†], Amjad Hudaib [2,†] and Nadim Obeid [2,†]

1. Department of Computer Science, King Abdullah II School for Information Technology, The University of Jordan, Amman 11942, Jordan
2. Department of Computer Information Systems, King Abdullah II School for Information Technology, The University of Jordan, Amman 11942, Jordan; ahudaib@ju.edu.jo (A.H.); obein@ju.edu.jo (N.O.)
* Correspondence: nby9170078@ju.edu.jo
† These authors contributed equally to this work.

Abstract: Exploration and exploitation are the two main concepts of success for searching algorithms. Controlling exploration and exploitation while executing the search algorithm will enhance the overall performance of the searching algorithm. Exploration and exploitation are usually controlled offline by proper settings of parameters that affect the population-based algorithm performance. In this paper, we proposed a dynamic controller for one of the most well-known search algorithms, which is the Genetic Algorithm (GA). Population Diversity Controller-GA (PDC-GA) is proposed as a novel feature-selection algorithm to reduce the search space while building a machine-learning classifier. The PDC-GA is proposed by combining GA with k-mean clustering to control population diversity through the exploration process. An injection method is proposed to redistribute the population once 90% of the solutions are located in one cluster. A real case study of a bankruptcy problem obtained from UCI Machine Learning Repository is used in this paper as a binary classification problem. The obtained results show the ability of the proposed approach to enhance the performance of the machine learning classifiers in the range of 1% to 4%.

Keywords: diversity control; genetic algorithm; bankruptcy problem; classification

1. Introduction

The concept of exploration and exploitation plays a vital role in the effectiveness of any searching algorithm. In general, exploration is the process of visiting a new area in the search space, while exploitation is visiting a specific area within the neighborhood of previously visited points [1]. In order to find a good search algorithm, a good ratio between exploration and exploitation should be defined. The performance of population-based algorithms such as genetic algorithms, genetic programming, particle swarm optimization, brainstorm optimization algorithms, and other algorithms, is determined by the relationship between exploitation and exploration processes. Many researchers believe that population-based algorithms are effective due to the good ratio between exploitation and exploration [2,3].

In general, exploration and exploitation are usually controlled offline by proper settings of parameters that affect the population-based algorithm performance. However, the development process of such algorithms depends on the problem itself, which needs different amounts of exploration and exploitation. Since the problem nature is not known in advance, it is needed to find a controller to determine the amount of exploration and exploitation that are to be dynamically changed during a run. This motivates us to propose a novel approach that is able to control the exploration and exploitation processes inside GA. We employed GA as a binary feature selection approach to select the most valuable features for the bankruptcy prediction problem. The bankruptcy problem is one of the most critical economic problems in the world, which economic decision makers have to

face. Small or large companies that are related to the local community play a vital role in decisions for policymakers and the directions for global economic growth. Therefore, the performance of these companies and the prediction of bankruptcies attract researchers due to its effluence on social economics [4,5]. Moreover, bankruptcy reflects on companies and individuals based on several factors such as reputation, potential loss of customer base, and credit history.

In the field of risk management for the bank's sector, the bankruptcy problem is one of the most critical problems that face the financial department every day. Predicting bankruptcy in advance plays a key role in evaluating credit loan systems since it helps banks to prevent them from insolvency due to bad loans. Moreover, predicting corporate bankruptcy in an accurate way can contribute to the community by giving the right loan to successful companies. In general, a bankruptcy problem depends on a set of input features that are collected carefully from customer information (see Table 1). Moreover, this problem is a binary classification problem, where the main objective is to predict if the customer will have bankruptcy or not. Due to a large number of input features, finding a rigid wrapper algorithm is needed to enhance the overall performance of machine learning classifiers. As a result, we propose an enhanced version of the GA (i.e., Population Diversity Controller-GA (PDC-GA)) to explore the search space in a good manner and prevent premature convergence of the standard GA.

Table 1. Dataset features descriptions.

ID	Description	ID	Description
X1	net profit / total assets	X33	operating expenses / short-term liabilities
X2	total liabilities / total assets	X34	operating expenses / total liabilities
X3	working capital / total assets	X35	profit on sales / total assets
X4	current assets / short-term liabilities	X36	total sales / total assets
X5	[(cash + short-term securities + receivables- short-term liabilities) / (operating expenses - depreciation)] * 365,	X37	(current assets - inventories) / long-termliabilities
X6	retained earnings / total assets	X38	constant capital / total assets
X7	EBIT / total assets	X39	profit on sales / sales
X8	book value of equity / total liabilities	X40	(current assets - inventory - receivables) /short-term liabilities
X9	sales / total assets	X41	total liabilities / ((profit on operating activities+ depreciation) * (12/365))
X10	equity / total assets	X42	profit on operating activities / sales
X11	(gross profit + extraordinary items + financial expenses) / total assets	X43	rotation receivables + inventory turnover in days
X12	gross profit / short-term liabilities	X44	(receivables * 365) / sales
X13	(gross profit + depreciation) / sales	X45	net profit / inventory
X14	(gross profit + interest) / total assets	X46	(current assets - inventory) / short-term liabilities
X15	(total liabilities * 365) / (gross profit +depreciation)	X47	(inventory * 365) / cost of products sold
X16	(gross profit + depreciation) / total liabilities	X48	EBITDA (profit on operating activities -depreciation) / total assets
X17	total assets / total liabilities	X49	EBITDA (profit on operating activities -depreciation) / sales
X18	gross profit / total assets	X50	current assets / total liabilities
X19	gross profit / sales	X51	short-term liabilities / total assets
X20	(inventory * 365) / sales	X52	(short-term liabilities * 365) / cost of products sold)
X21	sales (n) / sales (n-1)	X53	equity / fixed assets
X22	profit on operating activities / total assets	X54	constant capital / fixed assets
X23	net profit / sales	X55	working capital
X24	gross profit (in 3 years) / total assets	X56	(sales - cost of products sold) / sales
X25	(equity - share capital) / total assets	X57	(current assets - inventory - short-term liabilities)/ (sales - gross profit - depreciation)
X26	(net profit + depreciation) / total liabilities	X58	total costs /total sales
X27	profit on operating activities / financial expenses	X59	long-term liabilities / equity
X28	working capital / fixed assets	X60	sales / inventory
X29	logarithm of total assets	X61	sales / receivables
X30	(total liabilities - cash) / sales	X62	(short-term liabilities *365) / sales
X31	(gross profit + interest) / sales	X63	sales / short-term liabilities
X32	(current liabilities * 365) / cost of products sold	X64	sales / fixed assets

In general, the bankruptcy problem is still a complex hard problem due to two main factors: (i) a complex relationship between a large number of variables, and (ii) imbalanced datasets, where the collected data for this problem usually are imbalanced (i.e., number of bankrupt cases are less than positive cases). The first factor makes the search space of this problem very high, where optimization methods try to reduce it as a feature selection problem, while the second factor affects the performance of ML methods. As a result, it is important to examine the collected data before building the prediction method and solving the imbalanced data problem first.

Recently, machine learning (ML) and optimization methods have shown promising performance while tackling the bankruptcy prediction problem [6,7]. ML solved the bankruptcy prediction problem as a binary classification problem (i.e., bankrupt and non-bankrupt), while optimization methods addressed this problem as a feature selection (FS) problem [8,9]. ML and optimization methods are able to solve the bankruptcy prediction problem due to their ability to handle big data and a large number of features (i.e., input data) [7]. In contrast, statistical methods have a set of limitations while solving this problem, such as solving the problem as a linear one, shortage in exploring the hidden relations between all input data, and sensitivity to outliers [10].

Many research papers have been proposed using ML to solve the bankruptcy prediction problem. For example, artificial neural networks (ANN) [11], genetic programming [12], support vector machines [13], and ensemble learning [6], while optimization methods such as genetic algorithm (GA)[14], ant colony optimization (ACO) [9], particle swarm optimization (PSO) [15], and Grey wolf optimization (GWO) [16] are employed successfully for the bankruptcy prediction problem. In general, optimization algorithms should have a balance between exploration and exploitation processes while solving complex problems to reach the optimal or near-optimal solutions. This balance can be achieved by controlling the diversity of the optimization algorithm [17].

The main contribution of this paper involves employing an enhanced version of GA that is able to control the population diversity during the search process. The main idea works by injecting the search space with new solutions to enhance the balance between exploration and exploitation processes to avoid being trapped in local optima.

In what follows, Section 2 presents the related works of bankruptcy prediction problem and diversity control for optimization algorithms. Section 3 presents the dataset used in this research, which is a public dataset obtained from the UCI Machine Learning Repository. Section 4 explores the proposed PDC-GA method in detail. Section 5 investigates the obtained results and our findings. Finally, Section 6 presents the conclusion and future works of this research.

2. Related Works
2.1. Diversity Control

All searching algorithms have two main concepts, which are *Exploration* and *Exploitation* processes [3]. *Exploration* refers to exploring or visiting all points in the search space, while *Exploitation* means visiting the surrounding points for a specific area. In general, finding a balance between *Exploration* and *Exploitation* is considered the main criterion to evaluate the overall performance of optimization algorithms. Many research papers highlight the concept of exploration and exploitation to gain a good ratio between both criteria. Sun et al. [18] propose a clustering method (i.e., k-mean method) to achieve a good ratio between exploration and exploitation for multi-objective optimization problems. Mittal et al. [19] enhanced the overall performance of Grey Wolf Optimizer (GWO) to prevent the premature convergence of the GWO algorithm. In the original GWO, half of the iterations work on exploration, while the second half work on exploitation, which is considered as a weak point of the GWO algorithm. The authors address this problem by finding a proper ratio between exploration and exploitation. Lynn and Suganthan [20] modified the Particle Swarm Optimization (PSO) by generating two subpopulations from the main population pool to enhance the exploration and exploitation processes. The main idea is that each sub-population focuses on exploration or exploitation based on the current distribution of solutions in the search space. Chen et al. [21] employed a sigmoid function that controls the velocity update process for PSO to overcome the premature convergence.

Shojaedini et al. [22] enhanced the performance of GA based on an adaptive genetic operator for selection and mutation. The proposed approach tries to find a good ratio between exploration and exploitation based on the number of outliers. Kelly et al. [23] control the exploration and exploitation processes for genetic programming (GP) by proposing a new selection method called *knobelty*. Mirsaleh and Meybodi [24] studied the premature convergence for GA and memetic algorithms.

2.2. Bankruptcy Prediction Problem

The bankruptcy prediction problem is a challenging problem that is related to companies' evaluation and organizations' solvency. Bankruptcy prediction was initially solved based on statistical methods [25] and Multiple Discriminant Analysis [26]. However, statistical methods are not able to explore all factors of the financial data or discover the hidden information inside it.

To overcome the weaknesses of statistical methods, machine learning methods show a great ability to address this problem [6,27]. Several methods have been adopted, such as support vector machines (SVMs) [28], artificial neural networks (ANNs) [29], Decision Tree [8], ensemble models [6], and deep learning methods [30]. Recently, many research papers have focused on ensemble models due to their ability to model bankruptcy problem in a good manner due to their ability to handle a large amount of financial data and produce lower error frequency [31]. Moreover, ensemble models are able to handle imbalance data [32].

2.3. Classification Models for Bankruptcy Prediction Problem

There are many machine learning methods that have been employed to predict the bankruptcy problem. We will list the methods that have been employed in the literature for this problem. Linear discriminant analysis (LDA) is a well-known classifier that has been employed successfully for many real-world applications. LDA was firstly employed for feature extraction, feature selection, and classification [33]. Altman [34] applied LDA in 1968 to predict the bankruptcy problem. Multilayer perceptron (MLP) is the most well-known model of artificial neural networks (ANNs), which consists of three layers (i.e., input, hidden, and output layers) [35]. MLP has been widely employed to predict the bankruptcy problem [29,36]. The JRip algorithm refers to Repeated Incremental Pruning to Produce Error Reduction (RIPPER), which enhances the correctness of a set of rules by tuning individual rules [37]. JRip algorithm works as an incremental learner that builds a set of rules in the training process. The error is evaluated based on the number of misclassified records for the training dataset [38]. JRip algorithm has been employed successfully in financial prediction problems [39,40].

The decision tree classifier (J48) is a well-known classifier constructed as a hierarchical model that consists of three main components (i.e., internal nodes, leaf nodes, and branches). The J48 model is very simple to implement, and the obtained trees are readily interpretable, while the weakness of this model is that it does not support multiple outputs and is susceptible to noisy data [41]. Logistic regression (LR) is a statistical classification model in ML, which is suitable for a binary class (bankruptcy and non-bankruptcy). LR has been employed successfully in different fields, such as medical, engineering, and social science [42]. Support vector machine (SVM) is an excellent classification method for data classification and regression [43]. However, the main weakness of SVM is high computational complexity.

Boosting is an ensemble learning method on ML for binary classification problems. In general, boosting comes in many flavors, such as bagging and random forest. There are many versions of boosting methods that have been employed successfully for binary classification problems (e.g., XGB and EXGB [4]). AdaBoost (Adaptive Boosting) is a simple

ML classifier based on boosting algorithms. In AdaBoost, the decision trees are employed for weak learners with trees with a single split termed as decision stumps. The samples are weighted in AdaBoost. The weights are increased for samples that are difficult to classify, while the weights are decreased for weights that are simple to classify [44]. Random forest (RF) is an ensemble machine learning method that involves construction (growing) of multiple decision trees via bootstrap aggregation [45]. In simple terms, the input of RF is passed down each of the constituent decision trees. Each tree classifies the input values and "votes" for the corresponding class. The majority of the votes decide the overall RF prediction [46].

Recently, many research studies have focused on reducing the number of variables of bankruptcy prediction to enhance the overall performance of machine learning classifiers. Wrapper feature selection methods are one of the most applicable methods that are able to reduce the search space [47,48]. Therefore, we believe employing feature selection methods will help decision-makers to explore the financial data and extract valuable features that play an important role in bankruptcy cases.

3. Dataset

The dataset used in this research is adopted from [4], which represents several of Polish companies. Since 2004, many Poland companies in the manufacturing sector went to bankruptcy state, which makes this problem valuable to study and analyze. The dataset was extracted from the Emerging Markets Information Service (EMIS) database. In general, the dataset has information on emerging markets around the world, including the Polish one, such as financial information, political, macroeconomic, and companies' news in Polish and English for 540 publications.

In this dataset, there are two periods of time. One is for bankruptcy, which is between 2007 and 2013, and the other period, which is between 2000 and 2012, without bankruptcy, for companies that are still working based on the reported data in EMIS. Therefore, this dataset has imbalanced samples. The dataset presents a binary classification problem with 64 features (inputs) and a single output (i.e., 0 means bankruptcy or 1 means no bankruptcy). Table 1 demonstrates the 64 features for this dataset.

4. Proposed Approach: PDC-GA

The proposed method of this work is depicted in Figure 1. The PDC-GA enhances the performance of GA by controlling the population diversity of GA. In this work, GA works as a feature selection method. After collecting the financial data for companies based on historical data, eliminating the weak features will enhance the overall performance of ML classifier(s) by reducing the search space of this problem. The population diversity is controlled by clustering the population based on k-means method and redistributing the solutions once most of the solutions are related to one cluster. An adaptive injection process is proposed to balance the exploration and exploitation of GA. Five different ML classifiers are employed to evaluate the selected features. Moreover, we applied the K-fold cross-validation with k-fold = 10 to avoid the overfitting problem. The following subsections explore the proposed method in more detail.

4.1. Preprocessing

The nature of financial data is very complex and may suffer from several problems due to several reasons, such as missing data, imbalanced cases, human error, or corrupted data. Therefore, building an accurate model is a challenging task. In this paper, to build an accurate model, we examine the collected data, and we found that there are some records that have missing data, as shown in Table 2. Figure 2 demonstrates the missing data patterns for the dataset. It is clear that the distribution of missing data overall are input features. Moreover, we performed two types of experiments: without missing data and with imputting missing data. We employed Synthetic Minority Oversampling Technique

(SMOTE) to address imbalanced data. To ensure that all the data were consistent, we normalized all features to be within the range of [0,1].

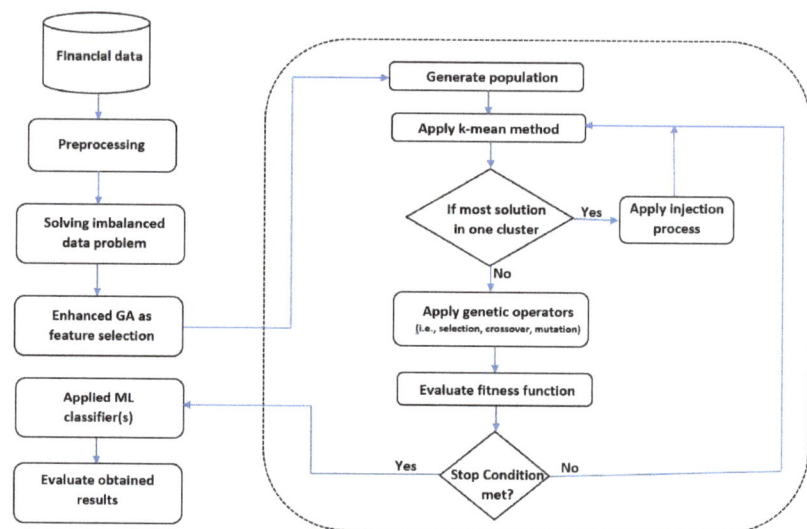

Figure 1. Population Diversity Controller–Genetic Algorithm (PDC-GA) method.

Table 2. Missing data percentage.

	Missing Data Percentage
Year 1	1.2775
Year 2	1.8385
Year 3	1.4484
Year 4	1.3788
Year 5	1.2146

4.2. Imbalanced Dataset

The performance of ML classifiers and optimization methods is affected by two factors: number of class type and number of samples. The major problem is called an imbalance problem when the class of interest is quite small compared to the normal one. In general, ML performance is based on a dataset skewed toward the normal class (i.e., majority class). In reality, financial data suffer from imbalanced data problem, which reduces the overall performance of ML classifiers [49].

There are two main approaches to handle the imbalanced data problem: algorithm perspective and data perspective [50]. The data perspective works based on re-sampling the data space. Resampling works using either over-sampling for the minority class or under-sampling instances for the majority class. Moreover, the re-sampling method address the imbalanced data problem randomly or deterministically.

SMOTE is a well-known method used to address imbalanced data, which creates synthetic samples between positive samples and the closest samples to it [50]. Adaptive synthetic sampling (ADASYN) generates a weighted distribution for several minority class based on their difficulty during the learning process [49]. ADASYN is able to reduce the bias toward the minority class and adaptively learn. In this work, we employed the ADASYN method to handle the imbalanced financial data.

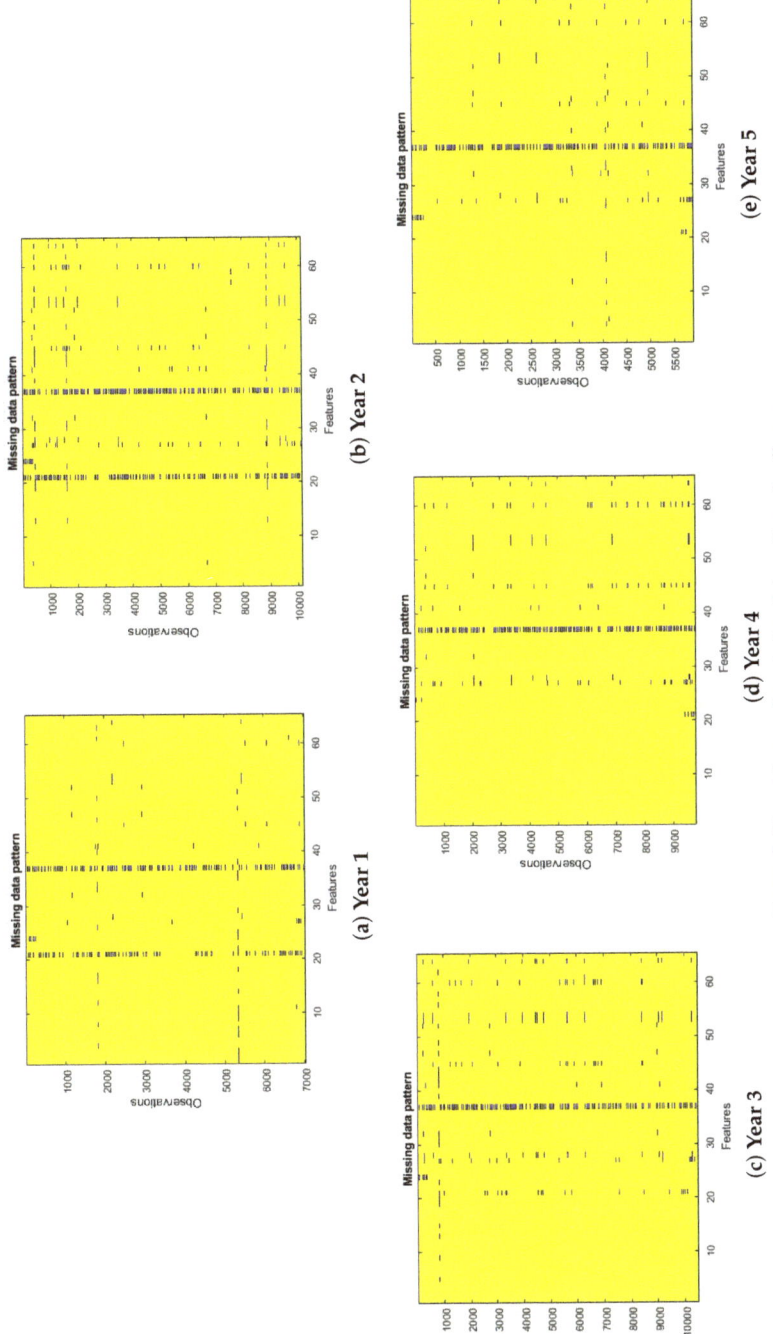

Figure 2. Pictorial maps for missing data patterns in all years.

4.3. Enhanced Genetic Algorithm

In this paper, we proposed a clustered genetic algorithm, as shown in Figure 3. The proposed approach works based on hybridization between k-means algorithm and the genetic algorithm. To prevent the local optima problem, at each iteration, we performed k-means clustering with k = 3. The process of injecting is demonstrated in Figure 4. The initial population of GA is clustered using k-means method, as shown in Figure 4a. At iteration *t*, all the solutions are converged (i.e., most solutions are located in one cluster), as shown in Figure 4b. Once most of the solutions are located in one cluster, an injection process of new solutions is obtained from a pool of solutions (i.e., worst solutions from previous genetic operations).

Given:
    -sP: size of the population.
    -nI: number of iterations.
    -rC: rate of crossover.
    -rM: rate of mutation.
Generate initial population of size sP.
Evaluate initial population based on the fitness function.
While ($current\_iteration \leq nI$)
Cluster the population using K-mean clustering algorithm.
    while(90% of solutions located in one cluster)
      Apply injection process
    Select two parent solutions from current population.
    Form offspring's solutions via crossover.
IF($rand(0.0, 1.0) < $ rM)
    Mutate the offspring's solutions.
end IF
Evaluate each child solution based on the fitness function.
Add offspring's to population.
//population size is now MaxPop=sP$\times$ (1+rC).
Remove the rC$\times$ sP least-fit solutions from population.
end While
Output the global best solution.

Figure 3. The pseudo-code for enhanced Genetic Algorithm.

Figure 5 presents the pseudo-code for the injection process, which presents the redistribution process of the solutions in the search space once 90% of the solutions are located in one cluster. Since the population size is fixed in GA, the number of injected solutions should be equal to the number of removed solutions. This number is determined based on Equation (2), where Max_c presents the highest number of solutions located in clusters, Min_c presents the lowest number of solutions located in clusters, *Iter* presents the maximum number of iterations, and $Population_{size}$ presents the population size. All injected solutions are obtained from a pool of solutions (i.e., generated randomly) to make sure they are distributed over the search space.

We employed an internal classifier based on the k-Nearest Neighbors (kNN) algorithm to evaluate the selected features. The reason behind selecting kNN as an internal classifier is that the time complexity of kNN is $O(n \times s)$, where *n* means the number of training samples, while *s* means the number of selected features. The fitness function of kNN is presented in Equation (1), where the accuracy of kNN classification is used as a fitness function, where TP, FP, TN, and FN are the calculated values of true positive, false positive, and true negative, and false negative, respectively.

$$Accuracy = \frac{TP + TN}{TP + TN + FP + FN} \tag{1}$$

$$Injection_{percentage} = \frac{Max_c - Min_c}{Iter} \times Population_{size} \tag{2}$$

4.4. Machine Learning Classifier(S)

There are many classification methods that can be employed on binary classification problems. However, in this work, the 10-fold cross-validation technique was employed using five classification methods, namely K nearest neighbor (kNN), linear discriminant (LD), decision tree (DT), support vector machine (SVM), and logistic regression (LR). Each classifier works in a different manner, so examining the performance of these classifiers over the bankruptcy problem using the 10-fold cross-validation technique is investigated in this work. Moreover, the cross-validation method reduces the possibility of over-fitting and redundancy of the data, so it enables us to generate a robust model.

4.5. Evaluation Process

Since the bankruptcy problem is a binary classification, we employed the area under the receiver operating characteristics curve (AUC) as an evaluation criterion. The AUC value reports the ratio between the number of correctly classified samples to the total number of testing samples. Equation (3) presents an over average for the k-fold cross-validation method.

$$CV = \frac{1}{k} \sum_{i=1}^{k} A_i \tag{3}$$

where CV represents the cross-validation accuracy, k represents the number of folds (i.e., k-fold = 10), and A represents the AUC value for each fold. Figure 6 demonstrates a pictorial diagram for AUC curves with its meanings.

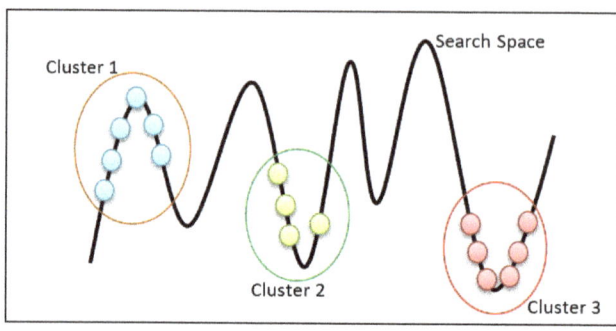

(a) Initial population

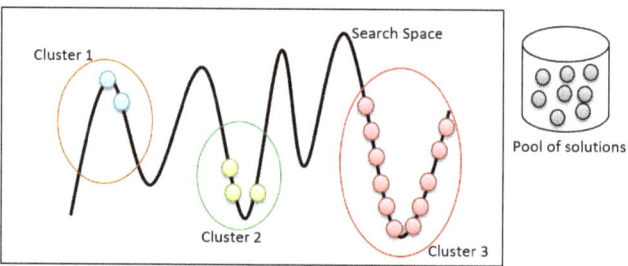

(b) Population state at iteration t.

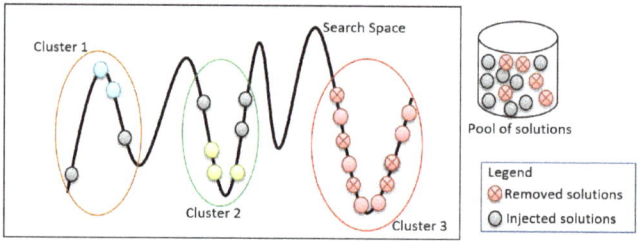

(c) Injection process of new solutions.

Figure 4. Injection process for new solutions.

Given:
Current Population at iteration t
Cluster the population using K-mean clustering algorithm.
If 90% *of the population located in one cluster*
    Determine the percentage of injection (D) based on Equation (2)
    Remove the worst solutions from the population based on D
    Inject new solutions to the population based on D

Figure 5. The pseudo-code for injection process.

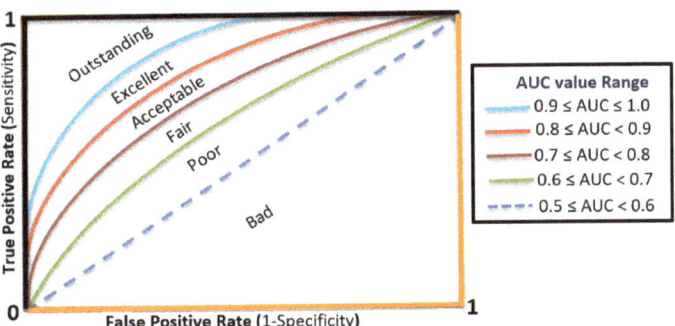

Figure 6. A pictorial diagram for the ROC curves and AUC values.

5. Results and Analysis

The proposed approach was implemented in MATLAB R2019b environment, and the simulations were executed on an Intel Core i5- M40 2.4-GHz processor with 4 GB RAM. The parameters adopted in all experiments, either for standard GA or the proposed enhanced GA, are listed in Table 3. In this work, we performed two types of experiments: (i) standard genetic algorithm based on the pseudo-code, and (ii) enhanced genetic algorithm. Both experiments employed the same fitness function as shown in Equation (1). To evaluate the overall performance of selected features either from standard or enhanced genetic algorithm, we executed four different classifiers (i.e., KNN, LD, DT, SVM, and LR). Each classifier is evaluated based on the average AUC value as shown in Equation (3). The following subsections demonstrate the obtained results and analysis.

Table 3. Parameters setting of standard GA and enhanced GA.

Parameters	Value
Population size	100
Crossover rate	0.7
Crossover type	Uniform
Mutation Rate	0.14
Mutation type	Multiple points
Selection type	Roulette wheel Selection (RWS)
Number of iterations	500

5.1. Feature Selection Results

Finding the most important features for the bankruptcy problem gives us more information about this problem. We executed 21 different independent runs for each year. Table 4 explores the obtained results for standard and enhanced genetic algorithm without imputing missing data (i.e., removing missing data). It is obvious that the proposed approach outperforms the standard genetic algorithm based on fitness value and the number of selected features. For example, the year 1 results show that the performance of enhanced GA is 0.9573, which outperforms the standard GA, which is 0.9391. Moreover, the number of selected features is improved for all datasets. For example, there are 18 features selected using standard GA for year 5, while enhanced GA selects 14 features. The reported selected features are the best features out of 21 independent runs. Reducing the number of features will reduce the computational time for machine learning and remove the redundant/irrelevant features. Boxplot diagrams for enhanced GA and standard GA are presented in Figure 7. It is obvious that the performance of enhanced GA for all years has more robust performance compared to the standard GA during the 21 independent runs and able to overcome the premature convergence during the search space. Figure 8 demonstrates the performance of the exploration process for both methods. It is clear that an enhanced genetic algorithm is able to explore the search space better than standard GA.

Table 4. Obtained results for standard and enhanced GA as a feature selection.

Dataset	Approach	Fitness Value (Accuracy)	Selected Features
Year 1	Standard GA	0.9391	[5,7,11,14,15,23,44,48,50,54,57,60,62]
	Enhanced GA	**0.9573**	[5,8,11,14,15,20,27,35,41,52,60]
Year 2	Standard GA	0.9054	[2,8,16,17,19,22,27,29,30,33,34,35,38,42,55,59,61,62]
	Enhanced GA	**0.9379**	[7,8,13,15,18,21,22,27,29,34,38,39,42,62]
Year 3	Standard GA	0.8687	[5,6,8,9,25,27,34,54,56,58,61,63]
	Enhanced GA	**0.885**	[14,33,35,38,46,47,49,57,63]
Year 4	Standard GA	0.9338	[7,11,17,29,33,37,41,47,55,56,59,62]
	Enhanced GA	**0.9559**	[4,12,14,18,19,20,33,44,59,63]
Year 5	Standard GA	0.9188	[2,7,8,12,14,16,17,22,24,27,35,39,40,47,49,57,59,63]
	Enhanced GA	**0.9666**	[1,5,9,12,17,18,19,20,31,37,39,48,56,61]

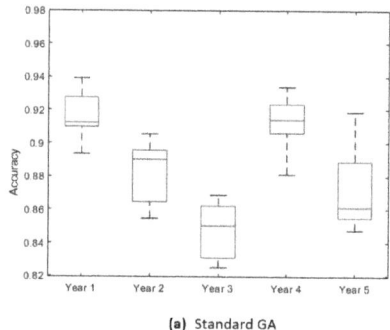 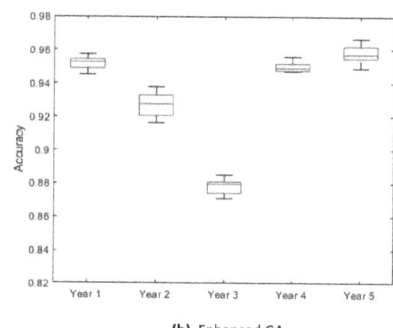

Figure 7. Boxplot diagrams for 21 independent runs.

5.2. Classifiers Models Results

To evaluate the best feature selected in the previous section, we employed five different binary classifiers (i.e., KNN, LD, DT, SVM, and LR), Since the datasets used in this work have missing data, we employed two types of experiments: (i) without imputing missing data, and (ii) with imputing missing data. We executed each classifier 21 times using a 10-fold cross-validation method. All reported results represent the mean value of AUC. Table 5 shows the obtained results for all years without imputing. It is clear that SVM with enhanced GA outperforms other methods in three datasets (i.e., year 2, year 3, and year 4), while KNN with enhanced GA outperforms in two datasets (i.e., year 1 and year 5). It is clear that our proposed approach is able to enhance the performance of machine learning classifiers.

Table 5. Obtained mean results with feature selection without imputing missing data.

	Enhanced GA					Standard GA				
	KNN	LD	DT	SVM	LR	KNN	LD	DT	SVM	LR
Year 1	**0.9573**	0.9236	0.8742	0.9563	0.9221	0.9391	0.8821	0.8452	0.8893	0.8996
Year 2	0.9379	0.9072	0.8236	**0.9453**	0.9235	0.9054	0.7996	0.9236	0.8437	0.9351
Year 3	0.885	0.9136	0.8337	**0.9447**	0.9436	0.8687	0.8863	0.9108	0.7834	0.9228
Year 4	0.9559	0.9440	0.8906	**0.9750**	0.9412	0.9338	0.9327	0.9004	0.8961	0.9183
Year 5	**0.9666**	0.9571	0.8813	0.9636	0.9336	0.9188	0.8526	0.8821	0.8379	0.9037

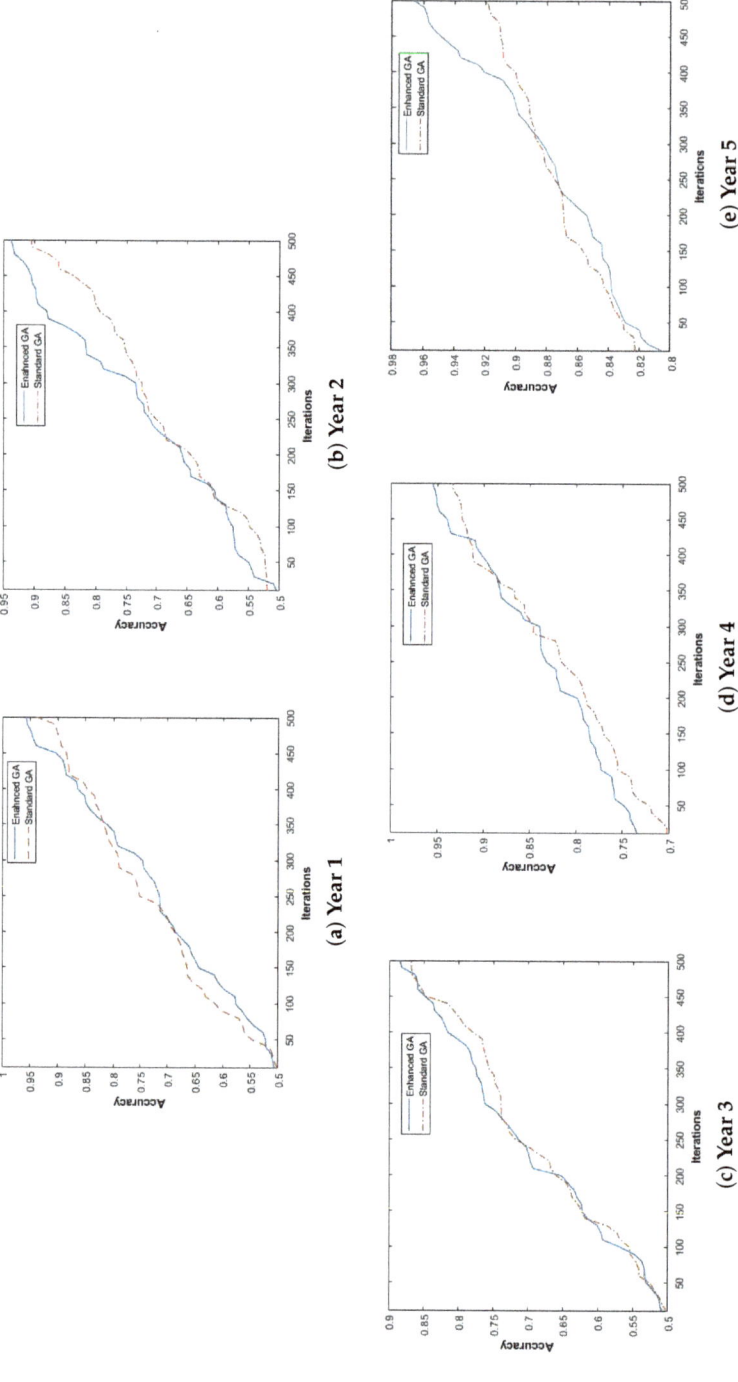

Figure 8. The exploration performance for enhanced and standard GA.

Table 6 shows the obtained results with imputing missing data. We employed the average method to impute the missing data. The obtained results show that the enhanced genetic algorithm outperforms the standard GA in four datasets (i.e., year 1, year 2, year 4, and year 5), while the performance of standard GA outperforms the enhanced GA only in one dataset (i.e., year 3). Moreover, from comparing the obtained results from Tables 5 and 6, it is clear that imputing missing data reduces the overall performance of classifiers. Four datasets (i.e., year 1, year 2, year 4, and year 5) gain less accurate values after imputing missing data. We believe that imputing missing is not valuable in our case for two reasons: (i) percentage of missing data is small (less than 2%), and (ii) imputing missing data based on average value is not a sufficient method.

Table 6. Obtained mean results with feature selection with imputing missing data.

Dataset	Enhanced GA					Standard GA				
	KNN	LD	DT	SVM	LR	KNN	LD	DT	SVM	LR
Year 1	0.9123	0.8794	**0.9299**	0.9171	0.9396	0.8457	0.9066	0.8594	0.8800	0.9058
Year 2	**0.9400**	0.9339	0.8734	0.8769	0.9020	0.8627	0.8678	0.9102	0.8645	0.9372
Year 3	0.9231	0.9354	0.9108	0.9520	0.9418	**0.9577**	0.9249	0.8747	0.9059	0.8440
Year 4	**0.9669**	0.9231	0.9025	0.8806	0.9311	0.9478	0.9444	0.9363	0.8639	0.9073
Year 5	**0.9479**	0.9123	0.8791	0.8966	0.8854	0.8329	0.8853	0.8707	0.8510	0.8532

Table 7 shows the obtained results without a feature selection algorithm. We executed all classifiers in two different ways: without imputing missing data, and with imputing missing data. The algorithm with imputing missing data outperforms that without imputing missing data in all datasets except the year 5 dataset. Moreover, the reported results in Table 6 show that applying feature selection improves the overall performance of machine learning classifiers since the search space is reduced.

Table 7. Obtained mean results without feature selection.

Dataset	Without Imputing Missing Data					With Imputing Missing Data				
	KNN	LD	DT	SVM	LR	KNN	LD	DT	SVM	LR
Year 1	0.8266	0.8950	0.8247	0.8279	0.8774	0.8452	**0.9052**	0.9000	0.8809	0.9047
Year 2	0.8306	0.9100	0.9099	0.8995	0.8365	0.9000	0.8267	0.8192	**0.9499**	0.8257
Year 3	0.8926	0.8770	0.9270	0.8914	0.9080	0.8049	0.8842	**0.9323**	0.9004	0.8286
Year 4	0.8699	0.8676	0.9108	0.8292	0.8346	0.8553	0.8691	**0.9472**	0.8235	0.9283
Year 5	0.8391	0.8630	**0.9115**	0.9084	0.8267	0.8967	0.8564	0.8286	0.8642	0.8723

Table 8 compares the obtained results with a set of algorithms from the literature. It is clear that the proposed feature selection is able to enhance the obtained results in three datasets (i.e., year 3, year 4, and year 5), while imputing missing data without a feature selection approach (i.e., Approach 4) outperforms all other methods in Year 2 using SVM method. However, the EXGB method that is proposed by Zięba et al. [4] outperforms all other methods in Year 1.

Table 8. Comparison results of the literature.

Author	Approach	Year 1	Year 2	Year 3	Year 4	Year 5
Zięba et al. [4]	LDA	0.6390	0.6600	0.6880	0.7140	0.7960
	MLP	0.5430	0.5140	0.5480	0.5960	0.6990
	Jrip	0.5230	0.5400	0.5350	0.5380	0.6540
	CJRip	0.7450	0.7740	0.8040	0.7990	0.7780
	J48	0.7170	0.6530	0.7010	0.6910	0.6100
	CJ48	0.6580	0.6520	0.6180	0.6110	0.7190
	LR	0.6200	0.5130	0.5000	0.5000	0.6320
	CLR	0.7040	0.6710	0.7140	0.7240	0.8210
	AB	0.9160	0.8500	0.8610	0.8850	0.9250
	AC	0.9160	0.8490	0.8590	0.8860	0.9280
	SVM	0.5020	0.5020	0.5000	0.5000	0.5050
	CSVM	0.5780	0.5170	0.6140	0.6150	0.7160
	RF	0.8510	0.8420	0.8310	0.8480	0.8980
	XGB	0.9450	0.9170	0.9220	0.9350	0.9510
	EXGB	**0.9590**	0.9440	0.9400	0.9410	0.9550
Our Approach	Approach 1	0.9573 (kNN)	0.9453 (SVM)	0.9447 (SVM)	**0.9750 (SVM)**	**0.9666 (kNN)**
	Approach 2	0.9299 (DT)	0.9400 (kNN)	**0.9577 (kNN)**	0.9669 (kNN)	0.9479 (kNN)
	Approach 3	0.8774 (LR)	0.9100 (LD)	0.9270 (DT)	0.9108 (DT)	0.9115 (DT)
	Approach 4	0.9052 (LD)	**0.9499 (SVM)**	0.9323 (DT)	0.9472 (DT)	0.8723 (LR)

where

- LDA: linear discriminant analysis.
- MLP: multilayer perceptron with a hidden layer.
- JRip: decision rules inducer.
- CJRip: cost-sensitive variation of JRip.
- J48: decision tree model.
- CJ48: cost-sensitive variation of J48.
- LR: Logistic Regression.
- CLR: cost-sensitive variation of Logistic Regression.
- AB: AdaBoost.
- AC: AdaCost.
- SVM: Support Vector Machines.
- CSVM: Cost-sensitive Support Vector Machines (CSVM)
- RF: Random Forest.
- XGB: Boosted trees trained with Extreme Gradient Boosting.
- EXGB: Ensemble of boosted trees.
- Approach 1: With feature selection and without imputing missing data.
- Approach 2: With feature selection and with imputing missing data.
- Approach 3: Without feature selection without imputing missing data.
- Approach 4: Without feature selection with imputing missing data.

Finally, employing feature selection based on enhanced GA is able to enhance the overall performance of machine learning classifiers. Moreover, controlling the population diversity of GA will help the searching algorithm to overcome the local optima problem.

6. Conclusions and Future Works

In this work, we present a novel approach to control the population diversity for genetic algorithm while searching for optimal solutions. The PDC-GA method is presented as a feature selection method. The genetic algorithm is hybridized with k-means clustering method to cluster the population while performing genetic operators. An injection method is proposed to redistribute the population once 90% of the solutions are located in one cluster. Predicting bankruptcy cases in advance is a challenging task for banks and companies, due to its importance for financial sectors in all countries. As a result, finding a robust method to predict bankruptcy cases is needed. PDC-GA method has been employed successfully in this domain. The obtained results of the PDC-GA outperform the standard GA as a feature selection. Moreover, we employed five machine learning classifiers (i.e., KNN, LD, DT, SVM, and LR) to evaluate the obtained results of the proposed algorithm. The performance of the proposed PDC-GA enhanced the original performance of standard GA between 1 and 4%. In our future work, we will apply the proposed method to optimize several real problems in industrial and research fields.

Author Contributions: Methodology, N.A.-M.; Software, N.A.-M.; Supervision, A.H. and N.O.; Validation, N.A.-M., A.H. and N.O; Writing—original draft, N.A.-M., A.H. and N.O. All authors have read and agreed to the published version of the manuscript.

Funding: This research received no external funding.

Acknowledgments: We would like to acknowledge the anonymous reviewers for the valuable comments that improved the quality of this paper. Moreover, we would also like to thank the Editors for their generous comments and support during the review process.

Conflicts of Interest: The authors declare no conflict of interest.

References

1. Lin, L.; Gen, M. Auto-tuning strategy for evolutionary algorithms: balancing between exploration and exploitation. *Soft Comput.* **2009**, *13*, 157–168. [CrossRef]
2. Alba, E.; Dorronsoro, B. The exploration/exploitation tradeoff in dynamic cellular genetic algorithms. *IEEE Trans. Evol. Comput.* **2005**, *9*, 126–142. [CrossRef]
3. Črepinšek, M.; Liu, S.H.; Mernik, M. Exploration and Exploitation in Evolutionary Algorithms: A Survey. *ACM Comput. Surv.* **2013**, *45*, 35:1–35:33. [CrossRef]
4. Zięba, M.; Tomczak, S.K.; Tomczak, J.M. Ensemble boosted trees with synthetic features generation in application to bankruptcy prediction. *Expert Syst. Appl.* **2016**, *58*, 93–101. [CrossRef]
5. Soui, M.; Smiti, S.; Mkaouer, M.W.; Ejbali, R. Bankruptcy Prediction Using Stacked Auto-Encoders. *Appl. Artif. Intell.* **2020**, *34*, 80–100. [CrossRef]
6. Chen, Z.; Chen, W.; Shi, Y. Ensemble learning with label proportions for bankruptcy prediction. *Expert Syst. Appl.* **2020**, *146*, 113155. [CrossRef]
7. Zoričák, M.; Gnip, P.; Drotár, P.; Gazda, V. Bankruptcy prediction for small- and medium-sized companies using severely imbalanced datasets. *Econ. Model.* **2020**, *84*, 165–176. [CrossRef]
8. Boucher, T.R.; Msabaeka, T. Trained Synthetic Features in Boosted Decision Trees with an Application to Polish Bankruptcy Data. In *Advances in Information and Communication*; Arai, K., Kapoor, S., Bhatia, R., Eds.; Springer International Publishing: Cham, Switzerland, 2020; pp. 295–309.
9. Uthayakumar, J.; Metawa, N.; Shankar, K.; Lakshmanaprabu, S. Financial crisis prediction model using ant colony optimization. *Int. J. Inf. Manag.* **2020**, *50*, 538–556. [CrossRef]
10. Chen, N.; Chen, A.; Ribeiro, B. Influence of Class Distribution on Cost-Sensitive Learning: A Case Study of Bankruptcy Analysis. *Intell. Data Anal.* **2013**, *17*, 423–437. [CrossRef]
11. Wilson, R.L.; Sharda, R. Bankruptcy prediction using neural networks. *Decis. Support Syst.* **1994**, *11*, 545–557. [CrossRef]
12. Lee, W.C. Genetic Programming Decision Tree for Bankruptcy Prediction. In *9th Joint International Conference on Information Sciences (JCIS-06)*; Atlantis Press: Paris, France, 2006. [CrossRef]
13. Min, S.H.; Lee, J.; Han, I. Hybrid genetic algorithms and support vector machines for bankruptcy prediction. *Expert Syst. Appl.* **2006**, *31*, 652–660. [CrossRef]
14. Chou, C.H.; Hsieh, S.C.; Qiu, C.J. Hybrid genetic algorithm and fuzzy clustering for bankruptcy prediction. *Appl. Soft Comput.* **2017**, *56*, 298–316. [CrossRef]
15. Azayite, F.Z.; Achchab, S. Topology Design of Bankruptcy Prediction Neural Networks Using Particle Swarm Optimization and Backpropagation. In Proceedings of the International Conference on Learning and Optimization Algorithms: Theory and Applications, LOPAL '18, Rabat, Morocco, 2–5 May 2018; Association for Computing Machinery: New York, NY, USA, 2018; [CrossRef]
16. Wang, M.; Chen, H.; Li, H.; Cai, Z.; Zhao, X.; Tong, C.; Li, J.; Xu, X. Grey wolf optimization evolving kernel extreme learning machine: Application to bankruptcy prediction. *Eng. Appl. Artif. Intell.* **2017**, *63*, 54–68. [CrossRef]
17. Kimura, S.; Konagaya, A. A Genetic Algorithm with Distance Independent Diversity Control for High Dimensional Function Optimization. *Trans. Jpn. Soc. Artif. Intell.* **2003**, *18*, 193–202. [CrossRef]
18. Sun, J.; Zhang, H.; Zhang, Q.; Chen, H. Balancing Exploration and Exploitation in Multiobjective Evolutionary Optimization. In Proceedings of the Genetic and Evolutionary Computation Conference Companion, GECCO '18, Kyoto, Japan, 15–19 July 2018; ACM: New York, NY, USA, 2018; pp. 199–200.

19. Mittal, N.; Singh, U.; Sohi, B.S. Modified Grey Wolf Optimizer for Global Engineering Optimization. *Appl. Comp. Intell. Soft Comput.* **2016**, *2016*, 8. [CrossRef]
20. Lynn, N.; Suganthan, P.N. Heterogeneous comprehensive learning particle swarm optimization with enhanced exploration and exploitation. *Swarm Evol. Comput.* **2015**, *24*, 11–24. [CrossRef]
21. Chen, F.; Sun, X.; Wei, D.; Tang, Y. Tradeoff strategy between exploration and exploitation for PSO. In Proceedings of the 2011 Seventh International Conference on Natural Computation, Shanghai, China, 26–28 July 2011; Volume 3, pp. 1216–1222. [CrossRef]
22. Shojaedini, E.; Majd, M.; Safabakhsh, R. Novel adaptive genetic algorithm sample consensus. *Appl. Soft Comput.* **2019**, *77*, 635–642. [CrossRef]
23. Kelly, J.; Hemberg, E.; O'Reilly, U.M. Improving Genetic Programming with Novel Exploration - Exploitation Control. In *Genetic Programming*; Sekanina, L., Hu, T., Lourenço, N., Richter, H., García-Sánchez, P., Eds.; Springer International Publishing: Cham, Switzerland, 2019; pp. 64–80.
24. Rezapoor Mirsaleh, M.; Meybodi, M.R. Balancing exploration and exploitation in memetic algorithms: A learning automata approach. *Comput. Intell.* **2018**, *34*, 282–309. [CrossRef]
25. Collins, R.A.; Green, R.D. Statistical methods for bankruptcy forecasting. *J. Econ. Bus.* **1982**, *34*, 349–354. [CrossRef]
26. Lee, S.; Choi, W.S. A multi-industry bankruptcy prediction model using back-propagation neural network and multivariate discriminant analysis. *Expert Syst. Appl.* **2013**, *40*, 2941–2946. [CrossRef]
27. Barboza, F.; Kimura, H.; Altman, E. Machine learning models and bankruptcy prediction. *Expert Syst. Appl.* **2017**, *83*, 405–417. [CrossRef]
28. Eygi Erdogan, B.; Özöğür Akyüz, S.; Karadayı Ataş, P. A novel approach for panel data: An ensemble of weighted functional margin SVM models. *Inf. Sci.* **2019**. [CrossRef]
29. Hosaka, T. Bankruptcy prediction using imaged financial ratios and convolutional neural networks. *Expert Syst. Appl.* **2019**, *117*, 287–299. [CrossRef]
30. Qu, Y.; Quan, P.; Lei, M.; Shi, Y. Review of bankruptcy prediction using machine learning and deep learning techniques. *Procedia Comput. Sci.* **2019**, *162*, 895–899. [CrossRef]
31. Son, H.; Hyun, C.; Phan, D.; Hwang, H. Data analytic approach for bankruptcy prediction. *Expert Syst. Appl.* **2019**, *138*, 112816. [CrossRef]
32. Brown, I.; Mues, C. An experimental comparison of classification algorithms for imbalanced credit scoring data sets. *Expert Syst. Appl.* **2012**, *39*, 3446–3453. [CrossRef]
33. Rathi, V.P.G.P.; Palani, S. Brain tumor MRI image classification with feature selection and extraction using linear discriminant analysis. *CoRR* **2012**, abs/1208.2128.
34. Altman, E.I. Financial ratios, discriminant analysis and the prediction of corporate bankruptcy. *J. Financ.* **1968**, *23*, 589–609. [CrossRef]
35. Taud, H.; Mas, J. Multilayer Perceptron (MLP). In *Geomatic Approaches for Modeling Land Change Scenarios*; Camacho Olmedo, M.T., Paegelow, M., Mas, J.F., Escobar, F., Eds.; Springer International Publishing: Cham, Switzerland, 2018; pp. 451–455. [CrossRef]
36. Inam, F.; Inam, A.; Mian, M.A.; Sheikh, A.A.; Awan, H.M. Forecasting Bankruptcy for organizational sustainability in Pakistan. *J. Econ. Adm. Sci.* **2019**, *35*, 183–201. [CrossRef]
37. Kalmegh, S.; Ghogare, M.S. Performance comparison of rule based classifier: Jrip and decisiontable using weka data mining tool on car reviews. *Int. Eng. J. Res. Dev.* **2019**, *4*, 5.
38. Cohen, W.W. Fast Effective Rule Induction. In *Machine Learning Proceedings 1995*; Prieditis, A., Russell, S., Eds.; Morgan Kaufmann: San Francisco, CA, USA, 1995; pp. 115–123. [CrossRef]
39. Boytcheva, S.; Tagarev, A. Company Investment Recommendation Based on Data Mining Techniques. In *Business Information Systems Workshops*; Abramowicz, W., Corchuelo, R., Eds.; Springer International Publishing: Cham, Switzerland, 2019; pp. 73–84.
40. Amuda, K.A.; Adeyemo, A.B. Customers Churn Prediction in Financial Institution Using Artificial Neural Network. *arXiv* **2019**, arXiv:1912.11346.
41. Zhao, Y.; Zhang, Y. Comparison of decision tree methods for finding active objects. *Adv. Space Res.* **2008**, *41*, 1955–1959. [CrossRef]
42. Briceño-Arias, L.M.; Chierchia, G.; Chouzenoux, E.; Pesquet, J.C. A random block-coordinate Douglas–Rachford splitting method with low computational complexity for binary logistic regression. *Comput. Optim. Appl.* **2019**, *72*, 707–726. [CrossRef]
43. Vapnik, V. *The Nature of Statistical Learning Theory*; Springer: Berlin/Heidelberg, Germany, 2013.
44. Bahad, P.; Saxena, P. Study of AdaBoost and Gradient Boosting Algorithms for Predictive Analytics. In *International Conference on Intelligent Computing and Smart Communication 2019*; Singh Tomar, G., Chaudhari, N.S., Barbosa, J.L.V., Aghwariya, M.K., Eds.; Springer: Singapore, 2020; pp. 235–244.
45. Opitz, D.; Maclin, R. Popular ensemble methods: An empirical study. *J. Artif. Intell. Res.* **1999**, *11*, 169–198. [CrossRef]
46. Weed, N.; Bakken, T.; Graddis, N.; Gouwens, N.; Millman, D.; Hawrylycz, M.; Waters, J. Identification of genetic markers for cortical areas using a Random Forest classification routine and the Allen Mouse Brain Atlas. *PLoS ONE* **2019**, *14*, e0212898. [CrossRef]
47. Farooq, U.; Qamar, M.A.J. Predicting multistage financial distress: Reflections on sampling, feature and model selection criteria. *J. Forecast.* **2019**, *38*, 632–648. [CrossRef]

48. Xiaomao, X.; Xudong, Z.; Yuanfang, W. A Comparison of Feature Selection Methodology for Solving Classification Problems in Finance. *J. Physics: Conf. Ser.* **2019**, *1284*, 012026. [CrossRef]
49. He, H.; Bai, Y.; Garcia, E.A.; Li, S. ADASYN: Adaptive synthetic sampling approach for imbalanced learning. In Proceedings of the 2008 IEEE International Joint Conference on Neural Networks (IEEE World Congress on Computational Intelligence), Hong Kong, China, 1–8 June 2008; pp. 1322–1328. [CrossRef]
50. Bowyer, K.W.; Chawla, N.V.; Hall, L.O.; Kegelmeyer, W.P. SMOTE: Synthetic Minority Over-sampling Technique. *CoRR* **2011**, *16*, 321–357.

Article

The Real-Life Application of Differential Evolution with a Distance-Based Mutation-Selection

Petr Bujok

Department of Informatics and Computers, Faculty of Science, University of Ostrava, 30. Dubna 22, 70103 Ostrava, Czech Republic; petr.bujok@osu.cz; Tel.: +420-553462176

Abstract: This paper proposes the real-world application of the Differential Evolution (DE) algorithm using, distance-based mutation-selection, population size adaptation, and an archive for solutions (DEDMNA). This simple framework uses three widely-used mutation types with the application of binomial crossover. For each solution, the most proper position prior to evaluation is selected using the Euclidean distances of three newly generated positions. Moreover, an efficient linear population-size reduction mechanism is employed. Furthermore, an archive of older efficient solutions is used. The DEDMNA algorithm is applied to three real-life engineering problems and 13 constrained problems. Seven well-known state-of-the-art DE algorithms are used to compare the efficiency of DEDMNA. The performance of DEDMNA and other algorithms are comparatively assessed using statistical methods. The results obtained show that DEDMNA is a very comparable optimiser compared to the best performing DE variants. The simple idea of measuring the distance of the mutant solutions increases the performance of DE significantly.

Keywords: differential evolution; distance-based; mutation-selection; real application; experimental study; global optimisation

1. Introduction

The solving of global optimisation problems is frequently needed in many areas of research, industry, and engineering where minimal or maximal cost values are required. In general, a global optimisation problem is specified in the search space Ω which is limited by its boundary constraints, $\Omega = \prod_{j=1}^{D}[a_j, b_j]$, $a_j < b_j$. The objective function f is defined in all $x \in \Omega$ and the point x^* for $f(x^*) \leq f(x), \forall x \in \Omega$ is the solution of the global optimisation problem.

In this study, several engineering optimisation problems are used to illustrate the performance of both existing (well-known) and newly proposed optimisation methods. The motivation and aim is to show the efficiency of the newly proposed optimisation algorithms. Achieving an optimal solution for engineering problems is a very popular research area [1]. Generally, the real-life application of optimisation methods is extremely important in many fields of industry, energy, and scheduling, etc. [2].

In addition to the area of engineering optimisation problems, the field of industrial economics is also very popular. In 2019, Dosi et al. introduced a comprehensive theoretical survey of the history of agent-based macroeconomics [3]. The authors critically discussed the issues found in macroeconomics from different points of view. The authors recommended the direct cooperation of agent-based macroeconomics with financial institutions. In 2020, Bellomo et al. introduced a theoretical cooperation between evolutionary theory and the theory of active particles [4]. The authors deeply analysed areas of evolutionary landscapes and the interactions found in endogenous systems which resulted in a model of differential equations. The results of the simulations showed the potential of their proposed approach in aiding the cooperation between states and private companies.

There are various optimisation approaches to find the minimal (maximal) function value of objective functions. The biggest group of optimisation methods is called the

Evolutionary Algorithms (EAs), which are inspired by natural systems. One of the most frequently used EA is the Differential Evolution (DE) algorithm [5]. The high popularity of the DE algorithm is based on its simplicity and efficiency. Over more than 15 years, a lot of powerful DE variants have been developed and studied very intensively [6–8]. Despite the efficiency of DE, there is still not one specific variant of the optimisation algorithm which is possible to solve all global optimisation problems in the most efficient way (No Free Lunch theorem [9]).

Differential Evolution

Differential evolution was introduced by Storn and Price as a simple and efficient optimisation algorithm in 1996 [5]. DE is a population-based optimisation algorithm that uses three numerical control parameters. The main idea of DE is as follows. In the beginning, the population of N individuals (D-dimensional vectors) is generated randomly in Ω and evaluated by the objective function f. After initialisation, the development of the population is performed from generation to generation until the stopping condition is met. The development of the individuals in the population is controlled by evolutionary operators—mutation, crossover, and selection. A new trial individual (offspring) y_i is derived from the current point x_i as follows. A mutated individual u_i is constructed from the current individual using mutation. There are several well-known mutation variants, the most widely-used mutation variant in DE is denoted rand/1 (1), where $r1, r2, r3$ are randomly selected mutual indices from $[1, N]$, different from i. The parameter $F \in (0, 2]$ is called a scale factor.

$$u = x_{r1} + F \cdot (x_{r2} - x_{r3}) \quad (1)$$

After mutation, a crossover operation is performed. Here, elements of the original x_i and the mutated individuals u_i are used for a new offspring solution—y_i. The most widely used crossover variant is known as binomial crossover (2), where the crossover ratio $CR \in (0, 1)$ controls the number of elements from a mutated individual selected for a trial solution.

$$y_{i,j} = \begin{cases} u_{i,j}, & \text{if } rand_j(0,1) \leq CR \text{ or } j = rand_j(1, D) \\ x_{i,j}, & \text{otherwise}. \end{cases} \quad (2)$$

A new individual y_i is evaluated by a cost function and it replaces the parent individual x_i in the population if it is better, $f(y_i) \leq f(x_i)$. This evolutionary operation is known as selection. When standard canonical DE is used for solving complex optimisation problems or large scale problems with high dimensions D, its efficiency is worse. The issue is mainly caused by the fixed values of the control parameters—N, CR, F. Then, the adaptive approach of the DE control parameters' values helps to solve various optimisation tasks. A lot of successful adaptive DE variants have been introduced and applied to real-world problems [6–8].

In this paper, a new DE variant based on a distance-based selection of mutation individuals, using an archive of old-good solutions and a population-size reduction mechanism, was applied to real-world problems. The main motivation for using the new algorithm was derived from an attempt to control the speed of convergence in the DE by the proper selection of a mutation individual [10,11]. Euclidean distance is employed to select the correct mutation individual from a triplet based on the current stage of the algorithm. Additionally, using historical yet correct, solutions in a reproduction process can enhance the ability to avoid the local minimal area. Finally, changing the population size during the search (from a bigger value to a smaller value) enables the support of exploration in early generations and exploitation in later generations. The most important aspect of the research is the practical use of the proposed optimisation methods for real-world problems. Therefore, three real-world engineering problems and 13 constrained problems were used to evaluate the proposed DE and compare the results with other state-of-the-art DE variants.

The rest of the paper is organised as follows. The newly proposed DE variant is presented in Section 2. The real-world problems and experimental settings are represented in Section 3. The results obtained from the experimental study are presented and discussed in Section 4. The paper is briefly concluded in Section 5.

2. A Novel DE with Distance-Based Mutation-Selection (DEDMNA)

In this section, a DE variant with Distance-based Mutation-selection, population size (N) reduction, and the use of an archive of old-good solutions (DEDMNA) is introduced. The main motivation for this approach is to manage the speed of convergence in the DE algorithm because the selection of the proper mutation operation significantly influences the ability to increase or decrease the population diversity.

In 2012, Liang et al. proposed a new DE variant with a distance-based selection approach [12]. Here, newly generated solutions are based on the Euclidean distance of an individuals' cost functions. A weakness of this approach is found in the necessity for the evaluation of the individuals. This is because it is typically the most time-consuming operation during the optimisation process.

In 2017, Gosh et al. proposed a DE variant with a distance-based mutation scheme using the central tendency of the population [13]. The Manhattan distance of the parent and offspring solution was applied to prioritise newly generated solutions with worse quality. The mechanism proposed a higher level of population diversity during the search.

In 2020, Liang et al. presented a novel DE algorithm based on the function value of the Euclidean-distance ratio [14]. This ratio reflects the function value and distance between two individuals in the population, and it is computed for the whole population. Therefore, the parent individuals are selected by roulette using the ratio values. The results of their experiments showed an increased efficiency in some classification problems.

2.1. Proper Mutation Variants for Convergence-Control

Standard DE uses mutation and crossover operations to generate new solutions to produce the next generation. There are many mutation variants, and preliminary results show that various mutation variants perform significantly differently [15]. Therefore, a couple of well-performing mutation variants which provide a variety of convergence-speeds were selected. Preliminary experimental results [10] and a theoretical analysis [11] provide an evaluation of DE mutation based on the speed of convergence. Based on preliminary experiments, the DE mutation variants *rand/1* (1), *best/2* were assessed as a balanced set of fast-converging and diversity-keeping mutation variants.

$$u = x_{best} + F \cdot (x_{r_1} - x_{r_2}) + F \cdot (x_{r_3} - x_{r_4}) \tag{3}$$

$$u = x_{r_1} + F \cdot (x_{best} - x_{r_1}) + F \cdot (x_{r_2} - x_{r_3}) \tag{4}$$

where $x_{r_1}, x_{r_2}, x_{r_3}, x_{r_4}$ are mutually different points $r_1 \neq r_2 \neq r_3 \neq r_4 \neq i$ and x_{best} is best point of P.

2.2. Distance-Based Mutation-Selection Mechanism

The newly proposed DE with a distance-based approach is based on a previously designed DEMD variant [16]. The original DEMD uses only a distance-based mutation-selection approach and a control parameter adaptation approach. To improve the original DEMD, a linear population size reduction approach and an archive for old solutions were employed in our research. Here, more details of the original DEMD and its new enhanced variant are provided.

The main motivation for using the original DEMD was the control of the convergence ability (speed) of the DE algorithm. For each individual x_i in population P, three mutation individuals are generated using the three mutation variants as discussed above. Then, the most proper mutation individual is selected for the crossover and selection, using the standard Euclidean distance, with respect to the current stage of the search process (exploration or exploitation). Note that where the CoDE variant [17] selects one of the three trial individuals evaluated by the cost function, the proposed DEDMNA uses the Euclidean distance between the coordinates of the points in the population. Therefore, the computational costs of the DEDMNA approach are substantially lower because the function evaluation of the individuals is a computationally expensive operation.

At the beginning of DEDMNA, a population of P of N individuals x_i, $i = 1, 2, \ldots, N$ is generated randomly in Ω and evaluated by objective function. Next, for each individual x_i from P, a new solution y_i is generated. The reproduction process of DEDMNA is divided into two phases—exploration and exploitation. The exploration phase is performed in the early generations of DEDMNA, and it keeps the coarse detection of potentially good regions of Ω. In this phase, for each x_i three new mutant vectors u_1, u_2, and u_3 are produced using (1), (3) and (4) mutations. Subsequently, a mutation point with the least Euclidean distance between the mutation individuals and the current position x_i is selected to choose the proper mutation individual and to achieve a better exploration of Ω. The second, exploitation, phase is controlled by (5), and the mutation point of the triplet of mutation individuals u_1, u_2, and u_3 with the least Euclidean distance between the mutation individuals and the best individual x_{best} is selected to choose the proper mutation individuals and maintain a better exploitation ability.

$$dist = \begin{cases} \sqrt{\sum_{j=1}^{D} (u_{k,j} - x_{i,j})^2}, & \text{if FES/maxFES} < \text{rand} \\ \sqrt{\sum_{j=1}^{D} (u_{k,j} - x_{\text{best},j})^2} & \text{otherwise}, \quad k = 1, 2, 3. \end{cases} \quad (5)$$

where *FES* is the current number of depleted function evaluations and *maxFES* the maximum *FES* for one run. Next, a new trial individual y_i is developed using a standard binomial crossover (2).

It is clear that setting the control parameters F and CR are crucial for the efficiency of the DEDMNA algorithm. In DEDMNA, an adaptive approach to changing the values of F and CR during the search process is employed. Simply, the values of CR are generated randomly, uniformly from the interval $(0,1)$, and independently for each point in P. Furthermore, the value of CR_i is randomly re-sampled if it has a small probability of 0.1. The adaptive mechanism for the values of F depend on the current phase. In the early exploration phase, the values of F_i are computed as a random permutation of length N divided by N for each point from P. Such values equidistantly cover the interval $(0, 1)$. In the late exploitation phase, values of F_i are sampled as a random number from the uniform interval $(0, 1)$. In both phases, the F_i values are randomly assigned to individuals of P and modified by $F_i = F_i + 0.1 * rand$. Such a modification guarantees slightly varying values in each generation. Similar adaptation mechanisms for the DE control parameters were also used in the original algorithms [18,19].

2.3. Archive of Historically Good Solutions

To simplify the use of archived historical solutions in A, point x_{r3} (see (1), (3) and (4)) is randomly selected from $P \bigcup A$ (the remaining points for mutation are selected solely from P). It means that when the archive is fully written, the randomly selected individual x_{r3} has a 50% chance from being from P and a 50% chance from A.

2.4. Population Size Adaptation

Preliminary experiments showed that varying the population size during the search significantly increases the performance of the DE algorithm [20–22]. The population size N of the DEDMNA algorithm is linearly reduced during the search process from a bigger value at the beginning to a smaller value at the end. After each generation, the current proper population size (based on linear dependency) is computed (6). When the current population size N differs from the needed value, the population size is reduced:

$$N = round[(\frac{N_{\min} - N_{\text{init}}}{maxFES})FES + N_{\text{init}}], \tag{6}$$

where FES is the current number of function evaluations, N_{init} is the initial population size, $N_{\min}$ represents the size of population at the end of the search process (counted by the total number of $maxFES$ function evaluations).

3. Experimental Settings

The proposed DEDMNA algorithm was applied to three engineering problems and 13 constrained problems. The results from DEDMNA were compared with six state-of-the-art DE variants.

3.1. State-of-the-Art Variants in Comparison

Six state-of-the-art DE variants were selected for an experimental comparison to assess the performance of the proposed DEDMNA variant. A brief description of the methods in a chronological manner follows.

In 2006, Brest et al. proposed a simple and efficient adaptive DE variant (jDE) [18]. jDE uses a DE/rand/1/bin strategy with an adaptive approach of F and CR. Each individual has separate values of F and CR, and in each generation, it is regenerated with a probability of 0.1. More details of the efficient jDE method can be found in [18].

In 2009, Qin et al. proposed a DE algorithm with strategy adaptation (SaDE) [23]. In Sade, four mutation strategies (rand/1/bin, rand/2/bin, rand-to-best/2/bin, and current-to-rand/1) are used for generating new trial solutions. The strategy to be applied is selected by roulette based on the success and failure of previous LP generations. Each strategy has the same probability set to 1/4, i.e., all the strategies have an equal probability of being selected.

In 2013, Tanabe and Fukunaga introduced the Success-History Based Parameter Adaptation for Differential Evolution (SHADE) [24] which was the best performing DE variant in the CEC 2013 competition. SHADE is derived from JADE [25], where the main difference between SHADE and the original JADE is a different history-based adaptation of the control parameters F and CR. Both algorithms use a current-to-$p$best mutation strategy where one parent individual is selected from $P \cup A$. The SHADE algorithm is abbreviated in the results of this paper as SHA.

In 2014, Wang et al. proposed a new DE variant using covariance-matrix learning and bimodal parameter settings (CoBiDE and CoBi in results) [26]. CoBiDE advances the canonical DE in two new aspects–the covariance-matrix crossover (based on Eigenvectors of the population) and bimodal sampling of the control parameters, which distinguishes between exploration and exploitation. The authors of CoBiDE supposed a higher performance in problems defined by rotated objective functions. The Eigenvector crossover is controlled by two control parameters $pb = 0.4$ is the probability of using the Eigenvector crossover (instead of the classic binomial crossover) in the whole population, and $ps = 0.5$ is the portion of the population used to determine the Eigenvectors. More details are provided in the original paper.

In 2015, Tang et al. introduced a DE with an Individual-Dependent Mechanism [19]. The search process in IDE is divided into explorative and exploitative phases. The dynamic setting of the F and CR values using the quality of the individuals is employed. Better individuals with lesser objective function values have smaller values of F and CR and vice versa. In 2017, an advanced IDE variant was proposed with a novel mutation variant and diversity-based population size control (IDEbd) [27]. The details of the IDEbd method can be found in the original paper, and it is labelled simply by 'IDE' in the results section of this paper.

In 2017, Brest et al. introduced an adaptive DE variant derived from the successful JADE, SHADE, and L-SHADE called jSO [21]. The jSO algorithm achieved second position in the CEC 2017 competition. jSO uses historical circle memories of length 5 containing the mean values for generating F and CR. In the first half of the jSO search process, higher values of CR are used. In the first 60% of evaluations, the values of F are kept under 0.7. jSO uses an advanced weighted current-to-$p$best mutation. Finally, jSO uses a linear adaptation of the population size where the initial population size is $N = 25 \times \sqrt{D} \times \log D$. More details are available in [21].

In 2019, Brest et al. proposed a very efficient adaptive DE variant called jDE100 [28], In 2019, jDE100 was the optimisation algorithm with the best results in the CEC competition. The jDE100 algorithm is derived from jDE. In jDE100, two independent populations are used—one big and one small. Also, the initial values of the mutation and crossover are set to $F = 0.5$ and $CR = 0.9$ for each individual in both populations. After one generation of the big population, if the best solution for the jDE100 is in the big population, it is copied to the small population. Then, when the condition for the re-initialisation of the big population is satisfied, it is reset. Then several generations of the small population are performed (equally to the number of function evaluations of the big population), and also the reset condition is verified, and the best solutions are stored. More details regarding jDE100 can be found in the original paper.

3.2. Well-Known Engineering Problems

The experimental comparison found here is based on three well-known engineering problems [29]. All the problems are related to minimisation, i.e., the global minimum point is the solution. The computational complexity of the problems are varied, and the dimensionality of the search space is ($D \in \{3, 4\}$). For each algorithm and problem, 25 independent runs were performed. Each algorithm stops when it achieves a predefined number of function evaluation, i.e., *MaxFES* = 150,000. A better insight into the results of the algorithms is provided by results achieved at *MaxFES* = 50,000 and *MaxFES* = 100,000. Finally, the individual of the final population with the least function value is the solution of the algorithm for the given problem.

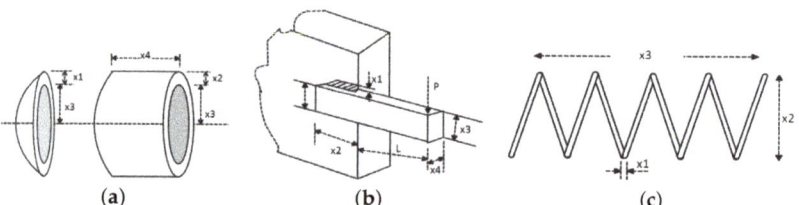

Figure 1. (a) Pressure vessel design problem, (b) Welded beam design problem, and (c) Tension-compression string problem.

In the pressure vessel design problem (labelled preved in results), the production costs represented by four parameters and constraints are minimised. The decision space area is represented by a four-dimensional real-valued space: x_1 defines the thickness of the head, x_2 is the thickness of the cylinder, x_3 is the inner radius, and x_4 is the length of the cylinder part (see Figure 1a)). The objective function is defined:

$$f(y) = 0.6224 x_1 x_3 x_4 + 1.7781 x_2 x_3^2 + 3.1661 x_1^2 x_4 + 19.84 x_1^2 x_3 \tag{7}$$

with constraints:

$$y_1 = -x_1 + 0.0193 x_3 \leq 0$$
$$y_2 = -x_2 + 0.00954 x_3 \leq 0$$
$$y_3 = -\pi x_3^2 x_4 - \frac{4}{3}\pi x_3^3 + 1{,}296{,}000 \leq 0 \quad (8)$$
$$y_4 = x_4 - 240 \leq 0$$

The purpose of the second Welded Beam Design problem (labelled welded in results) is to achieve the best production cost regarding a set of project constraints. An illustration of this problem is depicted in Figure 1b). The problem variables are—the weld thickness ($x1$) length ($x2$), height ($x3$), and thickness of the bar ($x4$).

$$f(y) = 1.10471 x_1^2 x_2 + 0.04811 x_3 x_4 (14.0 + x_2) \quad (9)$$

with settings:

$$t_{max} = 13{,}600$$
$$s_{max} = 30{,}000$$
$$d_{max} = 0.25$$
$$M = P(L + \frac{x_2}{2})$$
$$R = \sqrt{0.25(x_2^2 + (x_1 + x_3)^2)}$$
$$P = 6000$$
$$L = 14$$
$$E = 3 \times 10^6 \quad (10)$$
$$G = 12 \times 10^6$$
$$J = 2\sqrt{2} x_1 x_2 (x_2^2/12 + 0.25(x_1 + x_3)^2)$$
$$P_c = (\frac{4.013 E}{(6L^2)}) x_3 x_4^3 (1 - 0.25 x_3 \sqrt{E/G}/L)$$
$$t_1 = P/(\sqrt{2} x_1 x_2), \quad t_2 = MR/J$$
$$t = \sqrt{1^2 + t_1 t_2 x_2 / R + t_2^2}$$
$$s = 6PL/(x_4 x_3^2), \quad d = 4PL^3/(E x_4 x_3^3)$$

and constraints:

$$y_1 = t - t_{max}$$
$$y_2 = s - s_{max}$$
$$y_3 = x_1 - x_4 \quad (11)$$
$$y_4 = 0.10471 x_1^2 + 0.04811 x_3 x_4 (14.0 + x_1) - 5.0$$

In the Tension-Compression String problem (labelled tecost in results), the weight of the spring is minimised. The problem variables of the tecost problem are the wire diameter ($x1$), the mean coil diameter ($x2$), and the number of active coils ($x3$). The tecost problem is restricted by the constraints of shear stress, surge frequency, and minimum deflection (Figure 1c)). The objective function is:

$$f(y) = x_1^2 x_2 (x_3 + 2) \quad (12)$$

with constraints:

$$y_1 = 1 - \frac{x_2^3 x_3}{71{,}785 x_1^4}$$

$$y_2 = \frac{4x_2^2 - x_1 x_2}{12{,}566 x_1^3 (x_2 - x_1)} + \frac{1}{5108 x_1^2} - 1 \qquad (13)$$

$$y_3 = 1 - 140.45 x_1 / (x_3 x_2^2)$$

$$y_4 = \frac{x_1 + x_2}{1.5} - 1$$

3.3. Constrained Optimisation Problems

Real-world problems are very often defined as constrained optimisation problems. The constrained conditions (based on equality or inequality) specify more accurate areas for the allowed values of optimised variables. Therefore, a set of 13 minimisation constrained problems are used in experiments to distinguish more and less efficient methods. Details and definitions of the objective functions of the constrained problems are available in [29]. The constrained problems are labelled $p1$–$p13$ following the order of the original report. The dimensionality of the search space is $D \in (2, 20)$.

All algorithms and problems are implemented and experimentally compared in a Matlab 2020b environment. All computations were carried out on a standard PC with Windows 10, Intel(R) Core(TM)i7-9700 CPU 3.0 GHz, 16 GB RAM. For each algorithm and problem $maxFES = 100{,}000$ and is the stopping condition of the search, and 25 independent runs were performed to achieve statistically significant results. The population size of all algorithms is $N = 90$. The control parameters are minimal population size ($N_{min} = 5, 20$), initial population size ($N_{init} = round(25 * log(D) * \sqrt{D}$ [21]), and the size of the archive is equal to the population size N. Based on the final population size values, two different DEDMNA variants are labelled in the results as $DDMA_5$ and $DDMA_{20}$. The control parameters for the state-of-the-art algorithms used in this comparison, follow the recommended settings from the original papers.

4. Results

In this paper, two variants of the novel DEDMNA algorithm are compared with six state-of-the-art DE variants when solving three engineering and 13 constrained problems. At first, the performance of all nine algorithms is compared using the Friedman test. This method provides the mean ranks of the algorithms in comparison using the median values of the best-achieved function values. The best-achieved solution for each algorithm was recorded in ten phases of the search. The mean ranks for each algorithm and problem for the ten phases are in Table 1. The mean ranks represent the overall performance of the algorithm, including all 16 problems. The algorithms are ordered based on the mean rank in the final 10th phase ($MR_{st} = 10$). The mean rank of the best algorithm is printed bold and underlined, the second-best is printed bold, and the algorithm in the third position is underlined. In the last column, the achieved significance level of the Friedman tests is presented. If the null hypothesis is rejected, symbol of $***$ ($p < 0.001$), $**$ ($p < 0.01$), and $*$ ($p < 0.05$) is presented. Otherwise, symbol of $\approx$ demonstrate cases, where the null hypothesis is not rejected.

Table 1. Mean ranks of all algorithms from the Friedman tests computed for each stage independently.

MR$_{st}$	DDMA$_{20}$	DDMA$_5$	SHA	SaDE	jDE	jDE100	IDE	CoBi	Sig.
1	4.38	4.50	3.41	5.53	5.84	**2.66**	3.66	6.03	***
2	4.84	<u>4.72</u>	3.97	4.91	5.16	**2.75**	3.97	5.69	*
3	4.91	4.34	<u>4.25</u>	4.44	4.91	**3.25**	4.03	5.88	≈
4	4.59	4.22	<u>4.16</u>	4.19	4.78	**3.78**	4.53	5.75	≈
5	4.28	4.22	<u>4.19</u>	4.06	4.75	**4.03**	4.94	5.53	≈
6	<u>4.19</u>	4.25	**4.03**	4.06	4.56	4.41	5.03	5.47	≈
7	4.50	<u>4.38</u>	**4.16**	4.25	4.44	4.25	4.81	5.22	≈
8	4.13	4.13	**4.09**	<u>4.44</u>	<u>4.44</u>	4.69	4.84	5.25	≈
9	**3.94**	<u>4.00</u>	4.22	4.44	4.63	4.69	4.84	5.25	≈
10	**3.81**	<u>4.19</u>	4.22	4.38	4.63	4.75	4.78	5.25	≈

The null hypothesis is rejected only in the first two phases; the performance of the algorithms in the remaining phases is rather similar. Very interesting information is provided by the development of the mean rank values for each algorithm during the progression of stages. In the early phases, jDE100 and SHADE variants are well-performing. The best results, including all 16 problems in the last two (final) phases, were achieved by the newly proposed DEDMNA$_{20}$ and DEDMNA$_5$. It highlights the effective performance of the proposed DEDMNA method. A better insight into the mean rank comparison is provided by the plots of the mean ranks in Figure 2. The performance of jDE100 decreases during the search, whereas the efficiency of the DEDMNA algorithm increases (especially for DEDMNA$_{20}$).

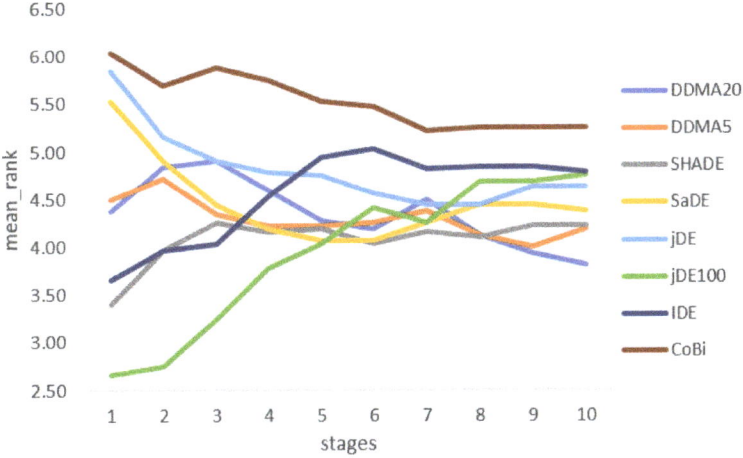

Figure 2. Illustration of the algorithms' mean ranks from the Friedman tests.

A more detailed comparison is provided by the Wilcoxon rank-sum tests. The test is applied to compare the results of two algorithms with one problem. The reference method is DEDMNA$_{20}$ (best mean rank from the Friedman test), and it is compared with the seven remaining counterparts. In Tables 2 and 3, the median values of all algorithms and problems are shown, including the significance from the Wilcoxon rank-sum tests ('−' denotes the better performance of a counterpart method, '+' shows the better performance of DEDMNA$_{20}$, and '≈' is for similar results). Mostly, the median values of all the compared algorithms are very similar to the achieved true solution.

Table 2. Median values for each algorithm and problem, with significance from the Wilcoxon rank-sum tests.

Fun	DDMA$_{20}$	IDE	CoBi	jDE	SaDE
preved	5885.333	5885.3328 ($\approx$)	5885.333 ($\approx$)	5885.333 ($\approx$)	5885.33 ($---$)
welded	2.218151	2.218151 ($\approx$)	2.218151 ($\approx$)	2.218151 ($\approx$)	2.21815 ($---$)
tecost	0.012665	0.012665 ($---$)	0.012665 ($---$)	0.012665 ($\approx$)	0.012665 ($---$)
p1	-15	-14.99 (+++)	-15 ($\approx$)	-15 ($\approx$)	-15 ($\approx$)
p2	-0.8049	-0.792 (+++)	-0.754 (+++)	-0.803 (+)	-0.804 ($\approx$)
p3	-0.02377	-0.25338 ($---$)	-1.04×10^4 (+++)	-3.00×10^6 (+++)	-5.00×10^6 (+++)
p4	-30665.5	-30665.5 ($\approx$)	-30665.5 (+++)	-30665.5 ($\approx$)	-30665.5 (+++)
p5	1.19×10^{12}	1.19×10^{12} ($\approx$)	1.19×10^{12} ($---$)	1.19×10^{12} ($\approx$)	1.19×10^{12} ($---$)
p6	-6961.81	-6961.81 ($\approx$)	-6961.81 (+++)	-6961.81 ($\approx$)	-6961.81 (+++)
p7	24.30697	24.35218 (+++)	24.307 ($\approx$)	24.30798 (++)	24.3064 ($--$)
p8	-0.09583	-0.09583 ($\approx$)	-0.095825 (+++)	-0.09583 ($\approx$)	-0.095825 (+++)
p9	680.6301	680.63007 (+++)	680.63 ($\approx$)	680.6301 ($\approx$)	680.63 ($---$)
p10	7049.42	7059.31 (+++)	7054.68 (+++)	7049.43 ($\approx$)	7049.41 ($\approx$)
p11	0.7499	0.7499 ($\approx$)	0.7499 ($\approx$)	0.7499 ($\approx$)	0.9656 (+++)
p12	-1	-1 ($\approx$)	-1 ($\approx$)	-1 ($\approx$)	-1 ($\approx$)
p13	4.1×10^8	0.95456 ($--$)	3.25×10^9 (+)	1.44×10^{10} (++)	8.95×10^{11} (+++)
Σ		5/8/3	7/6/3	4/12/0	6/4/6

For a better comparison of the algorithms, the counts of better, similar, and worse results for the reference DEDMNA$_{20}$ algorithm are depicted in the last row of the tables.

Compared to IDEbd (labelled IDE), DEDMNA$_{20}$ performs better in five constrained problems and is worse in one constrained and one engineering problem. CoBiDE is outperformed by the reference method in seven problems, and it performs better in three problems. DEDMNA$_{20}$ outperforms jDE in four constrained problems and never performs worse. DEDMNA$_{20}$ is better in six constrained problems and worse in three constrained problems and three engineering problems, compared to SaDE. SHADE is able to outperform the reference method in three constrained problems, and it performs worse in three constrained and one engineering problem. The results of the two DEDMNA variants are very similar, and each is better in one problem. DEDMNA$_{20}$ outperforms jDE100 in eight constrained problems, and it is worse in two engineering problems and two constrained problems.

More insight into the algorithms' performance is provided by convergence plots for all 16 problems (see Figures 3–6). It is clear that in constrained problems 8 and 12, all the algorithms converge in the first phase. In the remaining problems, the convergence process takes some time. An interesting observation is the convergence of constrained problem 2, where the curves of the best algorithms' solutions differ to the last phase. The worst convergence is with CoBiDE, whereas very good results are provided by DEDMNA$_{20}$.

Table 3. Median values for each algorithm and problem, with significance from the Wilcoxon rank-sum tests.

Fun	DDMA$_{20}$	SHADE	DDMA$_5$	jDE100
preved	5885.333	5885.3328($\approx$)	5885.3328($\approx$)	5885.330($---$)
welded	2.218151	2.2181509($\approx$)	2.2181509($\approx$)	2.21815($---$)
tecost	0.012665	0.012666(+++)	0.012665(+)	0.012665($\approx$)
p1	-15	-15($\approx$)	-15($\approx$)	-15($\approx$)
p2	-0.80359	-0.8036($---$)	-0.8036($\approx$)	-0.79256(+++)
p3	-0.02377	-0.0004899(+++)	-0.02249($\approx$)	-0.00268(+++)
p4	-30665.5	-30665.5387($\approx$)	-30665.5387($\approx$)	-30665.5(+++)
p5	1.19×10^{12}	1.19×10^{12}($\approx$)	1.19×10^{12}($\approx$)	1.19×10^{12}($---$)
p6	-6961.81	-6961.81388($\approx$)	-6961.81388($\approx$)	-6961.81(+++)
p7	24.30697	24.30625($---$)	24.30699($\approx$)	24.3269(+++)
p8	-0.09583	-0.095823($\approx$)	-0.09583($\approx$)	-0.095825(+++)
p9	680.6301	680.630057($\approx$)	680.630057($\approx$)	680.633(+++)
p10	7049.42	7049.29($--$)	7049.55($\approx$)	7071.57(+++)
p11	0.7499	0.9401(+++)	0.7499($\approx$)	0.7499($\approx$)
p12	-1	-1($\approx$)	-1($\approx$)	-1($\approx$)
p13	4.1×10^8	5.07×10^{11}(+++)	0.97026($-$)	0.902($--$)
Σ		4/9/3	1/14/1	8/4/4

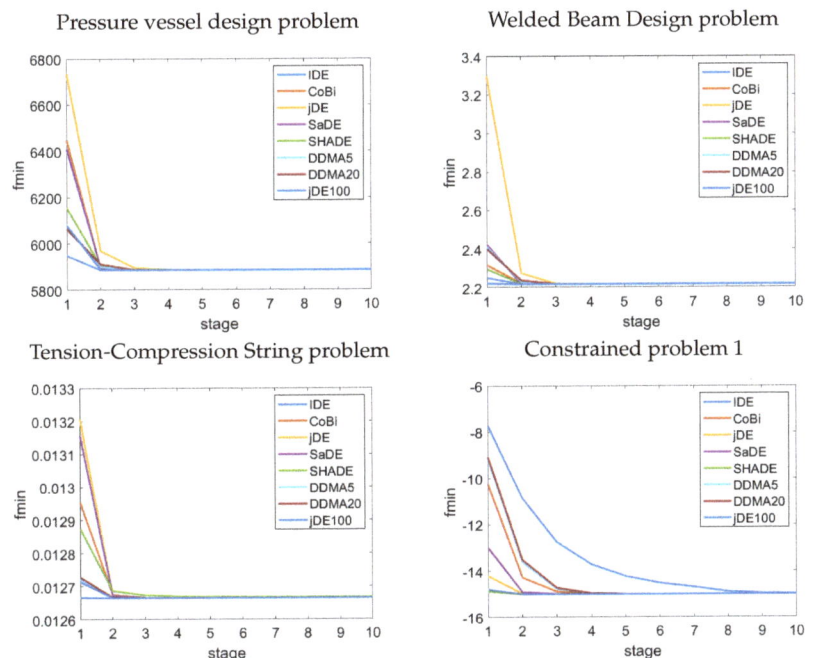

Figure 3. Convergence plots of the algorithms in comparison.

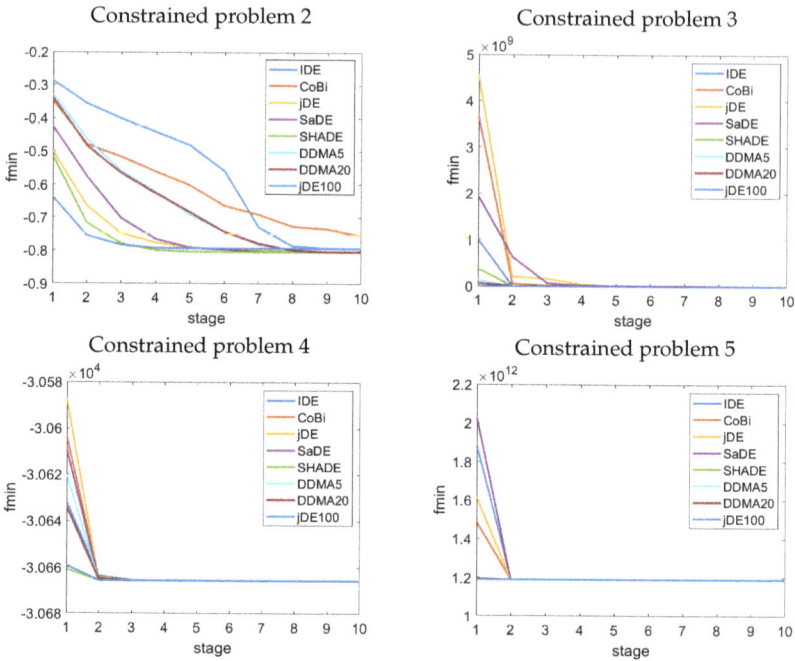

Figure 4. Convergence plots of the algorithms in comparison.

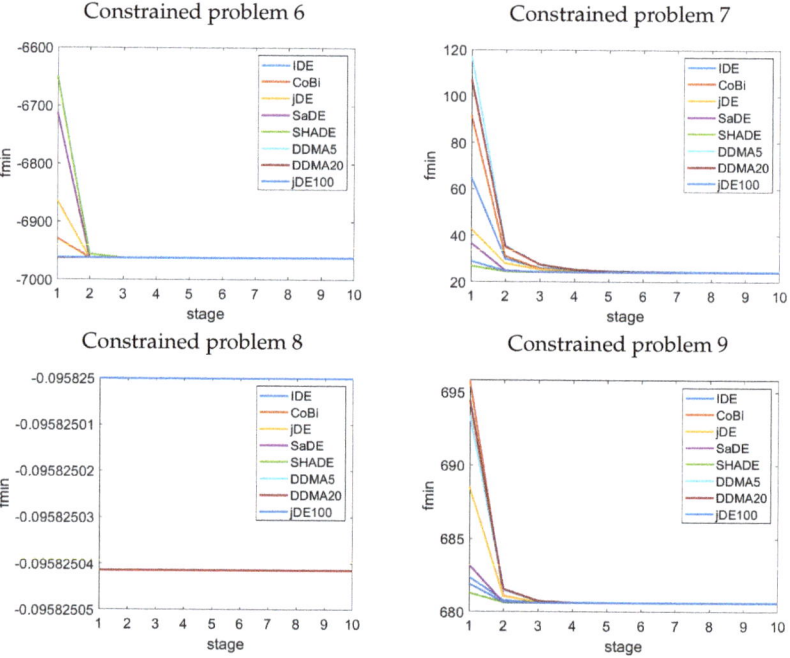

Figure 5. Convergence plots of the algorithms in comparison.

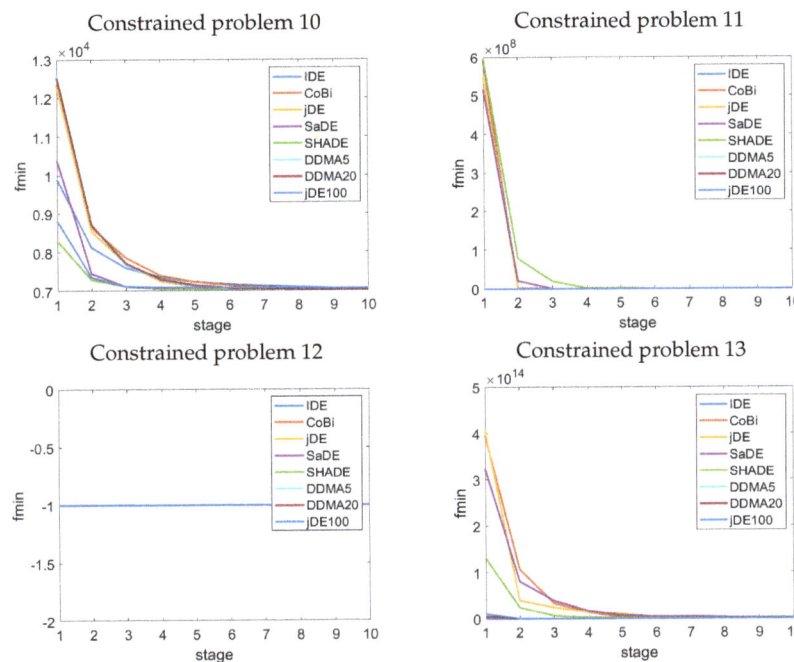

Figure 6. Convergence plots of the algorithms in comparison.

5. Conclusions

In this experimental comparison, two newly proposed DEDMNA variants are compared with six state-of-the-art DE variants when solving three engineering problems and 13 constrained problems. The results of the Friedman tests show that the DEDMNA variant provides the best performance in the last phase of the search, whereas the successful jDE100 variant performs better in the early phases of the search. The better results for DEDMNA, with a bigger final population size, indicates the necessity for higher diversity during the search process.

The counts of better and worse results from the Wilcoxon rank-sum test show that the new DEDMNA variant is able to be comparable with optimised state-of-the-art methods when applied to real-world problems. Despite this, all algorithms achieved mostly quite similar results, which illustrate the ability of the methods to determine the area of the true solution. The Proposed DEDMNA variant was successfully applied to the current CEC 2021 competition, and it achieves a very promising performance compared to the state-of-the-art DE algorithms from the preliminary experiments. This finding is very promising for the future development of new optimisation methods. The performance of DEDMNA will be studied and further tuned in future research.

Funding: This research was funded by Internal Grant Agency of University of Ostrava grant number SGS17/PrF-MF/2021.

Institutional Review Board Statement: Not applicable.

Informed Consent Statement: Not applicable.

Data Availability Statement: All data was measured in Matlab during the experiments.

Conflicts of Interest: The funders had no role in the design of the study; in the collection, analyses, or interpretation of data; in the writing of the manuscript, or in the decision to publish the results.

References

1. Rhinehart, R.R. *Engineering Optimization: Applications, Methods and Analysis*; John Wiley & Sons: Hoboken, NJ, USA, 2018.
2. Fujisawa, K.; Shinano, Y.; Waki, H. *Optimization in the Real World: Toward Solving Real-World Optimization Problems*; Mathematics for Industry, Springer: Berlin/Heidelberg, Germany, 2016.
3. Dosi, G.; Roventini, A. More is Different ... and Complex! The Case for Agent-Based Macroeconomics. *J. Evol. Econ.* **2019**, *29*, 1–37. [CrossRef]
4. Bellomo, N.; Dosi, G.; Knopoff, D.A.; Virgillito, M.E. From particles to firms: on the kinetic theory of climbing up evolutionary landscapes. *Math. Model. Methods Appl. Sci.* **2020**, *30*, 1441–1460. [CrossRef]
5. Storn, R.; Price, K.V. Differential evolution—A Simple and Efficient Heuristic for Global Optimization over Continuous Spaces. *J. Glob. Optim.* **1997**, *11*, 341–359. [CrossRef]
6. Das, S.; Mullick, S.; Suganthan, P. Recent advances in differential evolution—An updated survey. *Swarm Evol. Comput.* **2016**, *27*, 1–30. [CrossRef]
7. Das, S.; Suganthan, P.N. Differential Evolution: A Survey of the State-of-the-Art. *IEEE Trans. Evol. Comput.* **2011**, *15*, 27–54. [CrossRef]
8. Neri, F.; Tirronen, V. Recent advances in differential evolution: A survey and experimental analysis. *Artif. Intell. Rev.* **2010**, *33*, 61–106. [CrossRef]
9. Wolpert, D.H.; Macready, W.G. No Free Lunch Theorems for Optimization. *IEEE Trans. Evol. Comput.* **1997**, *1*, 67–82. [CrossRef]
10. Jeyakumar, G.; Shanmugavelayutham, C. Convergence analysis of differential evolution variants on unconstrained global optimization functions. *Int. J. Artif. Intell. Appl. (IJAIA)* **2011**, *2*, 116–127. [CrossRef]
11. Zaharie, D. Differential Evolution: From Theoretical Analysis to Practical Insights. In *MENDEL 2012, 18th International Conference on Soft Computing, 27–29 June 2012*; University of Technology: Brno, Czech Republic, 2013; pp. 126–131.
12. Liang, J.; Qu, B.; Mao, X.; Chen, T. Differential Evolution Based on Fitness Euclidean-Distance Ratio for Multimodal Optimization. In *Emerging Intelligent Computing Technology and Applications*; Huang, D.S., Gupta, P., Zhang, X., Premaratne, P., Eds.; Springer Berlin Heidelberg: Berlin/Heidelberg, Germany, 2012; pp. 495–500.
13. Ghosh, A.; Das, S.; Mallipeddi, R.; Das, A.K.; Dash, S.S. A Modified Differential Evolution With Distance-based Selection for Continuous Optimization in Presence of Noise. *IEEE Access* **2017**, *5*, 26944–26964. [CrossRef]
14. Liang, J.; Wei, Y.; Qu, B.; Yue, C.; Song, H. Ensemble learning based on fitness Euclidean-distance ratio differential evolution for classification. *Nat. Comput.* **2021**, *20*, 77–87. [CrossRef]
15. Bujok, P.; Tvrdík, J. A Comparison of Various Strategies in Differential Evolution. In Proceedings of the MENDEL, 17th International Conference on Soft Computing, Brno, Czech Republic, 14–17 June 2011; pp. 48–55.
16. Bujok, P. Improving the Convergence of Differential Evolution. In *Numerical Analysis and Applications*; Lecture Notes in Computer Science; Springer: Berlin/Heidelberg, Germany, 2016; pp. 248–255.
17. Wang, Y.; Cai, Z.; Zhang, Q. Differential Evolution with Composite Trial Vector Generation Strategies and Control Parameters. *IEEE Trans. Evol. Comput.* **2011**, *15*, 55–66. [CrossRef]
18. Brest, J.; Greiner, S.; Bošković, B.; Mernik, M.; Žumer, V. Self-adapting Control Parameters in Differential Evolution: A Comparative Study on Numerical Benchmark Problems. *IEEE Trans. Evol. Comput.* **2006**, *10*, 646–657. [CrossRef]
19. Tang, L.; Dong, Y.; Liu, J. Differential Evolution With an Individual-Dependent Mechanism. *IEEE Trans. Evol. Comput.* **2015**, *19*, 560–574. [CrossRef]
20. Tanabe, R.; Fukunaga, A.S. Improving the search performance of SHADE using linear population size reduction. In Proceedings of the IEEE Congress on Evolutionary Computation (CEC), Beijing, China, 6–11 July 2014; pp. 1658–1665.
21. Brest, J.; Maučec, M.S.; Bošković, B. Single Objective Real-Parameter Optimization: Algorithm jSO. In Proceedings of the 2017 IEEE Congress on Evolutionary Computation (CEC), Donostia, Spain, 5–8 June 2017; pp. 1311–1318.
22. Polakova, R.; Tvrdik, J.; Bujok, P. Differential evolution with adaptive mechanism of population size according to current population diversity. *Swarm Evol. Comput.* **2019**, *50*, 100519. [CrossRef]
23. Qin, A.K.; Huang, V.L.; Suganthan, P.N. Differential Evolution Algorithm With Strategy Adaptation for Global Numerical Optimization. *IEEE Trans. Evol. Comput.* **2009**, *13*, 398–417. [CrossRef]
24. Tanabe, R.; Fukunaga, A.S. Success-history based parameter adaptation for Differential Evolution. In Proceedings of the IEEE Congress on Evolutionary Computation (CEC), Cancun, Mexico, 20–23 June 2013; pp. 71–78.
25. Zhang, J.; Sanderson, A.C. JADE: Adaptive Differential Evolution With Optional External Archive. *IEEE Trans. Evol. Comput.* **2009**, *13*, 945–958. [CrossRef]
26. Wang, Y.; Li, H.X.; Huang, T.; Li, L. Differential evolution based on covariance matrix learning and bimodal distribution parameter setting. *Appl. Soft Comput.* **2014**, *18*, 232–247. [CrossRef]
27. Bujok, P.; Tvrdík, J. Enhanced individual-dependent differential evolution with population size adaptation. In Proceedings of the 2017 IEEE Congress on Evolutionary Computation (CEC), Donostia, Spain, 5–8 June 2017; pp. 1358–1365.
28. Brest, J.; Maučec, M.S.; Bošković, B. The 100-Digit Challenge: Algorithm jDE100. In Proceedings of the 2019 IEEE Congress on Evolutionary Computation (CEC), Wellington, New Zealand, 10–13 June 2019; pp. 19–26. [CrossRef]
29. Hedar, A.R. Global Optimization Test Problems. Available online: http://www-optima.amp.i.kyoto-u.ac.jp/member/student/hedar/Hedar_files/TestGO_files/Page422.htm (accessed on 30 June 2021).

MDPI
St. Alban-Anlage 66
4052 Basel
Switzerland
Tel. +41 61 683 77 34
Fax +41 61 302 89 18
www.mdpi.com

Mathematics Editorial Office
E-mail: mathematics@mdpi.com
www.mdpi.com/journal/mathematics

www.ingramcontent.com/pod-product-compliance
Lightning Source LLC
LaVergne TN
LVHW070203100526
838202LV00015B/1988